«Uncertainty Quantification and Stochastic Modeling with Matlab®»

Series Editor
Piotr Breitkopf

Uncertainty Quantification and Stochastic Modeling with Matlab®

Eduardo Souza de Cursi
Rubens Sampaio

First published 2015 in Great Britain and the United States by ISTE Press Ltd and Elsevier Ltd

Apart from any fair dealing for the purposes of research or private study, or criticism or review, as permitted under the Copyright, Designs and Patents Act 1988, this publication may only be reproduced, stored or transmitted, in any form or by any means, with the prior permission in writing of the publishers, or in the case of reprographic reproduction in accordance with the terms and licenses issued by the CLA. Enquiries concerning reproduction outside these terms should be sent to the publishers at the undermentioned address:

ISTE Press Ltd
27-37 St George's Road
London SW19 4EU
UK

www.iste.co.uk

Elsevier Ltd
The Boulevard, Langford Lane
Kidlington, Oxford, OX5 1GB
UK

www.elsevier.com

Notices
Knowledge and best practice in this field are constantly changing. As new research and experience broaden our understanding, changes in research methods, professional practices, or medical treatment may become necessary.

Practitioners and researchers must always rely on their own experience and knowledge in evaluating and using any information, methods, compounds, or experiments described herein. In using such information or methods they should be mindful of their own safety and the safety of others, including parties for whom they have a professional responsibility.

To the fullest extent of the law, neither the Publisher nor the authors, contributors, or editors, assume any liability for any injury and/or damage to persons or property as a matter of products liability, negligence or otherwise, or from any use or operation of any methods, products, instructions, or ideas contained in the material herein.

For information on all Elsevier publications visit our website at
http://store.elsevier.com/

© ISTE Press Ltd 2015
The rights of Eduardo Souza de Cursi and Rubens Sampaio to be identified as the authors of this work have been asserted by them in accordance with the Copyright, Designs and Patents Act 1988.

MATLAB® is a trademark of The MathWorks, Inc. and is used with permission.

The MathWorks does not warrant the accuracy of the text or exercises in this book.

This book's use or discussion of MATLAB® software or related products does not constitute endorsement or sponsorship by The MathWorks of a particular pedagogical approach or particular use of the MATLAB® software.

British Library Cataloguing in Publication Data
A CIP record for this book is available from the British Library
Library of Congress Cataloging in Publication Data
A catalog record for this book is available from the Library of Congress
ISBN 978-1-78548-005-8

Printed and bound in the UK and US

Contents

INTRODUCTION xi

CHAPTER 1. ELEMENTS OF PROBABILITY THEORY
AND STOCHASTIC PROCESSES 1

   1.1. Notation 2
   1.2. Numerical characteristics of finite populations 3
   1.3. Matlab implementation 7
      1.3.1. Using a data vector 7
      1.3.2. Using a frequency table 10
   1.4. Couples of numerical characteristics 11
   1.5. Matlab implementation 15
   1.6. Hilbertian properties of the numerical characteristics 21
      1.6.1. Conditional probability and approximation 23
      1.6.2. Expectation and approximation of a constant 25
      1.6.3. Linear correlation and affine approximation 26
      1.6.4. Matlab implementation 27
      1.6.5. Conditional mean and best approximation 31
   1.7. Measure and probability 33
   1.8. Construction of measures 36
      1.8.1. Measurable sets 36
      1.8.2. Lebesgue measure on $\mathbb{R}^p$ 41
      1.8.3. Equivalent functions 43
      1.8.4. Integrals 47
      1.8.5. Measures defined by densities 52

1.9. Measures, probability and integrals in infinite
    dimensional spaces   53
    1.9.1. Finite measures on infinite dimensional Hilbert spaces . . .   57
    1.9.2. Integration involving a measure on an infinite
        dimensional Hilbert space   62
1.10. Random variables   65
    1.10.1. Matlab implementation   73
    1.10.2. Couples of random variables   76
1.11. Hilbertian properties of random variables   80
    1.11.1. Approximation   82
1.12. Sequences of random variables   88
    1.12.1. Quadratic mean convergence   88
    1.12.2. Convergence in the mean   89
    1.12.3. Convergence in probability   89
    1.12.4. Almost sure convergence   90
    1.12.5. Convergence in distribution   90
    1.12.6. Relations among different types of convergence   91
1.13. Some usual distributions   91
    1.13.1. Poisson distribution   92
    1.13.2. Uniform distribution   92
    1.13.3. Normal or Gaussian distribution   92
    1.13.4. Gaussian vectors   94
1.14. Samples of random variables   99
1.15. Gaussian samples 101
1.16. Stochastic processes 107
1.17. Hilbertian structure 108
1.18. Wiener process 112
1.19. Ito integrals 114
    1.19.1. Integrals with respect to time 115
    1.19.2. Integrals with respect to a process 117
    1.19.3. Integrals with respect to a Wiener process 119
1.20. Ito Calculus 122
    1.20.1. Ito's formula 122
    1.20.2. Ito stochastic diffusions 127

CHAPTER 2. MAXIMUM ENTROPY AND INFORMATION 133

2.1. Construction of a stochastic model 133
2.2. The principle of maximum entropy 135

2.2.1. Discrete random variables 136
2.2.2. Continuous random variables 143
2.2.3. Random vectors 153
2.2.4. Random matrices 156
2.3. Generating samples of random variables, random
vectors and stochastic processes 159
2.4. Karhunen–Loève expansions and numerical generation
of variates from stochastic processes 161
2.4.1. Karhunen–Loève expansions 162
2.4.2. Numerical determination of Karhunen–Loève
expansions 165

CHAPTER 3. REPRESENTATION OF RANDOM VARIABLES 177

3.1. Approximations based on Hilbertian properties 178
3.1.1. Using the conditional expectation in order to generate
a representation 181
3.1.2. Using the mean for the approximation by a constant 184
3.1.3. Using the linear correlation in order to
construct a representation 186
3.1.4. Polynomial approximation 189
3.1.5. General finite-dimensional approximations 193
3.1.6. Approximation using a total family 202
3.2. Approximations based on statistical properties
(moment matching method) 215
3.3. Interpolation-based approximations (collocation) 222

**CHAPTER 4. LINEAR ALGEBRAIC EQUATIONS
UNDER UNCERTAINTY** 227

4.1. Representation of the solution of uncertain linear systems . . . 228
4.1.1. Case where the distributions are known 229
4.1.2. Case where the distributions are unknown 237
4.2. Representation of eigenvalues and eigenvectors
of uncertain matrices 243
4.2.1. Determination of the distribution of eigenvalues
and eigenvectors by collocation 245
4.2.2. Determination of the distribution of eigenvalues
and eigenvectors by moment fitting 249

4.2.3. Representation of extreme eigenvalues by
   optimization techniques 249
4.2.4. Power iterations 252
4.2.5. Subspace iterations and Krylov iterations 255
4.3. Stochastic methods for deterministic linear systems 260

CHAPTER 5. NONLINEAR ALGEBRAIC EQUATIONS INVOLVING RANDOM PARAMETERS 265

5.1. Nonlinear systems of algebraic equations 265
   5.1.1. Collocation 266
   5.1.2. Moment fitting 271
   5.1.3. Variational approximation 273
   5.1.4. Adaptation of iterative methods 279
5.2. Numerical solution of noisy deterministic systems
   of nonlinear equations 288

CHAPTER 6. DIFFERENTIAL EQUATIONS UNDER UNCERTAINTY 297

6.1. The case of linear differential equations 298
6.2. The case of nonlinear differential equations 306
6.3. The case of partial differential equations 310
   6.3.1. Linear equations 310
   6.3.2. Nonlinear equations 311
   6.3.3. Evolution equations 312
6.4. Reduction of Hamiltonian systems 314
   6.4.1. Hamiltonian systems 314
   6.4.2. Reduction of autonomous Hamiltonian systems 317
6.5. Local solution of deterministic differential equations
   by stochastic simulation 319
   6.5.1. Ordinary differential equations 319
   6.5.2. Elliptic partial differential equations 325
   6.5.3. Parabolic partial differential equations 328
   6.5.4. Finite difference schemes 333
6.6. Statistics of dynamical systems 336
   6.6.1. Evaluation of statistics of the limit cycle 340

CHAPTER 7. OPTIMIZATION UNDER UNCERTAINTY 345

   7.1. Representation of the solutions in unconstrained
       optimization 346
       7.1.1. Collocation 347
       7.1.2. Moment matching 350
       7.1.3. Stochastic optimization 350
       7.1.4. Adaptation of iterative methods 351
       7.1.5. Optimal criteria 362
   7.2. Stochastic methods in deterministic continuous
       optimization 364
       7.2.1. Version 1: Stochastic descent 370
       7.2.2. Version 2: Dynamics of Metropolis 381
       7.2.3. Version 3: Hybrid methods 391
   7.3. Population-based methods 403
   7.4. Determination of starting points 407

CHAPTER 8. RELIABILITY-BASED OPTIMIZATION 421

   8.1. The model situation 422
   8.2. Reliability index 424
   8.3. FORM 426
   8.4. The bi-level or double-loop method 427
   8.5. One-level or single-loop approach 428
   8.6. Safety factors 430

BIBLIOGRAPHY 435

INDEX 441

Introduction

Until recently, the deterministic view of Nature prevailed in Physics. Starting from the Enlightenment, there was the belief that Mechanics was completely described by Newton's laws. If the initial conditions of a mechanical systems were known, then its state could be completely determined in the future as well as in the past. This was precisely the viewpoint of Lagrange and it ruled from the 18th Century to around the middle of the 20th Century.

However this panglossian view of Nature was disputed, especially with the advent of Thermodynamics that challenged the belief in reversibility and started the study of complex random systems. In addition, the discovery of simple systems with chaotic behavior, that despite being deterministic, showed a very complex behavior, that the mechanical problems are far from simple. Quantum Theory, however, showed that the processes in Nature are not deterministic, but stochastic. This change in paradigm is not yet accepted, as is well described in the book of Thomas Kuhn that discusses sociological aspects of sciences.

The main ideas of probability were slow to develop. Perhaps, because chance events were interpreted as the wish of the gods and hence were not believed to be random. The first theoretical works on probability were connected with games of chance and since the set of possibilities are finite, the studies are combinatorial, but the big breakthrough being when the continuous case is studied.

Stochastic modeling is, of course, harder than deterministic modeling and the implementation of the model is more costly. Let us look at this in a simple example. In the simplest deterministic continuous model, the parameters of the model are constants. The equivalent of this in the stochastic case is to transform the constants in random variables. For each realization of the random variable, its value is also a constant, but the constant may change with the realization following a certain distribution of probability. So, the most simple object from a stochastic model – random variables – is formed of functions, hence objects of infinite dimension. Also, in a coherent deterministic model, we expect a solution if we fix suitable conditions. In a stochastic model this is not the case. For each realization of the stochastic data, the parameters are chosen and a deterministic problem is solved. Then, with the set of solutions obtained, statistics are made and the main result of the problem is the distribution of probability of the results. Thus, the main element to be determined is the distribution of the possible solutions. The values obtained from a single realization are of no importance, the distribution of the values is the important result. The understanding of this fact came very slowly for the manufacturing industry, for example, and only after long years of application of quality control in the manufacturing came the realization that they were dealing with a process stochastic in nature. Since then, stochastic modeling has been essential in Engineering. Today, reliability represents a new way of design and it takes into account the inevitable uncertainties of the processes.

The solution of a stochastic model can be decomposed into three steps: generation of the stochastic data following their distribution of probability, solution of a deterministic problem for each sample generated, and finally computation of statistics with the results (for example, to construct a histogram) until they show a certain persistence (do not change an error criterion accordingly). Histograms represent the approximation of the solution of a stochastic problem. When we want an approximation of the solution of a stochastic problem, a histogram is the best approximation. If they are hard to find, or costly to compute, we make do with mean and dispersion, or a certain number of moments. In some situations, such as the ones described in Chapter 8, we compute only a particular probability since computation of moments is too expensive.

This book presents the main ideas of Stochastic Modeling and Uncertainty Quantification using Functional Analysis as the main tool. More specifically,

Hilbert Spaces and orthogonal projections are the basic objects leading to the methods exposed. As presented in the text, some ideas often considered as complex, such as Conditional Expectations, may be developed in a systematic way by considering their definition as orthogonal projections in convenient Hilbert spaces.

Chapter 1 discusses, the main ideas of Probability Theory in a Hilbert context. This chapter is designed as a reference, but the main concepts of random variables and random processes are developed. The reader, having a previous knowledge of probability theory and orthogonal projections, may start reading from Chapter 2, which presents the construction of a stochastic model by using the Principle of the Maximal Entropy. When the data is formed by independent random variables, the generation of a model by this procedure may be considered as simple. But in the case where dependent random variables or a random process has to be considered, the generation difficulties appear and new tools have to be introduced – the main ideas concerning the generation of samples from random vectors and processes are discussed in this chapter. Chapter 3 presents the problem of approximation of a random variable by a function of another random variable and the general methods which may be used in order to numerically determine this approximation. The following chapters deal with applications of these methods in order to solve particular problems. Chapter 4 considers linear systems with uncertainties; Chapter 5 presents methods for nonlinear systems with uncertainties; Chapter 6 deals with differential equations and the next two chapters, i.e. Chapters 7 and 8 present methods for optimization under uncertainties.

In order to help the reader interested in practical applications, listings of Matlab® programs implementing the main methods are included in the text. In each chapter, the reader will find these listings which complete the methodological presentation by a practical implementation. Of course, these programs are to be considered as examples of implementation and not as optimized ones. The reader may construct their own implementations – we expect that this work will be facilitated by the examples of implementation given in the text.

The authors would like thank the ISTE for the reception of this project and the publication of this work.

1

Elements of Probability Theory and Stochastic Processes

The element which constitutes the foundation of the construction of stochastic algorithms is the concept of *random variable*, i.e. a *function* $X : \Omega \longrightarrow \mathbb{R}$ for which individual values $X(\omega)$ ($\omega \in \Omega$) are *not available or simply not interesting* and we are looking for global information connected to X.

For instance, we may consider Ω as the population of a geographic region (country, town, etc) and numerical quantities connected to each individual ω: age, distance or transportation time from a residence to work, level of studies, revenue in the past year, etc. Each of these characteristics may be considered as deterministic, since being perfectly determined for a given individual ω. But to obtain the global information for all the individuals may become expensive (recall the cost of a census) and errors may occur in the process. In addition, maybe we are not interested in a particular individual, but only in groups of individuals or in global quantities such as "how many individuals are more than 60 years old?", "what is the fraction of individuals needing more than one hour of transportation time?", "how many individuals have finished university", "how many families have an income lower than ... ?". In this case, we may look to the quantities under consideration as random variables.

These examples show that random variables may be obtained by considering *numerical characteristics of finite populations*. This kind of variable is introduced in section 1.2, and gives a comprehensive introduction to random variables and illustrates their practical use, namely for the numerical calculation of statistics. In the general situation, random variables may be defined on general abstract sets Ω by using the notions of *measure* and *probability* (see section 1.7).

1.1. Notation

Let us denote by $\mathbb{N}$ the set of the natural numbers, and by $\mathbb{N}^*$ the set of the strictly positive real numbers. $\mathbb{R}$ denotes the set of the real numbers $(-\infty, +\infty)$ and the notation $\mathbb{R}_e$ refers to the extended real numbers: $\mathbb{R} \cup \{-\infty, +\infty\}$. $(a,b) = \{x \in \mathbb{R}: a < x < b\}$ denotes an open interval of real numbers.

For $k \in \mathbb{N}^*$, $\mathbb{R}^k = \{\boldsymbol{x} = (x_1, ..., x_k) : x_n \in \mathbb{R}, 1 \leq n \leq k\}$ is the set of the k-tuples formed by real numbers. Analogous notation is used for $\mathbb{R}^k_e$. We denote by $|\bullet|_p$ the standard p-norm of $\mathbb{R}^k$. The standard scalar product of two elements $\boldsymbol{x}$ and $\boldsymbol{y}$ of $\mathbb{R}^k$ is:

$$(\boldsymbol{x}, \boldsymbol{y})_k = \sum_{i=1}^{k} x_i y_i.$$

When the context is clear enough in order to avoid any confusion, the scalar product on $\mathbb{R}^k$ is simply denote by a point:

$$\boldsymbol{x}.\boldsymbol{y} = (\boldsymbol{x}, \boldsymbol{y})_k .$$

We will use analogous notation for matrices: $\mathbf{A} = (A_{ij})_{1 \leq i \leq m, 1 \leq j \leq n}$ denotes an $m \times n$-matrix formed by real numbers. The set formed by all the $m \times n$-matrices formed by real numbers is denoted by $\mathcal{M}(m,n)$. Usually, the elements of $\mathbb{R}^k$ may be seen either as elements of $\mathcal{M}(k,1)$ (column vectors) or elements of $\mathcal{M}(1,k)$ (row vectors). *In this text, we consider the elements of $\mathbb{R}^k$ as column vectors, i.e. elements of $\mathcal{M}(k,1)$.* This allow us to write the product of an $m \times k$ matrix $\mathbf{A} \in \mathcal{M}(m,k)$ and an element of $\mathbb{R}^k$, considered as an element of $\mathbf{x} \in \mathbb{R}^k \equiv \mathcal{M}(k,1)$:

$$\mathbf{y} = \mathbf{A}\mathbf{x} \in \mathcal{M}(m,1) \equiv \mathbb{R}^m.$$

As usual, we have $\boldsymbol{y} = (y_1, ..., y_m)$, with

$$y_i = \sum_{j=1}^{k} A_{ij} x_j.$$

Sometimes, in order to simplify the expressions, we use the convention on the implicit sum on repeated indexes (called Einstein's convention) by simply writing $y_i = A_{ij} x_j$ in this case.

1.2. Numerical characteristics of finite populations

Counts or surveys of the numerical characteristics of finite populations are often generated by censuses, generally made at regular time intervals. For instance, the number of inhabitants, the family income, the type of the house or employment, the school or professional level of the members of the family are some examples of characteristics that are periodically verified.

From the mathematical standpoint, a numerical characteristic X defined on a finite population Ω is a numerical function $X : \Omega \longrightarrow \mathbb{R}$. As previously observed, an important feature in censuses is that, usually, we are not interested in the particular value $X(\omega)$ of the numerical characteristic for a particular individual $\omega \in \Omega$, but in *global quantities* connected to X, i.e. in the *global behavior* of X on the population Ω: what the fraction is of the population having an inferior or superior age to given bounds, what part of the population has revenues inferior or superior to given bounds, etc. Thus, a frequent framework is the following:

1) $\Omega = \{\omega^1, ..., \omega^N\}$ is a finite population, non-empty set ($N \geq 1$)

2) $X : \Omega \longrightarrow \mathbb{R}$ is a numerical characteristic defined on Ω, having as an image a set of k distinct values: $X(\Omega) = \{X_1, ..., X_k\}$, $X_i \neq X_j$ for $i \neq j$. Since these values are distinct, they may be arranged in a strictly crescent order: if necessary, we may assume that $X_i < X_j$ for $i < j$ without loss of generality.

3) The inverse image of the value $X_i \in X(\Omega)$ is the subpopulation $H_i = X^{-1}(\{X_i\}) = \{\omega \in \Omega : X(\omega) = X_i\}$. The number of elements of H_i is $\#(H_i) = n_i$ and the *relative frequency* or *probability* of X_i is $P(X = X_i) = p_i = n_i/N$.

4) Let $A \subset \mathbb{R}$: we define $P(A) = n_A/N$, where $n_A = \#(X^{-1}(A))$ is the number of elements of A.

We have:
$$\sum_{i=1}^{k} p_i = 1 \text{ and } \sum_{i=1}^{k} n_i = N.$$

In addition, since the number of elements of $H_i \subset \Omega$ does not exceed the number of elements N of the global population Ω, we have $n_i \leq N$. Thus, $0 \leq p_i \leq 1$.

DEFINITION 1.1 (mean on a finite population).– The mean, or expectation, of X is:

$$E(X) = \frac{1}{N} \sum_{n=1}^{N} X(\omega^n). \blacksquare$$

PROPOSITION 1.1.– The mean is a linear operation (i.e. $E(\bullet)$ is linear): let $X : \Omega \longrightarrow \mathbb{R}$ and $Y : \Omega \longrightarrow \mathbb{R}$ be two numerical characteristics of the population Ω and α, β be two real numbers. Then:

$$E(\alpha X + \beta Y) = \alpha E(X) + \beta E(Y). \blacksquare$$

PROOF.– We have:

$$E(\alpha X + \beta Y) = \frac{1}{N} \sum_{n=1}^{N} (\alpha X(\omega^n) + \beta Y(\omega^n)),$$

that shows that:

$$E(\alpha X + \beta Y) = \frac{\alpha}{N} \sum_{n=1}^{N} X(\omega^n) + \frac{\beta}{N} \sum_{n=1}^{N} Y(\omega^n) = \alpha E(X) + \beta E(Y). \blacksquare$$

PROPOSITION 1.2.– If $X \geq 0$ then $E(X) \geq 0$. $\blacksquare$

PROOF.– Since $X \geq 0$, we have $X(\omega) \geq 0, \forall \omega \in \Omega$. Thus $E(X)$ is a sum of non-negative terms and we have $E(X) \geq 0$. $\blacksquare$

DEFINITION 1.2 (variance on a finite population).– The variance of X is:

$$V(X) = \frac{1}{N} \sum_{n=1}^{N} (X(\omega^i) - E(X))^2.$$

The standard deviation of X is the square root of the variance: $\sigma(X) = \sqrt{V(X)}$. $\blacksquare$

PROPOSITION 1.3.– We have:

$$V(X) = E(X^2) - [E(X)]^2. \blacksquare$$

PROOF.– Since:

$$(X(\omega^n) - E(X))^2 = \left(X(\omega^n)^2\right) - 2E(X)X(\omega^n) + [E(X)]^2,$$

we have:

$$V(X) = \frac{1}{N}\left[\sum_{n=1}^{N}\left(X(\omega^n)^2\right) - 2E(X)\sum_{n=1}^{N}X(\omega^n) + [E(X)]^2\sum_{n=1}^{N}1\right]$$

and

$$V(X) = E(X^2) - 2[E(X)]^2 + [E(X)]^2 = E(X^2) - [E(X)]^2. \blacksquare$$

PROPOSITION 1.4.– $V(X) \geq 0$. In addition, $V(X) = 0$ if and only if X is a constant on Ω. $\blacksquare$

PROOF.– The first assertion is immediate, since $V(X)$ is a sum of non-negative terms (sum of squares). For the second, we observe that a sum of squares is null if and only if each term in the sum is null. Thus, $V(X) = 0$ if and only if $X(\omega^n) = E(X), \forall n \in \mathbb{N}$. $\blacksquare$

COROLLARY 1.1.– $E(X^2) \geq 0$ and $E(X^2) = 0$ if and only if $X = 0$. $\blacksquare$

PROOF.– We have $E(X^2) = V(X) + [E(X)]^2$. Thus, on the one hand, $E(X^2) \geq 0$ and, on the other hand, $E(X^2) = 0$ if and only if $V(X) = 0$ and $E(X) = 0$. $\blacksquare$

DEFINITION 1.3 (moments on a finite population).– The moment of order p, or p-moment, of X is:

$$M_p(X) = E(X^p) = \frac{1}{N}\sum_{n=1}^{N}(X(\omega^n))^p. \blacksquare$$

The whole body of information on the global behavior of a numerical characteristic X may be synthesized by a frequency table, which is a table giving the possible values of X and their relative frequencies:

X_1	X_2	...	X_k
p_1	p_2	...	p_k

or

X_1	X_2	...	X_k
n_1	n_2	...	n_k

Table 1.1. *Synthesizing information on X in a frequency table*

Frequency Tables do not contain any information about the value of X for a particular $\omega \in \Omega$, but only global information about X on Ω. Nevertheless, they permit the evaluation of global quantities involving X. For instance,

LEMMA 1.1.– Let $g : \mathbb{R} \longrightarrow \mathbb{R}$ be a function and $Z = g(X)$. Then:

$$E(Z) = \sum_{i=1}^{k} p_i g(X_i). \blacksquare \qquad [1.1]$$

PROOF.– Since $\bigcup_{i=1}^{k} H_i = \Omega$, we have:

$$E(Z) = \frac{1}{N} \sum_{i=1}^{k} \left(\sum_{\omega \in H_i} \underbrace{Z(\omega)}_{= g(X_i) \text{ on } H_i} \right) =$$
$$\frac{1}{N} \sum_{i=1}^{k} n_i g(X_i) = \sum_{i=1}^{k} p_i g(X_i). \blacksquare$$

PROPOSITION 1.5.– $M_p(X) = \sum_{i=1}^{k} p_i X_i^p$. Namely, $E(X) = \sum_{i=1}^{k} p_i X_i$ and $E(X^2) = \sum_{i=1}^{k} p_i X_i^2$. $\blacksquare$

PROOF.– The result follows straight from the preceding lemma. $\blacksquare$

The information contained in a frequency table may also be given by a *cumulative distribution function* – often referred to as *cumulative function*, or *distribution function*:

$$F(x) = P(X < x) = P(X \in (-\infty, x)).$$

x	$x < X_1$	$x < X_2$	$x < X_3$	...	$x < X_k$	$x < +\infty$
$F(x)$	0	p_1	$p_1 + p_2$	...	$p_1 + ... + p_{k-1}$	1

Table 1.2. *Synthesizing information on X in a cumulative distribution function*

We observe that F admits a distributional derivative (i.e. a weak derivative) and that $f = F'$ is a sum of k Dirac measures (i.e. Dirac masses): $f(x) = \sum_{i=1}^{k} p_i \delta(x - X_i)$. f is the *probability density* of X: in this situation, f is a distribution – we consider later situations where this probability density is a standard function.

1.3. Matlab implementation

1.3.1. *Using a data vector*

Matlab furnishes simple commands which evaluate statistics from data vectors. For instance, if the data is given by a vector $\mathbf{X} = (X_1, ..., X_N)$, where $X_n = X(\omega^n)$, $1 \leq n \leq N$, you may use one or several of the following commands:

Listing 1.1. *Statistics of a data vector*

```
ma = sum(X)/length(X); % mean of vector X
mb = mean(X); % mean of vector X
v1a = sum((X - ma).^2)/length(X); % variance of vector X
v1b = var(X,1); % variance of vector X
v2 = sum((X - ma).^2)/(length(X)-1); % unbiased estimative of
    the variance of vector X
v1c = var(X); % unbiased estimative of the variance of vector X
s1a = sqrt(v1a); % standard deviation of vector X
s1b = std(X,1); % standard deviation of vector X
s1c = std(X); % unbiased estimative of the standard deviation
    of vector X
s2 = sqrt(v2); % unbiased estimation of the standard deviation
    of vector X
m_p = mean(X.^p); % moment of order p of vector X
mc_p = mean((X - ma).^p); % centered moment of order p of
    vector X
```

The cumulative distribution may be obtained by using:

Listing 1.2. *Cdf of a data vector*

```
function [F_Y] = sample2cfd(ys,y)
%
% determines the empirical cumulative function associated to a
    sample
% ys from a random variable y. F_Y is determined at the
    abscissa y.
%
%
% IN:
% ys: sample of y - type array of double
% y: vector of the abscissa for evaluation of F_Y - type array
    of double
%
% SORTIE:
% F_Y: vector containing the values of the cfd - type array of
    double
% F_Y(i) = P( Y < y(i) ) is the cfd at y(i)
%
F_Y = zeros(size(y));
ns = max(size(ys));
n = max(size(y));
for i = 1: n
    ya = y(i);
    ind = find(ys < ya);
    if isempty(ind)
        na = 0;
    else
        na = max(size(ind));
    end;
    pa = na/ns;
    F_Y(i) = pa;
end;

return ;
end
```

The probability density associated may be determined by numerical derivation of the cumulative density

Listing 1.3. *Pdf of a data vector*

```
function fY = cdf2pdf(FY,y,h)
%
% determines the density function associated to the cumulative
   function
% FY evaluated at the equally spaced abscissa y.
% f_Y is determined by particle derivative.
%
% IN:
% FY: values of the cdf - type array of double
% y: vector of the abscissa - type array of double
% h: radius of influence - type double
%     defines how many neighbour points are to derivate
%     for equally spaced data, h = a small (1 to 5) multiple of
       the step
%                                  often 3 is used
%     the larger is h, the smoother is the pdf.
%
% OUT:
% fY: vector containing the values of the pdf - type array of
   double
%
g = @(y,x) exp(-0.5*((y-x)/h)^2)/(h*sqrt(2*pi));
dg = @(y,x) -((y-x)/h)*(1/h)*g(y,x);
%
ny = length(y);
v = zeros(size(y));
for i = 1: ny
    s1 = 0;
    for j = 1: ny
        aux = g(y(j),y(i));
        s1 = s1 + aux;
    end;
    v(i) = s1;
end;
fY = zeros(size(FY));
for i = 1: ny
    s1 = 0;
    s2 = 0;
    for j = 1: ny
        aux = dg(y(j),y(i))/v(j);
        s1 = s1 + aux*(y(j) - y(i));
        s2 = s2 + aux*(FY(j) - FY(i));
    end;
    fY(i) = s2/s1;
end;
return ;
end
```

An example is shown below (Figure 1.1): we generate a random sample of 250 variates from the Gaussian distribution $N(0,1)$ and we determine both the cdf and the pdf by using these programs

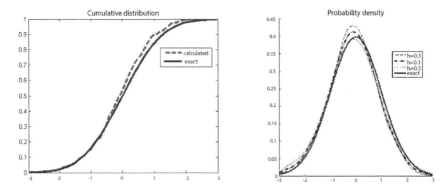

Figure 1.1. *Results for a sample of the Gaussian distribution. For a color version of the figure, see www.iste.co.uk/souzadecursi/quantification.zip*

1.3.2. *Using a frequency table*

Assuming that the data is summarized in two vectors $\mathbf{X} = (X_1, ..., X_k)$, and $\mathbf{nX} = (n_1, ..., n_k)$, you may use one or several of the following commands:

Listing 1.4. *Statistics of a frequency table*

```
ma = sum(nX.*X)/sum(nX); % mean of vector X
v1a = sum(nX.*(X - ma).^2)/sum(nX); % variance of vector X
v2 = sum(nX.*(X - ma).^2)/(sum(nX)-1); % unbiased estimative of
    the variance of vector X
s1a = sqrt(v1a); % standard deviation of vector X
s2 = sqrt(v2); % unbiased estimation of the standard deviation
    of vector X
m_p = mean(nX.*(X.^p)); % moment of order p of vector X
mc_p = mean(nX.*(X - ma).^p); % centered moment of order p of
    vector X
```

Notice that the relative frequencies may be obtained as follows:

Listing 1.5. *Relative frequencies from the absolute ones*

```
function pys = abs2relfreq(nys)
%
% generates the relative frequencies pys from the
% absolute frequencies nys
%
%IN:
% nys : table of absolute frequencies (numbers of occurences):
    type    array of integer
%
%OUT:
% pys : table of relative frequencies : type array of double
%
ns = sum(nys);
pys = nys/ns;
return;
end
```

Analogous to the preceding situation, the probability density associated may be determined by numerical derivation of the cumulative density.

1.4. Couples of numerical characteristics

Let us consider a pair (X, Y) of numerical characteristics $X, Y : \Omega \longrightarrow \mathbb{R}$ such that $X(\Omega) = \{X_1, ..., X_k\}$ and $Y(\Omega) = \{Y_1, ..., Y_m\}$. Since X and Y are connected by ω, we refer to (X, Y) as a couple of numerical characteristics on Ω.

Analogous to the preceding situation where a single numerical characteristic has been considered, the inverse image of (X_i, Y_j) is $H_{ij} = X^{-1}(\{X_i\}) \cap Y^{-1}(\{Y_j\}) = \{\omega \in \Omega : X(\omega) = X_i \text{ and } Y(\omega) = Y_j\}$. The number of elements of H_{ij} is $\#(H_{ij}) = n_{ij}$ and the relative frequency (i.e. the probability) of (X_i, Y_j) is $P(X = X_i, Y = Y_j) = p_{ij} = n_{ij}/N$. Here,

$$\sum_{i=1}^{k}\sum_{j=1}^{m} p_{ij} = 1 \text{ and } \sum_{i=1}^{k}\sum_{j=1}^{m} n_{ij} = N.$$

The information about the global behavior of the couple (X, Y) may be synthetized in a contingency table (or cross-tabulation table).

	Y_1	$\cdots$	Y_m
X_1	p_{11}	$\cdots$	p_{1m}
$\cdots$	$\cdots$	$\cdots$	$\cdots$
X_k	p_{k1}	$\cdots$	p_{km}

Table 1.3. *Synthesizing information on (X,Y) in a contingency table*

Analogous to frequency tables, contingency tables permit the evaluation of global quantities involving functions of the couple:

LEMMA 1.2.– Let $g : \mathbb{R}^2 \longrightarrow \mathbb{R}$ be a function and $Z = g(X,Y)$. Then:

$$E(Z) = \sum_{i=1}^{k} \sum_{j=1}^{m} p_{ij} g(X_i, Y_j). \blacksquare \qquad [1.2]$$

PROOF.– Since $\bigcup_{i=1}^{k} \bigcup_{j=1}^{m} H_{ij} = \Omega$, we have:

$$\sum_{n=1}^{N} g(X(\omega^n), Y(\omega^n)) = \sum_{i=1}^{k} \sum_{j=1}^{m} \left(\sum_{\omega \in H_{ij}} g \underbrace{\left(X(\omega), Y(\omega) \right)}_{=(X_i, Y_j) \text{ on } H_{ij}} \right),$$

so that:

$$\sum_{n=1}^{N} g(X(\omega^n), Y(\omega^n)) = \sum_{i=1}^{k} \sum_{j=1}^{m} p_{ij} g(X_i, Y_j). \blacksquare$$

Thus,

$$X^{-1}(\{X_i\}) = H_{i\bullet} = \bigcup_{j=1}^{m} H_{ij}; \quad Y^{-1}(\{Y_j\}) = H_{\bullet j} = \bigcup_{i=1}^{k} H_{ij},$$

and we have:

$$P(X = X_i) = p_{i\bullet} = \sum_{j=1}^{m} p_{ij}; \quad P(Y = Y_j) = p_{\bullet j} = \sum_{i=1}^{k} p_{ij}.$$

As a consequence:

$$E(X) = \sum_{i=1}^{k} p_{i\bullet} X_i = \sum_{i=1}^{k}\sum_{j=1}^{m} p_{ij} X_i;$$

$$E(Y) = \sum_{j=1}^{m} p_{\bullet j} Y_j = \sum_{i=1}^{k}\sum_{j=1}^{m} p_{ij} Y_j.$$

Into an analogous way,

$$V(X) = \sum_{i=1}^{k} p_{i\bullet}(X_i - E(X))^2 = \sum_{i=1}^{k}\sum_{j=1}^{m} p_{ij}(X_i - E(X))^2;$$

$$V(Y) = \sum_{j=1}^{m} p_{\bullet j}(Y_j - E(Y))^2 = \sum_{i=1}^{k}\sum_{j=1}^{m} p_{ij}(Y_j - E(Y))^2;$$

$$M_p(X) = \sum_{i=1}^{k} p_{i\bullet} X_i^p = \sum_{i=1}^{k}\sum_{j=1}^{m} p_{ij} X_i^p;$$

$$M_p(Y) = \sum_{j=1}^{m} p_{\bullet j} Y_j^p = \sum_{i=1}^{k}\sum_{j=1}^{m} p_{ij} Y_j^p.$$

When considering couples, the cumulative distribution function is:

$$F(x, y) = P(X < x, Y < y) = P((X, Y) \in (-\infty, x) \times (-\infty, y))$$

and the probability density is:

$$f(x, y) = \frac{\partial^2 F}{\partial x\, \partial y}(x, y).$$

DEFINITION 1.4 (covariance of a finite population).– The covariance of X and Y is:

$$Cov(X, Y) = \frac{1}{N}\sum_{n=1}^{N}(X(\omega^i) - E(X))(Y(\omega^i) - E(Y)).\ \blacksquare$$

PROPOSITION 1.6.– We have:

$$Cov(X, Y) = E(XY) - E(X) E(Y).\ \blacksquare$$

PROOF.– Since:

$$(X(\omega^n) - E(X))(Y(\omega^n) - E(Y)) = X(\omega^n)Y(\omega^n) - E(X)Y(\omega^n) - E(Y)X(\omega^n) + E(X)E(Y),$$

we have:

$$Cov(X,Y) = \frac{1}{N}\left[\sum_{n=1}^{N} X(\omega^n)Y(\omega^n) - E(X)\sum_{n=1}^{N} Y(\omega^n) - E(Y)\sum_{n=1}^{N} X(\omega^n) + E(X)E(Y)\sum_{n=1}^{N} 1\right]$$

and:

$$Cov(X,Y) = E(XY) - E(X)E(Y) - E(Y)E(X) + E(X)E(Y) = E(XY) - E(X)E(Y). \blacksquare$$

PROPOSITION 1.7.– We have:

$$V(\alpha X + \beta Y) = \alpha^2 V(X) + \beta^2 V(Y) + 2\alpha\beta Cov(X,Y). \blacksquare$$

PROOF.– Since:

$$(\alpha X + \beta Y)^2 = \alpha^2 X^2 + \beta^2 Y^2 + 2\alpha\beta XY,$$

we have:

$$E\left((\alpha X + \beta Y)^2\right) = \alpha^2 E(X^2) + \beta^2 E(Y^2) + 2\alpha\beta E(XY).$$

In addition,

$$E(\alpha X + \beta Y) = \alpha E(X) + \beta E(Y),$$

so that:

$$E[(\alpha X + \beta Y)]^2 = \alpha^2 [E(X)]^2 + \beta^2 [E(Y)]^2 + 2\alpha\beta E(X)E(Y).$$

and:

$$V(\alpha X + \beta Y) = \alpha^2\left(E(X^2) - [E(X)]^2\right) + \beta^2\left(E(Y^2) - [E(Y)]^2\right) + 2\alpha\beta\left(E(XY) - E(X)E(Y)\right),$$

which concludes the proof. $\blacksquare$

COROLLARY 1.2.– Let $X : \Omega \longrightarrow \mathbb{R}$ and $Y : \Omega \longrightarrow \mathbb{R}$ be two numerical characteristics on Ω. Then $|Cov(X,Y)| \leq \sqrt{V(X)}\sqrt{V(Y)}$. ∎

PROOF.– Let $\alpha \in \mathbb{R}$. Let us consider the second degree polynomial $f(\alpha) = \alpha^2 V(X) + 2\alpha Cov(X,Y) + V(Y)$. From the preceding proposition: $f(\alpha) = V(\alpha X + Y) \geq 0, \forall \alpha \in \mathbb{R}$. Thus,

$$\Delta = [2Cov(X,Y)]^2 - 4V(X)V(Y) \leq 0,$$

so that:

$$[Cov(X,Y)]^2 \leq V(X)V(Y)$$

and we have the result. ∎

1.5. Matlab implementation

Assuming that the data is summarized in two vectors $\mathbf{X} = (X_1, ..., X_k)$, $\mathbf{Y} = (Y_1, ..., Y_m)$, and the absolute number of occurrences is given by $\mathbf{nXY} = (n_{ij} : 1 \leq i \leq k, 1 \leq j \leq m)$, we obtain the contingency table by using:

Listing 1.6. *Generating the contingency table*
```
function pXY = abs2relfreq (nXY)
%
% generates the relative frequencies pXY from the
% absolute frequencies nXY
%
%IN :
% nXY : table of absolute frequencies (numbers of occurences):
     type integer
%
%OUT:
% pXY : table of relative frequencies : type double
%
ns = sum(sum(nXY));
pXY = nXY/ns;
return;
end
```

The marginal distributions may be obtained as follows (for instance, [pX, pY] = cont2marg(pXY,size(X), size(Y));):

Listing 1.7. *Generation of marginal values*

```
function [cX, cY] = cont2marg(cXY, sizeX, sizeY)
%
% generates the marginal values for each variable
% from the contingency table of the pair
%
%IN:
% cXY : contingency table of the pair : type array of integer
    or double
% sizeX : dimensions of X: type integer
% sizeY : dimensions of Y
%
%OUT:
% cX : table of marginal values for X: same type as cXY
% cY : table of marginal values for Y: same type as cXY
%
cX = reshape(sum(cXY, 2), sizeX);
cY = reshape(sum(cXY, 1), sizeY);
return;
end
```

and these results may be used in order to obtain statistics from X or Y by using the Matlab programs of section 1.3. The cdf of the pair may be obtained by using (the results may be visualized by using surf(x,y,FXY');)

Listing 1.8. *Cfd of a contingency table*

```
function FXY = cont2cdf(X,Y,pXY,x,y)
%
% determines the cdf of the pair (X,Y) from the
% contingency table pXY giving the relative frequencies
%
% IN:
% X: values of X — type array of double
% Y: values of Y — type array of double
% pXY : relative frequencies of the pair — type array of double
%       PXY(i,j) = frequency of (X(i),Y(j))
% x: abscissa for the calculation of the cdf — type array of
    double
% y: abscissa for the calculation of the cdf — type array of
    double
%
% OUT:
% FXY : table containg the values of the cdf — type array of
    double
%       FXY(i,j) = P( X < x(i) , Y < y(j) )
```

```
%
nx = length(x);
ny = length(y);
FXY = zeros(nx,ny);
for i = 1: nx
    indx = find(X < x(i));
    if isempty(indx)
        FXY(i,:) = 0;
    else
        for j = 1: ny
            indy = find(Y < y(j));
            if isempty(indy)
                FXY(i,j) = 0;
            else
                subpXY = pXY(indX,indY);
                FXY(i,j) = sum(sum(subpXY));
            end;
        end;
    end;
end;
return;
end
```

The statistics of the table may be obtained from the relative frequencies $\mathbf{pXY} = (p_{ij} : 1 \leq i \leq k, 1 \leq j \leq m)$ as follows:

Listing 1.9. *Statistics*

```
function [covXY,mX,mY,vX,vY] = cont2stat(pXY, X, Y)
%
% generates the statistics of the contingency table
%
% IN:
% X: values of X - type array of double
% Y: values of Y - type array of double
% pXY : relative frequencies of the pair - type array of double
%       PXY(i,j) = frequency of (X(i),Y(j))
%
% OUT:
% covXY : the covariance of the pair - type double
% mY : the mean of Y - type double
% mX : the mean of X - type double
% mY : the mean of Y - type double
% vX : the variance of X - type double
% vY : the variance of Y - type double
%
XX = reshape(X,[1,length(X)]);
```

```
YY = reshape(Y,[length(Y),1]);
mX = sum(XX*pXY); % mean of vector X
vX = sum(((XX - mX).^2)*pXY); % variance of vector X
mY = sum(pXY*YY); % mean of vector X
vY = sum(pXY*(YY - mY).^2); % variance of vector X
covXY = XX*pXY*YY - mX*mY;
```

In some situations, the contingency is not given, but only the values of the pair (X, Y). By assuming that the data is given by a $ns \times 2$ table of values $\mathbf{XYs} = (X_i, Y_i), 1 \leq i \leq ns$ (one line of $\mathbf{XYs}$ contains a value of the pair), the statistics may be obtained as follows:

Listing 1.10. *Statistics from a sample*

```
mX = mean(XYs(:,1)); % mean of X
mY = mean(XYs(:,2)); % mean of Y
C = cov(XYs,1); % covariance matrix of (X, Y)
vX = C(1,1); % variance of X
vY = C(2,2); % variance of Y
covXY = C(1,2); % covariance of X
C0 = cov(XYs); % unbiased estimative of covariance matrix of (X
    , Y)
vX0 = C0(1,1); % unbiased estimative of variance of X
vY0 = C0(2,2); % unbiased estimative of variance of Y
covXY0 = C0(1,2); % unbiased estimative of covariance of X
```

In this case, the cdf of the pair may be obtained by using

Listing 1.11. *Cdf from a sample*

```
function FXY = sample2cdf(XYs,x,y)
%
% determines the cdf of the pair (X,Y) from the
% sample values XYs
%
% IN:
% XYs : table ns x 2 of values of the pair - type array of
    double
%        XYs(i,:) contains one pair (X,Y)
%
% x: abscissa for the calculation of the cdf - type array of
    double
% y: abscissa for the calculation of the cdf - type array of
    double
%
% OUT:
```

```
% FXY : table containing the values of the cdf - type array of
    double
%        FXY(i,j) = P( X < x(i) , Y < y(j) )
%
nx = length(x);
ny = length(y);
FXY = zeros(nx,ny);
ns = size(XYs,1);
for i = 1: nx
    xx = x(i);
    for j = 1: ny
        yy = y(j);
%       s = 0;
%       for k = 1: ns
%           if (XYs(k,1) <= xx && XYs(k,2) <= yy)
%               s = s + 1;
%           end;
%       end;
%       FXY(i,j) = s;
        ind = find(XYs(:,1) < xx & XYs(:,2) < yy);
        FXY(i,j) = numel(ind);
    end;
end;
FXY = FXY/ns;
return;
end
```

In all the situations, the probability density may be evaluated by using a method analogous to the one used in section 1.3:

Listing 1.12. *Pdf from a cdf*

```
function fXY = cdf2pdf(FXY,x,y,h)
%
% determines the density function associated with the
    cumulative function
% FXY evaluated at the abscissa (x,y).
% fXY is determined by particle derivative.
%
% IN:
% FXY: values of the cdf - type array of double
% x: vector of the abscissa - type array of double
% y: vector of the abscissa - type array of double
% h: radius of influence - type double
%     defines how many neighbour points are to derivate
%     for equally spaced data, h = a small (1 to 5) multiple of
    the step
```

```
%                           often 3 is used
%     the larger h is, the smoother the pdf is.
%
% OUT:
% fXY: vector containing the values of the pdf - type array of
      double
%
g = @(y,x)  exp(-0.5*((y-x)/h)^2)/(h*sqrt(2*pi));
dg = @(y,x)  -((y-x)/h)*(1/h)*g(y,x);
%
ny = length(y);
v = zeros(size(y));
for i = 1: ny
    s1 = 0;
    for j = 1: ny
        aux = g(y(j),y(i));
        s1 = s1 + aux;
    end;
    v(i) = s1;
end;
%
fX = zeros(length(x),length(y));
for jjjc = 1: length(y)
    for i = 1: length(x)
        s1 = 0;
        s2 = 0;
        for j = 1: length(x)
            aux = dg(x(j),x(i))*v(j);
            s1 = s1 + aux*(x(j) - x(i));
            s2 = s2 + aux*(FXY(j,jjjc) - FXY(i,jjjc));
        end;
        fX(i,jjjc) = s2/s1;
    end;
end;
%
fXY = zeros(length(x),length(y));
for iiic = 1: length(x)
    for i = 1: length(y)
        s1 = 0;
        s2 = 0;
        for j = 1: length(y)
            aux = dg(y(j),y(i))*v(j);
            s1 = s1 + aux*(y(j) - y(i));
            s2 = s2 + aux*(fX(iiic,j) - fX(iiic,i));
        end;
        fXY(iiic,i) = s2/s1;
    end;
```

```
end ;
return ;
end
```

An example is provided below (Figure 1.5): we generate a random sample of 10000 variates from a Gaussian vector $N(0, \mathbf{Id})$ of dimension 2 and we determine both the cdf and the pdf by using these programs.

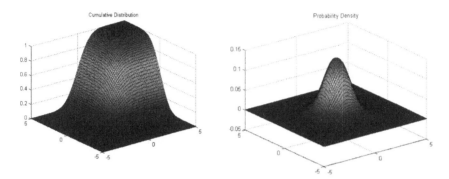

Figure 1.2. *Results for a sample of the Gaussian vector, with $h = 0.3$. For a color version of the figure, see www.iste.co.uk/souzadecursi/quantification.zip*

1.6. Hilbertian properties of the numerical characteristics

Let us introduce the set of the numerical characteristics on Ω:

$$\mathcal{C}(\Omega) = \{X : \Omega \longrightarrow \mathbb{R}\}.$$

The set of the *simple numerical characteristics* on Ω is:

$$\mathcal{V}(\Omega) = \{X \in \mathcal{C}(\Omega) : X(\Omega) \text{ is finite}\}.$$

Both $\mathcal{C}(\Omega)$ and $\mathcal{V}(\Omega)$ are linear spaces (i.e. vector spaces). Since Ω is finite, they coincide: we have $\mathcal{V}(\Omega) = \mathcal{C}(\Omega)$. Later, we will examine more general situations where these two sets do not coincide.

For $X, Y \in \mathcal{V}(\Omega)$, let us define:

$$(X, Y) = E(XY). \qquad [1.3]$$

Then:

LEMMA 1.3.– $(\bullet, \bullet)$ is a scalar product on $\mathcal{V}(\Omega)$. ∎

PROOF.– The definition above corresponds to a bilinear symmetric form on $\mathcal{V}(\Omega)$. Moreover, $(X, X) = E(X^2) \geq 0$ for any $X \in \mathcal{V}(\Omega)$. Finally, if $(X, X) = E(X^2) = 0$, then $X = 0$, and the definition corresponds to a definite-positive form. ∎

Let us denote by $L^2(\Omega)$ the completion of $\mathcal{V}(\Omega)$ for the scalar product given by equation [1.3]: $L^2(\Omega)$ is a Hilbert space for the scalar product [1.3]. The norm of an element $X \in L^2(\Omega)$ is:

$$\|X\| = \sqrt{E(X^2)}. \qquad [1.4]$$

The Hilbertian structure of $L^2(\Omega)$ makes possible, the use of results and methods from Hilbert space theory. We give particular attention to the concept of *orthogonal projection on a linear subspace*:

DEFINITION 1.5.– Let us consider a non-null linear subspace $S \subset L^2(\Omega)$ and $X \in L^2(\Omega)$. The orthogonal projection of X onto S is the element $PX \in S$ such that:

$$PX = \arg\min \{\|X - s\| : s \in S\}. \blacksquare$$

We have:

PROPOSITION 1.8.– If S is closed then PX exists and is uniquely defined. ∎

PROOF.– see [DE 08]. ∎

COROLLARY 1.3.– If S is a finite dimensional linear subspace then PX exists and is uniquely determined. ∎

PROOF.– Since S is finite dimensional, it is closed (see [DE 08]) and the result follows from the preceding one. ∎

PROPOSITION 1.9.– PX is the orthogonal projection of X onto S if and only if $PX \in S$ and $X - PX$ is orthogonal to S, i.e.,

$$PX \in S \quad \text{and} \quad (X - PX, s) = 0, \forall s \in S. \blacksquare$$

PROOF.– see [DE 08]. ■

If we are interested in *vectors of numerical characteristics*, the elements above extend straightly to this situation by considering product spaces. For instance, if we are interested in vectors $\mathbf{X} = (X_1, \ldots, X_k)$, we consider $[\mathcal{V}(\Omega)]^p$ and

$$(\mathbf{X}, \mathbf{Y}) = E(\mathbf{X}.\mathbf{Y}) = \sum_{i=1}^{k} E(X_i Y_i). \qquad [1.5]$$

In this case,

$$\|\mathbf{X}\| = \sqrt{E\left(|\mathbf{X}|^2\right)} = \sqrt{\sum_{i=1}^{k} E(X_i^2)}. \qquad [1.6]$$

All the approaches mentioned afterwards extend to vectors of numerical characteristics in this manner.

1.6.1. *Conditional probability and approximation*

Let $A \subset \Omega$ and the following numerical characteristic:

$$1_A(\omega) = 1, \text{ if } \omega \in A; \ 1_A(\omega) = 0, \text{ if } \omega \notin A.$$

1_A is the *characteristic function* of the subset A and we have:

$$P(1_A = 1) = P(A), \ P(1_A = 0) = P(\Omega - A).$$

Let us consider a second subset $B \subset \Omega$ and denote by 1_B, its characteristic function. We may consider the approximation of 1_B in terms of 1_A: for instance, we may look for the coefficient $\lambda \in \mathbb{R}$ such that $\lambda 1_A$ is the closest possible to 1_B. Thus, we may look for the orthogonal projection of 1_B onto a linear subspace having dimension 1 and given by:

$$S = \{Z \in L^2(\Omega, P) : Z \text{ is constant: } Z(\omega) = s1_A(\omega) \in \mathbb{R}, \forall \omega \in \Omega\}$$

The solution is the orthogonal projection of 1_B onto this linear subspace.

Assume that $P(A) = 0$ (i.e. that A is empty): in this case, 1_A is the null function, i.e. $1_A(\omega) = 0, \forall \omega \in \Omega$. Thus, S is formed by the single element 0: $S = \{0\}$ and the orthogonal projection is 0. We have $\lambda = 0$. Now, assume that $P(A) > 0$, (i.e. that A is non-empty). Recall that the orthogonal projection is:

$$(1_B - \lambda 1_A, s 1_A) = 0, \forall s \in \mathbb{R}.$$

Thus, by taking $s = 1$, we have:

$$\lambda (1_A, 1_A) = (1_B, 1_A).$$

We have:

$$1_A.1_B = 1, \text{ if } \omega \in A \cap B; 1_A(\omega) = 0, \text{ if } \omega \notin A \cap B.$$

Thus, on the one hand, $1_A.1_B = 1_{A \cap B}$ and, on the other hand $1_A.1_A = 1_A$. So, we obtain:

$$\lambda P(A) = P(A \cap B) \Longrightarrow \lambda = \frac{P(A \cap B)}{P(A)}.$$

This value of λ is called the *probability of B conditional to A* or *conditional probability of B with respect to A* and it is denoted by $P(B|A)$. Thus,

$$P(B|A) = \frac{P(A \cap B)}{P(A)} \text{ if } P(A) \neq 0; P(B|A) = 0 \text{ otherwise.} \quad [1.7]$$

In an analogous way, we have:

$$P(A|B) = \frac{P(A \cap B)}{P(B)} \text{ if } P(B) \neq 0; P(A|B) = 0 \text{ otherwise.} *$$

With these definitions, we have:

$$P(A \cap B) = P(B|A).P(A) = P(A|B).P(B). \quad [1.8]$$

A and B are said to be *independent* if and only if $P(A|B) = P(A)$ or $P(B|A) = P(B)$. If A and B are independent, then $P(A \cap B) = P(A).P(B)$.

1.6.2. *Expectation and approximation of a constant*

When we are looking for the best approximation of X of a constant, we may determine the value $m \in \mathbb{R}$ such that:

$$m = \arg\min \{\|X - \lambda\| : \lambda \in \mathbb{R}\}.$$

m is an orthogonal projection onto a linear subspace S having dimension 1:

$$S = \{Z \in L^2(\Omega) : Z \text{ is } constant : Z(\omega) = \lambda \in \mathbb{R}, \forall \omega \in \Omega\}$$

We have:

$$(X - m, \lambda) = 0, \forall \lambda \in \mathbb{R},$$

so that:

$$\lambda E(X) = \lambda m, \forall \lambda \in \mathbb{R} \iff m = E(X).$$

In addition,

$$\|X - m\| = \sqrt{E\left((X - E(X))^2\right)} = \sqrt{V(X)},$$

and the norm of the error of the approximation is the square root of the *variance* of X, i.e. the *standard deviation* of X.

For *vectors of numerical characteristics*, the best approximation of $\mathbf{X} = (X_1, \ldots, X_k)$ by a *constant vector* is $\mathbf{m} = E(\mathbf{X}) = (m_1, \ldots, m_k)$, where $m_i = E(X_i)$. The error in the approximation is $\|\mathbf{X} - \mathbf{m}\| = \sqrt{\sum_{i=1}^{k} E\left((X_i - m_i)^2\right)}$.

The values of the mean and of the variance of data vectors may be easily obtained by using Matlab. By assuming that the data is given by a table $\mathbf{Xs} = (Xs_{ij} : 1 \leq i \leq k, , 1 \leq j \leq ns)$, so that each column of $\mathbf{Xs}$ contains a variate from $\mathbf{X}$, we may use:

Listing 1.13. *Statistics of a data vector*

```
mX = mean(Xs,2); % vector of the means of the rows
vX = var(Xs,1,2); % vector of the variances of the rows
vX0 = var(Xs,0,2); % unbiased estimations of the variances of
     the rows
```

1.6.3. *Linear correlation and affine approximation*

Now, let us consider the best approximation of the numerical characteristic Y by an affine function of the numerical characteristic X. In this case, we may look for the parameters $a, b \in \mathbb{R}$ such that:

$$aX + b = \arg\min \ \{\|Y - Z\| : Z = \alpha X + \beta; \alpha, \beta \in \mathbb{R}\}.$$

Here, the solution is also the orthogonal projection onto a linear subspace S, having dimension 2 and given by:

$$S = \{s \in L^2(\Omega) : s = \alpha X + \beta; \alpha, \beta \in \mathbb{R}\},$$

We have:

$$(Y - aX - b, \alpha X + \beta) = 0, \forall \alpha, \beta \in \mathbb{R}.$$

Let us take successively $(\alpha, \beta) = (1, 0)$ and $(\alpha, \beta) = (0, 1)$ – we obtain:

$$aE(X^2) + bE(X) = E(XY); \quad aE(X) + b = E(Y).$$

The solution of this linear system of two variables is:

$$a = \frac{Cov(X, Y)}{V(X)}; \ b = E(Y) - aE(X).$$

In this case,

$$\|Y - aX - b\| = \sqrt{V(Y)\left(1 - [\rho(X, Y)]^2\right)},$$

where:

$$\rho(X, Y) = \frac{Cov(X, Y)}{\sqrt{V(X)V(Y)}}.$$

$\rho(X,Y)$ is the *linear correlation coefficient* between X and Y. We have $|\rho(X,Y)| \leq 1$ and the error is null if and only if $|\rho(X,Y)| = 1$.

For *vectors of numerical characteristics*, the best approximation of $\mathbf{Y} = (Y_1, \ldots, Y_k)$ by an affine function of $\mathbf{X} = (X_1, \ldots, X_m)$ is given by *multilinear regression*: $\mathbf{Y} = \mathbf{AX} + \mathbf{B}$, where $\mathbf{A} \in \mathcal{M}(k,m)$ and $\mathbf{B} \in \mathcal{M}(k,1)$ verify, for $1 \leq i \leq k$ and $1 \leq r \leq m$,

$$\sum_{j=1}^{m} A_{ij} E(X_j) + B_i = E(Y_i) \quad (k \text{ equations}),$$

$$\sum_{j=1}^{m} A_{ij} E(X_j X_r) + B_i E(X_r) = E(Y_i X_r) \quad (km \text{ equations}).$$

Thus, $\mathbf{A}$ is the solution of the linear system:

$$\sum_{j=1}^{m} A_{ij} Cov(X_j, X_r) = Cov(Y_i, X_r)$$

This system may be solved for a given i by considering $\mathbf{a} = (a_1, \ldots, a_m)^t \in \mathcal{M}(m,1)$ such that $a_j = A_{\bullet j}$, $\mathbf{C} \in \mathcal{M}(m,m)$ such that $C_{rj} = Cov(X_j, X_r)$ and $\mathbf{b} = (b_1, \ldots, b_m)^t \in \mathcal{M}(m,1)$ such that $b_r = Cov(Y_i, X_r)$. Then:

C.a = b.

Once $\mathbf{A}$ has been determined, $\mathbf{B}$ is obtained from the equation:

$$\mathbf{B} = \mathbf{Y} - \mathbf{A}^t.E(\mathbf{X}).$$

1.6.4. *Matlab implementation*

Assuming that the data is given by a $ns \times 2$ table of values $\mathbf{XYs} = (X_i, Y_i)$, $1 \leq i \leq ns$ (one line of $\mathbf{XYs}$ contains a value of the pair), the linear correlation coefficient and the coefficients of the linear approximation of Y as an *affine* function of x may be obtained as follows:

Listing 1.14. *Approximation by an affine function – univariate scalar*

```
mX = mean(XYs(:,1)); % mean of X
mY = mean(XYs(:,2)); % mean of Y
C = cov(XYs,1); % covariance matrix of (X, Y)
```

```
vX = C(1,1); % variance of X
vY = C(2,2); % variance of Y
covXY = C(1,2); % covariance of X
a = covXY/vX; % first coefficient
b = mY - a*mX; % second coefficient
rhoXY = covXY/sqrt(vX*vY); % linear correlation
err = sqrt(vY*(1 - rhoXY^2)); % error in the approximation
```

An example is given (Figure 1.3): we generate a random sample of 250 variates from a Gaussian vector $N(0, \mathbf{Id})$ of dimension 2 and we determine the coefficients. Since the variables X and Y are uncorrelated, the result is a constant equal to zero. In order to obtain correlated data, we consider a new table **XYsort**, generated by sorting the columns of **XYs** in an ascending order. In this case, the variables are correlated and the approximation is better.

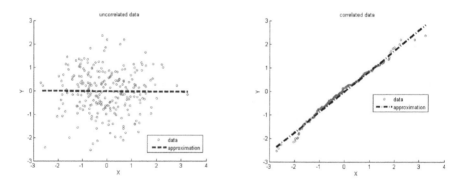

Figure 1.3. *Results for a sample of the Gaussian vector. On the left, the uncorrelated data* **XYs** *leads to a constant. On the right, the results for the correlated data* **XYsort**. *For a color version of the figure, see www.iste.co.uk/souzadecursi/quantification.zip*

Let us consider the situation where Y has to be approximated by an affine function of the vector $\mathbf{X} = (X_1, \ldots, X_m)$. We can say that the data is given by an $ns \times m+1$ table of values $\mathbf{XYs} = (XYs_{ij} : 1 \leq i \leq m+1, 1 \leq j \leq ns)$, so that, for $1 \leq i \leq m$, each column of **XYs** contains a variate from **X**, while the column $m+1$ contains the values of Y (one line of **XYs** contains a value of the pair $(\mathbf{X}, Y)$). In this case, the linear correlation coefficient and the coefficients of the linear approximation of Y by an affine function of **X** are given by the following Matlab program:

Listing 1.15. *Approximation by an affine function - multivariate scalar*

```
function   [a,b,Ye]  =  multivar_scalar(XYs,m)
aux  =  cov(XYs',1);  % covariance  matrix  of  (X,  Y)
C  =  aux(1:m,1:m);  % covariance  matrix  of  X
B  =  aux(1:m,m+1);  % second  member
a  =  C\B;  % vector  of  coefficients
aux  =  mean(XYs,2);  % vector  of  the  means  of  the  rows
mX  =  aux(1:m);
mY  =  aux(m+1);
b  =  mY  -  a'*mX;  % second  coefficient
Ye  =  a'*XYs(1:m,:)  +  b;  %estimation
return;
end
```

Let us illustrate this approach: we generate **X** as a random sample of 250 variates from a Gaussian vector $N(0, \mathbf{Id})$ of dimension 2 and we consider $Y = X_1 - X_2 + 2 + 0.01 * U$, where U is uniformly distributed on $(-1, 1)$. The results are shown in Figure 1.4.

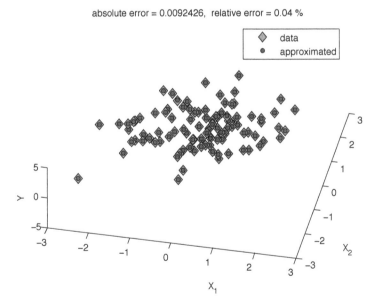

Figure 1.4. *Results for the multivariate approximation of a scalar random variable. For a color version of the figure, see www.iste.co.uk/souzadecursi/quantification.zip*

The approximation of a vector **Y** by an affine function of **X** is made by using this method for each component Y_i. For instance, we may use the following code:

Listing 1.16. *Approximation by an affine function – multivariate scalar*

```
function   [A,B,Ye] = multivar_vector(XYs,m,k)
Ye = zeros(k,ns);
A = zeros(m,k);
B = zeros(k,1);
for i = 1:k
    XXYY = [XYs(1:m,:) XYs(m+i,:)];
    [a,b,YYe] = multivar_scalar(XXYY,m);
    A(:,i) = a;
    B(i) = b;
    Ye(i,:) = YYe;
end;
YYe = A'*XYs(1:m,:);
for i = 1: k
    YYe(i,:) = YYe(i,:) + B(i);
end;
return;
end
```

An example is given in (Figure 1.5): data **X** is a random sample of 250 variates from a Gaussian vector $N(0, \mathbf{Id})$ of dimension 2 and $Y_1 = X_1 - X_2 + 2 + 0.01 * U_1$, $Y_2 = -2X_1 + 3X_2 + 1 + 0.01 * U_2$, where U_1 and U_2 are independent, uniformly distributed on $(-1, 1)$.

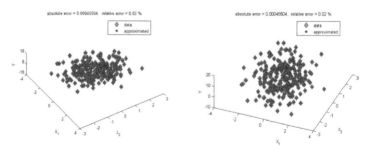

Figure 1.5. *Results for the multivariate approximation of a random vector. For a color version of the figure, see www.iste.co.uk/souzadecursi/quantification.zip*

1.6.5. *Conditional mean and best approximation*

Let us consider the situation where we are looking for an approximation of Y as a generic function of X – no prior expression of the approximation is introduced and we must determine:

$$g(X) = \arg\min \{\|Y - Z\| : Z = \varphi(X); \varphi : \mathbb{R} \longrightarrow \mathbb{R}\}.$$

The solution is the orthogonal projection onto:

$$S = \{s \in L^2(\Omega) : s = \varphi(X); \varphi : \mathbb{R} \longrightarrow \mathbb{R}\}.$$

In this case, we have:

$$(Y - g(X), \varphi(X)) = 0, \forall \varphi : \mathbb{R} \longrightarrow \mathbb{R},$$

i.e.,

$$\sum_{i=1}^{k}\sum_{j=1}^{m} p_{ij}(Y_j - g(X_i))\varphi(X_i) = 0, \forall \varphi : \mathbb{R} \longrightarrow \mathbb{R}. \qquad [1.9]$$

Let us introduce $g_i = g(X_i) \in \mathbb{R}$, $G = (g_1, ..., g_k) \in \mathbb{R}^k$, $\varphi_i = \varphi(X_i) \in \mathbb{R}$, $\Phi = (\varphi_1, ..., \varphi_m) \in \mathbb{R}^m$. Equation [1.9] becomes:

$$\sum_{i=1}^{k}\sum_{j=1}^{m} p_{ij}(Y_j - g_i)\varphi_i = 0, \forall \Phi \in \mathbb{R}^m.$$

Let us consider Φ such that $\varphi_i = 0$ for $i \neq \ell$ and $\varphi_\ell = 1$. We have:

$$\sum_{j=1}^{m} p_{\ell j}(Y_j - g_\ell) = 0 \Longrightarrow g_\ell = \sum_{j=1}^{m} p_{\ell j} Y_j / \sum_{j=1}^{m} p_{\ell j} = \sum_{j=1}^{m} p_{\ell j} Y_j / p_{\ell \bullet}.$$

The function g above defined is the *conditional mean of Y with respect to X* (or *conditional expectation of Y with respect to X*. We use the notations $E(Y|X)$ in order to refer to the numerical characteristic $Z = g(X)$ and $E(Y|X = X_\ell)$ in order to refer to the value g_ℓ. The error in the approximation is $\|Y - E(Y|X)\|$.

We observe that the expression of g_ℓ involves the term $p_{\ell j}/p_{\ell \bullet}$, which defines the *conditional distribution* of Y with respect to X (the expression *distribution of Y conditional to X* may also be found in the literature):

$$P(Y = Y_j \mid X = X_i) = \begin{cases} P(X = X_i, Y = Y_j)/P(X = X_i), \\ \quad \text{if } P(X = X_i) \neq 0 \\ 0, \text{ otherwise.} \end{cases}$$

We have:

$$E(Y \mid X = X_i) = \sum_{j=1}^{m} P(Y = Y_j \mid X = X_i) Y_j.$$

In an analogous way, the *conditional distribution* of X with respect to Y is:

$$P(X = X_i \mid Y = Y_j) = \begin{cases} P(X = X_i, Y = Y_j)/P(Y = Y_j), \\ \quad \text{if } P(Y = Y_j) \neq 0 \\ 0, \text{ otherwise.} \end{cases}$$

and we have:

$$E(X \mid Y = Y_j) = \sum_{i=1}^{k} P(X = X_i \mid Y = Y_j) X_i.$$

The two numerical characteristics are said to be *independent* if and only if:

$$\forall i,j : P(X = X_i \mid Y = Y_j) = P(X = X_i)$$
$$\text{or } P(Y = Y_j \mid X = X_i) = P(Y = Y_j),$$

i.e.,

$$\forall i,j : P(X = X_i, Y = Y_j) = P(X = X_i) P(Y = Y_j).$$

When X and Y are independent, we have $E(Y \mid X) = E(Y)$ and $E(X \mid Y) = E(X)$.

For *vectors of numerical characteristics*, the conditional means are analogously defined: we must simply take into account the existence of several components: the scalars g_i, X_i, Y_j become the vectors $\mathbf{g_i}$, $\mathbf{X_i}$, $\mathbf{Y_j}$, respectively.

1.7. Measure and probability

The preceding ideas generalize to more general universes Ω (for instance, infinite ones, eventually uncountable). The generalization is obtained by using the concept of measure: let us recall $\mathbb{R}_e = \mathbb{R} \cup \{-\infty, +\infty\}$ the set of the "extended real numbers". The elements of $\mathbb{R}_e$ are manipulated according to the following rules:

$$\forall x \in \mathbb{R} : x + (+\infty) = (+\infty) + x = +\infty$$
$$\text{and } x + (-\infty) = (-\infty) + x = -\infty$$

$$\forall x \in \mathbb{R} : x - (+\infty) = -(+\infty) + x = -\infty$$
$$\text{and } x - (-\infty) = -(-\infty) + x = +\infty$$

$$\forall x > 0 : x.(+\infty) = (+\infty).x = +\infty$$
$$\text{and } x.(-\infty) = (-\infty).x = -\infty$$

$$\forall x < 0 : x.(+\infty) = (+\infty).x = -\infty$$
$$\text{and } x.(-\infty) = (-\infty).x = +\infty$$

$$0.(+\infty) = 0.(-\infty) = (+\infty).0 = (-\infty).0 = 0$$
$$(+\infty).(+\infty) = +\infty; \quad (-\infty).(-\infty) = +\infty;$$
$$(-\infty).(+\infty) = (+\infty).(-\infty) = -\infty$$
$$(+\infty) + (+\infty) = +\infty; \quad (-\infty) + (-\infty) = -\infty$$

The expressions $(+\infty) - (+\infty)$, $(-\infty) - (-\infty)$, $(-\infty) + (+\infty)$, $(+\infty) + (-\infty)$ are undetermined. We also use the following order relations:

$$-\infty < +\infty; \quad \forall x \in \mathbb{R} : x < +\infty \text{ and } x > -\infty$$

DEFINITION 1.6 (measure).– Let Ω be a non-empty set and $\mathcal{P}(\Omega)$ be the set of the parts of Ω. A measure on Ω is an application $\mu : \mathcal{P}(\Omega) \longrightarrow \mathbb{R}_e$ such that:

i) μ is positive: $\mu(A) \geq 0, \forall A \subset \Omega$;

ii) μ is countably additive, i.e. additive for any countably disjoint family: $\mu\left(\bigcup_{n \in \mathbb{N}} A_n\right) = \sum_{n \in \mathbb{N}} \mu(A_n), \forall \{A_n\}_{n \in \mathbb{N}} \subset \mathcal{P}(\Omega)$ such that $A_i \cap A_j = \varnothing$ for $i \neq j, \forall i, j \in \mathbb{N}$;

iii) $\mu(\varnothing) = 0$.

In this case, the pair (Ω, μ) is referred to as *measure space*.

We say that μ is a finite measure on Ω when, in addition, $\mu(\Omega) \in \mathbb{R}$, i.e. $\mu(\Omega) < +\infty$. ∎

We have:

PROPOSITION 1.10.– $\mu\left(\bigcup_{i=1}^{n} B_i\right) = \sum_{i=1}^{n} \mu(B_i)$ for any finite disjoint family ($B_i \cap B_j = \varnothing$ for $i \neq j, 1 \leq i, j \leq n$). ∎

PROOF.– Let us consider the family $\{A_n\}_{n \in \mathbb{N}} \subset \mathcal{P}(\Omega)$ given by:

$$A_i = B_i, i \leq n; A_i = \varnothing, i > n.$$

Then $A_i \cap A_j = \varnothing$ for $i \neq j, \forall i, j \in \mathbb{N}$, so that:

$$\mu\left(\bigcup_{i=1}^{n} B_i\right) \underbrace{=}_{A_k = \varnothing, k > n} \mu\left(\bigcup_{k \in \mathbb{N}} A_k\right) = \sum_{k \in \mathbb{N}} \mu(A_k) \underbrace{=}_{\mu(A_k) = 0, k > n} \sum_{i=1}^{n} \mu(B_i). \ \blacksquare$$

COROLLARY 1.4.– Let μ be a measure on Ω. If $A \subset B \subset \Omega$, then $\mu(A) \leq \mu(B)$. ∎

PROOF.– Let us consider $B_1 = A$, $B_2 = B - A$. Then $B = B_1 \cup B_2$ and $B_1 \cap B_2 = \varnothing$, so that:

$$\mu\left(\bigcup_{i=1}^{n} B_i\right) \underbrace{=}_{A_k = \varnothing, k > n} \mu\left(\bigcup_{k \in \mathbb{N}} A_k\right) = \sum_{k \in \mathbb{N}} \mu(A_k) \underbrace{=}_{\mu(A_k) = 0, k > n} \sum_{i=1}^{n} \mu(i). \ \blacksquare$$

PROPOSITION 1.11.– Let μ be a measure on Ω. If $A \subset \Omega$ verifies $\mu(A) < +\infty$, then $\mu(A - B) = \mu(A) - \mu(A \cap B), \forall B \subset \Omega$. ∎

PROOF.– Let us consider $B_1 = A - B$, $B_2 = A \cap B$. Then $A = B_1 \cup B_2$ and $B_1 \cap B_2 = \emptyset$, so that

$$\mu(A) = \mu(B_1 \cup B_2) = \mu(B_1) + \mu(B_2) = \mu(A - B) + \mu(A \cap B).$$

or, $A - B \subset A$ and $A \cap B \subset A$, so that $\mu(A - B) \leq \mu(A) < +\infty$ and $\mu(A \cap B) \leq \mu(A) < +\infty$, that establishes the result. ∎

PROPOSITION 1.12.– Let μ be a measure on Ω and $A, B \subset \Omega$. If one of these sets has a finite measure, then ($\mu(A) < +\infty$ or $\mu(B) < +\infty$), then $\mu(A \cup B) = \mu(A) + \mu(B) - \mu(A \cap B)$. ∎

PROOF.– Let us assume, without loss of generality, that $\mu(A) < +\infty$. Let us also consider $B_1 = A - B$, $B_2 = B$. Then $A \cup B = B_1 \cup B_2$ and $B_1 \cap B_2 = \emptyset$, so that:

$$\mu(A \cup B) = \mu(B_1 \cup B_2) = \mu(B_1) + \mu(B_2) = \mu(A - B) + \mu(B)$$

and the result follows from the preceding proposition. The proof is analogous, if we assume that $\mu(B) < +\infty$ (in such a situation $B_1 = B - A$, $B_2 = A$). ∎

DEFINITION 1.7.– Let μ be a measure on Ω and $A \subset \Omega$. We say that A is a μ–null set if and only if $\mu(A) = 0$. In this case, we say that $x \in A$ holds μ–almost never on Ω or μ–almost nowhere on Ω. Analogously, we say that $x \in A$ holds μ–almost always on Ω or μ–almost everywhere on Ω if and only if $\Omega - A$ is a μ–null set. ∎

In the situations where no confusion is possible, the measure μ is not mentioned and we say simply "null set", "almost always", "almost everywhere", "almost nowhere", "almost never". These expressions are usually abbreviated by using "a.a." for "almost always" , "a.e." for "almost everywhere", "a.n." for "almost never" or "almost nowhere".

We focus on a particular family of measures:

DEFINITION 1.8 (probability).– Let μ be a measure on Ω. We say that μ is a probability on Ω if and only if $\mu(\Omega) = 1$. We also say that μ is the *probability distribution on* Ω and that the measure $\mu(A)$ associated with the subset $A \subset \Omega$ is the *probability* of A, which is denoted by $P(x \in A)$ or simply $P(A)$:

$$P(x \in A) = P(A) = \mu(A). \blacksquare$$

In this case, the pair (Ω, P) is referred to as *probability space*.

Any finite measure ν such that $\nu(\Omega) > 0$ generates a probability by the relation: $\mu(A) = \nu(A) / \nu(\Omega)$.

In probability, we use the terminology *event* to refer to the subsets of Ω that are possible to associate with a probability (the set of events form a σ-algebra, but this concept will be dealt with next): $A \subset \Omega$ is called an *event*. Analogously, the properties *almost everywhere* and *almost nowhere* are often mentioned – for instance, in theorems or proofs – under the following form:

DEFINITION 1.9.– Let P be a probability on Ω and $A \subset \Omega$ an event. We say that A is P–negligible or P–almost impossible if and only if $P(A) = 0$. Reciprocally, we say that A is P–almost sure if and only if $\Omega - A$ is P–negligible (or, equivalently, $P(A) = 1$). $\blacksquare$

As in the case of general measures, the probability P may be dropped and we may simply say "negligible", "almost impossible", "almost sure". These expressions may also be used in an abbreviated form: "a.i." for "almost impossible" and "a.s." for "almost sure".

1.8. Construction of measures

1.8.1. *Measurable sets*

In a finite population, a measure (or a probability) may be easily constructed by giving an individual value for each member of the population and, then, using the additivity property in order to evaluate the measure of an arbitrary subset. For instance, the measure may be equal for any member of the population. In the case of a probability measure on a population formed by N different elements, this leads to the value $1/N$ for any member – what corresponds to a uniform distribution.

When considering more general situations, namely those involving uncountable populations such as, for instance, the set of real numbers, the problem becomes more complex.

In practice, the definition given in the preceding section cannot be used for the constructions of measures, since it requests the definition (i.e. the attribution of the numerical value) of the measure for each subset of Ω: for instance, if we want to use the definition in order to define a measure corresponding to the area on $\mathbb{R}^2$, it becomes necessary to define *a priori* the area of any arbitrary region of $\mathbb{R}^2$ – i.e. we must evaluate the area of a geometrically arbitrary region prior to the definition of what the area of a region is.

For this reason, a more efficient procedure consists of defining the measure for a particular family of subsets $\mathfrak{B} \subset \mathcal{P}(\Omega)$ from Ω – generally of simple geometry, such as, for instance, rectangles – and extending the definition to other parts of Ω by using, on the one hand, the elementary set operations (reunion, intersection, difference and complement) applied to elements of the basic family $\mathfrak{B}$ and, on the other hand, the properties established in the preceding section. Formally, we use a σ–algebra:

DEFINITION 1.10.– $\mathcal{A} \subset \mathcal{P}(\Omega)$ is a σ–algebra on Ω if and only if:

i) $\varnothing \in \mathcal{A}$;

ii) $A \in \mathcal{A} \Longrightarrow \Omega - A \in \mathcal{A}$;

iii) the reunion of any countable family of elements of $\mathcal{A}$ is an element of $\mathcal{A}$: $\{A_n\}_{n \in \mathbb{N}} \subset \mathcal{A} \Longrightarrow \underset{n \in \mathbb{N}}{\cup} A_n \in \mathcal{A}$. ∎

This definition implies that: if $\mathcal{A}$ and $\mathcal{B}$ are σ–algebras on Ω, then $\mathcal{A} \cap \mathcal{B}$ is a σ–algebra on Ω. This property extends to arbitrary collections of $\sigma - algebras$ on Ω: the intersection of a collection of σ–algebras on Ω is a σ–algebra on Ω.

For any Ω, $\mathcal{A} = \mathcal{P}(\Omega)$ is a $\sigma - algebra$ on V. Thus, for any family $\mathfrak{B} \subset \mathcal{P}(\Omega)$ formed by subsets of Ω, there exists at least one σ–algebra on Ω containing $\mathfrak{B}$ (trivially, $\mathcal{A} = \mathcal{P}(\Omega)$ satisfies this condition). Consequently, we may consider the non-empty family formed by the σ–algebras on Ω containing $\mathfrak{B}$. Zorn's lemma (see [DE 08]) shows that this family has a minimal element $\sigma(\mathfrak{B})$: the smallest σ–algebra on Ω containing $\mathfrak{B}$, i.e. the *Borel algebra* $\Sigma(\Omega, \mathfrak{B})$, where the measure will be formally defined, by using an extension based on the concepts of *external measure* and *internal measure*:

DEFINITION 1.11.– Let $\mu : \mathfrak{B} \longrightarrow \mathbb{R}$ be such that:

i) μ is positive: $\mu(A) \geq 0, \forall A \in \mathfrak{B}$;

ii) $\mu(\emptyset) = 0$.

iii) $\mu(\Omega) < +\infty$.

Let $A \subset \Omega$. The external measure of A is:

$$\mu^e(A) = \inf \left\{ \sum_{n \in \mathbb{N}} \mu(B_n) : A \subset \bigcup_{n \in \mathbb{N}} B_n \text{ and } \forall n \in \mathbb{N} : B_n \in \mathfrak{B} \right\}$$

The internal measure of A is:

$$\mu^i(A) = \mu(\Omega) - \mu^e(\Omega - A)$$

We can say that A is measurable if and only if $\mu_i(A) = \mu_e(A)$ and, in this case, we can say that this common value is the measure of A, denoted by $\mu(A)$. ∎

We have:

PROPOSITION 1.13.– Let $\{A_n\}_{n \in \mathbb{N}} \subset \Sigma(\Omega, \mathfrak{B})$. Then $\bigcup_{n \in \mathbb{N}} A_n \in \Sigma(\Omega, \mathfrak{B})$ and $\bigcap_{n \in \mathbb{N}} A_n \in \Sigma(\Omega, \mathfrak{B})$. ∎

PROOF.– $\bigcup_{n \in \mathbb{N}} A_n \in \Sigma(\Omega, \mathfrak{B})$ from the definition of σ-algebra on Ω. Analogously, $\{\Omega - A_n\}_{n \in \mathbb{N}} \subset \Sigma(\Omega, \mathfrak{B})$, so that $\bigcup_{n \in \mathbb{N}} (\Omega - A_n) \in \Sigma(\Omega, \mathfrak{B})$ and, consequently, $\Omega - \bigcup_{n \in \mathbb{N}} (\Omega - A_n) \in \Sigma(\Omega, \mathfrak{B})$. Or,

$$\Omega - \bigcup_{n \in \mathbb{N}} (\Omega - A_n) = \bigcap_{n \in \mathbb{N}} [\Omega - (\Omega - A_n)] = \bigcap_{n \in \mathbb{N}} A_n,$$

which establishes the result. ∎

COROLLARY 1.5.– Let $\{B_i\}_{1 \leq i \leq n} \subset \Sigma(\Omega, \mathfrak{B})$. Then $\bigcup_{i=1}^{n} B_i \in \Sigma(\Omega, \mathfrak{B})$ and $\bigcap_{i=1}^{n} B_i \in \Sigma(\Omega, \mathfrak{B})$. ∎

PROOF.– Let us consider $A_i = B_i$, for $1 \leq i \leq n$ and $A_i = \emptyset$, for $i > n$. Then $\{A_n\}_{n \in \mathbb{N}} \subset \Sigma(\Omega, \mathfrak{B})$, so that $\bigcup_{i=1}^{n} B_i = \bigcup_{n \in \mathbb{N}} A_n \in \Sigma(\Omega, \mathfrak{B})$. Let $A_i = B_i$, for $1 \leq i \leq n$ and $A_i = \Omega$, for $i > n$. Then $\{A_n\}_{n \in \mathbb{N}} \subset \Sigma(\Omega, \mathfrak{B})$, so that $\bigcap_{i=1}^{n} B_i = \bigcap_{n \in \mathbb{N}} A_n \in \Sigma(\Omega, \mathfrak{B})$. ∎

We must observe that this procedure *does not* lead to the attribution of a measure to each part of Ω, but only to those belonging to the "closure" of the family $\mathfrak{B}$ for the elementary set operations: some "pathological" subsets may not belong to those "closures" and are referred to as "not measurable", while the members of the "closure are " – the "measurable" sets. In other words, we ensure that $\Sigma(\Omega, \mathfrak{B}) \subset \mathcal{P}(\Omega)$, but it may happen that $\mathcal{P}(\Omega) \not\subset \Sigma(\Omega, \mathfrak{B})$.

In order to ensure that $\mathcal{P}(\Omega) = \Sigma(\Omega, \mathfrak{B})$, it is necessary to ensure that every open part $A \subset \Omega$ may be generated by a sequence of elementary set operations applied to elements of $\mathfrak{B}$: for instance, $\mathfrak{B}$ must contain a topological basis, i.e. a family of open sets which generates all the open sets contained in Ω by operations of reunion.

The practical construction of a topological basis may be equivalent to the construction of a measure by using its definition: for instance, let us consider $\mathbb{R}^2$: the construction of a topological basis requests the generation of a geometrically arbitrary region by using the elementary set operations of elements of $\mathfrak{B}$, this is equivalent to the definition of the area of a geometrically arbitrary region of $\mathbb{R}^2$.

So, the practical procedure which is most frequently used for the definition of measures consists of using a geometrically simple family $\mathfrak{B}$ (for instance, the set of the rectangles of $\mathbb{R}^2$), *even if this set is not a topological basis*. As previously observed, some "pathological" subsets of Ω are not members of $\Sigma(\Omega, \mathfrak{B})$. These subsets are excluded from the theory below, which is limited to the parts of Ω which are μ–*measurables*:

$$\mathfrak{M}(\mu, \Omega) = \{ A \subset \Omega : \mu^e(A) = \mu^i(A) \}.$$

This exclusion is generally used implicitly: we simply write "$A \subset \Omega$" instead of "$A \in \mathfrak{M}(\mu, \Omega)$". But the reader must keep in mind that the exclusion of some subsets has an impact on some definitions, such as, for instance, null sets, almost sure events and negligible events. Thus, this exclusion has an impact on the whole theory presented below.

Among the remarkable properties of the measures generated by the method above, we underline the following:

PROPOSITION 1.14.– Let $\Omega_1 \subset \Omega_2$, $\mathfrak{B}_1 \subset \mathcal{P}(\Omega_1)$, $\mathfrak{B}_2 \subset \mathcal{P}(\Omega_2)$, $\mathfrak{B}_1 \subset \mathfrak{B}_2$, $\mathcal{B} \in \mathfrak{B}_2 \implies \Omega_1 \cap \mathcal{B} \in \mathfrak{B}_1$, μ_1 associated to $\mathfrak{B}_1$, μ_2 associated to $\mathfrak{B}_2$ such that

$\mu_1 = \mu_2$ on $\mathfrak{B}_1$. Then $\mu_1^e(A) = \mu_2^e(A)$ and $\mu_1^i(A) = \mu_2^i(A)$ for all $A \subset \Omega_1$. ∎

PROOF.– Indeed, it is immediate that $\mu_1^e(A) \geq \mu_2^e(A)$, since $\mathfrak{B}_1 \subset \mathfrak{B}_2$. Moreover, since $A \subset \Omega_1 \subset \Omega_2$,

$$A \subset \bigcup_{n \in \mathbb{N}} \mathcal{B}_n \iff A \subset \bigcup_{n \in \mathbb{N}} (\Omega_1 \cap \mathcal{B}_n)$$

Thus,

$$\sum_{n \in \mathbb{N}} \mu(\mathcal{B}_n) \geq \sum_{n \in \mathbb{N}} \mu(\Omega_1 \cap \mathcal{B}_n) \geq \mu_1(A)$$

and we have also $\mu_2(A) \geq \mu_1(A)$, what establishes the equality between the two measures. In addition, $\Omega_2 - A = (\Omega_2 - \Omega_1) \cup (\Omega_1 - A)$ and we have:

$$\mu_2(\Omega_2 - A) = \mu_2(\Omega_2 - \Omega_1) + \mu_2(\Omega_1 - A)$$

and

$$\mu_2(\Omega_2 - A) = \mu_2(\Omega_2) - \mu_2(\Omega_2 \cap \Omega_1) + \mu_2(\Omega_1 - A)$$

i.e.,

$$\mu_2(\Omega_2 - A) = \mu_2(\Omega_2) - \mu_2(\Omega_1) + \mu_2(\Omega_1 - A)$$

and

$$\mu_2(\Omega_2) - \mu_2(\Omega_2 - A) = \mu_2(\Omega_1) - \mu_2(\Omega_1 - A)$$

Since $\mu_2(\Omega_1) = \mu_1(\Omega_1)$ and $\mu_2(\Omega_1 - A) = \mu_1(\Omega_1 - A)$, we obtain the result. ∎

The properties of the σ–algebras arise from $\mathfrak{M}(\mu, \Omega)$:

1) If $A \in \mathfrak{M}(\mu, \Omega)$ then $\Omega - A \in \mathfrak{M}(\mu, \Omega)$: $\mu(\Omega - A) = \mu(\Omega) - \mu(A)$;

2) If $\{A_n\}_{n \in \mathbb{N}} \subset \mathfrak{M}(\mu, \Omega)$ satisfies $A_i \cap A_j = \emptyset$ for $i \neq j$ then $\bigcup_{n \in \mathbb{N}} A_n \in \mathfrak{M}(\mu, \Omega)$: $\mu\left(\bigcup_{n \in \mathbb{N}} A_n\right) = \sum_{n \in \mathbb{N}} \mu(A_n)$.

3) If $\{A_n\}_{n \in \mathbb{N}} \subset \mathfrak{M}(\mu, \Omega)$ satisfies $A_i \subset A_j$ for $i \leq j$ then $\bigcup_{n \in \mathbb{N}} A_n \in \mathfrak{M}(\mu, \Omega)$: $\mu\left(\bigcup_{n \in \mathbb{N}} A_n\right) = \limsup \mu(A_n)$.

4) If $\{A_n\}_{n \in \mathbb{N}} \subset \mathfrak{M}(\mu, \Omega)$ satisfies $A_i \subset A_j$ for $i \geq j$ then $\bigcap_{n \in \mathbb{N}} A_n \in \mathfrak{M}(\mu, \Omega)$: $\mu\left(\bigcap_{n \in \mathbb{N}} A_n\right) = \liminf \mu(A_n)$.

We have:

PROPOSITION 1.15.– If $\forall\ n\ \in\ \mathbb{N}\ :\ A_n\ \in\ \mathfrak{M}(\mu, \Omega)$ then $\mu\left(\bigcup_{n \in \mathbb{N}} A_n\right) \leq \sum_{n \in \mathbb{N}} \mu(A_n)$. ∎

PROOF.– Since $A_n\ \in\ \mathfrak{M}(\mu, \Omega)$, $\forall\ n\ \in\ \mathbb{N}$, we have $A = \bigcup_{n \in \mathbb{N}} A_n \in \mathfrak{M}(\mu, \Omega)$. Let $B_0 = A_0$ and, for $n \geq 1$, $B_n = A_n - \bigcup_{i=1}^{n-1} B_i$. We have $B_n \in \mathfrak{M}(\mu, \Omega)$, $\forall\ n \in \mathbb{N}$. Moreover, $B_i \cap B_j$ for $i \neq j$ and $A = \bigcup_{n \in \mathbb{N}} B_n$. Thus,

$$\mu(A) = \mu\left(\bigcup_{n \in \mathbb{N}} B_n\right) = \sum_{n \in \mathbb{N}} \mu(B_n).$$

or,

$$\mu(B_n) = \mu\left(A_n - \bigcup_{i=1}^{n-1} B_i\right) \leq \mu(A_n),$$

this establishes the result. ∎

COROLLARY 1.6.– If $\forall n \in \mathbb{N} : \mu(A_n) = 0$ then $\mu\left(\bigcup_{n \in \mathbb{N}} A_n\right) = 0$. ∎

PROOF.– Since $\mu(A_n) = 0$, we have:

$$\mu\left(\bigcup_{n \in \mathbb{N}} A_n\right) \leq \sum_{n \in \mathbb{N}} \mu(A_n) = 0.\ \blacksquare$$

1.8.2. *Lebesgue measure on* $\mathbb{R}^p$

Let $a \in \mathbb{R}^p$ and $b \in \mathbb{R}^p$. The set:

$$\mathcal{R}(a,\ b) = \{x \in \mathbb{R}^p :\ a_n < x_n < b_n, 1 \leq n \leq p\} = \prod_{n=1}^{p} (a_n, b_n)$$

is an open "rectangle" from $\mathbb{R}^p$.

Let us consider $M \in \mathbb{R}$ and

$$\Omega_M = (-M, M)^p = \{x \in \mathbb{R}^p : -M < x_n < M, 1 \leq n \leq p\}.$$

The set of all the open "rectangles" from Ω_M is:

$$\mathfrak{R}(\Omega_M) = \{\mathcal{R}(a, b) : \mathcal{R}(a, b) \subset \Omega_M\}.$$

The measure of an element $\mathfrak{R}(\Omega_M)$ may be defined as (Borel measure):

$$\ell_M(\mathcal{R}(a, b)) = \prod_{n=1}^{p} (b_n - a_n).$$

ℓ_M may be extended to $\Sigma(\Omega_M, \mathfrak{R}(\Omega_M))$ by using the concepts of internal and external measures, as in the preceding section. One of the remarkable properties of ℓ_M is the following:

PROPOSITION 1.16.– Let $N \geq M > 0$ and $A \subset \Omega_M$. Then $\ell_M^e(A) = \ell_N^e(A)$ and $\ell_M^i(A) = \ell_N^i(A)$. ∎

PROOF.– The result follows straightly from the last proposition of the preceding section. ∎

The measure defined by this procedure is referred to as the *Lebesgue measure* and is denoted by ℓ, since it is invariant with respect to M. The set formed by the subsets of $\Omega \subset \mathbb{R}^p$ which are *measurable in the sense of Lebesgue* or *Lebesgue-measurable* is:

$$\mathfrak{M}(\ell, \Omega) = \bigcup_{n=1}^{+\infty} \mathfrak{M}(\ell_n, \Omega).$$

Among the interesting properties of the Lebesgue measure, we may stress the possibility of decomposition into a product of measures – the Borel measure may be written as a *product of interval measures*, such as:

$$\ell_M(\mathcal{R}(m, M)) = \prod_{n=1}^{p} \ell_M((m_n, M_n)).$$

So, the measure on $\mathbb{R}^p$ may be considered as the product of p measures on $\mathbb{R}$. This property is used in the *Fubini's theorem*.

1.8.3. *Equivalent functions*

It is now convenient to introduce one of the fundamental building blocks of the theory under construction: the concept of μ-equivalent functions.

DEFINITION 1.12 (μ-equivalent functions).– Let $f, g : \Omega \longrightarrow \mathbb{R}^p$ be two functions. We say that f and g are μ-equivalent if and only if

$$A = \{\omega \in \Omega : f(\omega) \neq g(\omega)\} \text{ is a } \mu\text{-null set.} \blacksquare$$

This definition introduces an equivalence relation between the functions

$$f \approx_\mu g \iff f \text{ and } g \text{ are } \mu\text{-equivalent.}$$

Indeed,

$$f \approx_\mu f; f \approx_\mu g \implies g \approx_\mu f; f \approx_\mu g \text{ and } g \approx_\mu h \implies f \approx_\mu h.$$

The class of equivalence of f is:

$$[f]_\mu = \{g : \Omega \longrightarrow \mathbb{R}^p : g \approx_\mu f\}.$$

The members of $[f]_\mu$ are usually identified to f, since they may differ only on a null set. For instance – as we can see in the following – all the members of $[f]_\mu$ have the same integral, and this common value becomes the integral of the class $[f]_\mu$ and not only the integral of f.

In the following, we will manipulate some particular sets of functions:

DEFINITION 1.13 (measurable function).– Let $f : \Omega \longrightarrow \mathbb{R}^p$ be a function. We can say that f is μ *–measurable on* Ω if and only if $f^{-1}(\mathfrak{M}(\ell, \mathbb{R}^p)) \subset \mathfrak{M}(\mu, \Omega)$, i.e. if f transforms measurable sets into measurable sets. The set formed by the functions μ–measurable on Ω taking their values $\mathbb{R}^p$ is denoted by:

$$\mathcal{M}(\mu, \Omega, \mathbb{R}^p) = \{f : \Omega \longrightarrow \mathbb{R}^p :$$
$$A \in \mathfrak{M}(\ell, \mathbb{R}^p) \implies f^{-1}(A) \in \mathfrak{M}(\mu, \Omega)\,\}.\blacksquare$$

DEFINITION 1.14 (simple function).– Let $f: \Omega \longrightarrow \mathbb{R}^p$ be a function. We can say that f is a *simple function* or *elementary function* on Ω if and only if:

$$f \in \mathcal{M}(\mu, \Omega, \mathbb{R}^p) \text{ and } f(\Omega) \text{ is finite,}$$

i.e. f is a measurable function taking a finite number of values. The set formed by the simple functions on Ω is:

$$\mathcal{E}(\mu, \Omega, \mathbb{R}^p) = \{f \in \mathcal{M}(\mu, \Omega, \mathbb{R}^p) : f(\Omega) \text{ is finite}\}. \blacksquare$$

For any $e \in \mathcal{E}(\mu, \Omega, \mathbb{R}^p)$, we define:

$$e^*(\Omega) = e(\Omega) - \{0\} = \{\alpha \in \mathbb{R}^p : \alpha \neq 0 \text{ and } \alpha \in e(\Omega)\}.$$

Since e is μ–measurable, we have:

$$\forall y \in e^*(\Omega) : e^{-1}(\{y\}) \text{ is } \mu - \text{measurable.}$$

Eliminating from $e^*(\Omega)$, the terms corresponding to null sets, we have:

$$e^*_\mu(\Omega) = \{\alpha \in e^*(\Omega) : \mu\left(e^{-1}(\{\alpha\})\right) > 0\}.$$

DEFINITION 1.15 (characteristic function of a subset).– Let $A \subset \Omega$. We denote 1_A the characteristic function of A, given by:

$$1_A(\omega) = 1, \text{ if } \omega \in A; \quad 1_A(\omega) = 0, \text{ if } \omega \notin A. \blacksquare$$

The characteristic function of a subset must be distinguished from the characteristic function of a random variable: the last one is the Fourier transform of the probability distribution of the random variable (see section 1.10). Characteristic functions of subsets may be used in order to represent the subsets themselves and have a strong connection to the notion of *conditional probability* (see section 1.11.1.1.). They have some useful properties, such as:

$$1_A.1_B = 1_{A \cap B}, \, \max\{1_A, 1_B\} = 1_A + 1_B - 1_A.1_B = 1_{A \cup B},$$
$$1_A(\Omega) = \{0, 1\}, \, (1_A)^{-1}(\{0\}) = \Omega - A, \, (1_A)^{-1}(\{1\}) = A.$$

PROPOSITION 1.17.– We have:

i) $A \in \mathfrak{M}(\mu, \Omega)$ if and only if $1_A \in \mathcal{M}(\mu, \Omega)$;

ii) If $e \in \mathcal{E}(\mu, \Omega, \mathbb{R}^p)$, $e^*(\Omega)$ has exactly n elements and $e^*(\Omega) = \{\alpha_1, ..., \alpha_n\}$, $A_i = e^{-1}(\{\alpha_i\})$, then $e = \sum_{i=1}^{n} \alpha_i 1_{A_i}$;

iii) If $e \in \mathcal{E}(\mu, \Omega, \mathbb{R}^p)$, $e^*_\mu(\Omega)$ has exactly m elements and $e^*_\mu(\Omega) = \{\beta_1, ..., \beta_m\}_{B_i}$, $B_i = e^{-1}(\{\beta_i\})$, then $e \approx_\mu \sum_{i=1}^{m} \beta_i 1_{B_i}$. ∎

PROOF.–

i) : is an immediate consequence of the properties of the characteristic function: $(1_A)^{-1}(\{0\}) = \Omega - A$ and $(1_A)^{-1}(\{1\}) = A$.

ii) : we have

$$\omega \in \Omega \Longrightarrow \text{either } 1_A(\omega) = 0 \text{ or } \omega \in \bigcup_{i=1}^{n} A_i.$$

By using that $A_i \cap A_j = \varnothing$ for $i \neq j$, we obtain the result.

iii) : is immediate. ∎

When $\Omega \subset \mathbb{R}^n$, we may consider rectangular simple functions:

DEFINITION 1.16 (simple rectangular function).– Let $\Omega \subset \mathbb{R}^n$ and $f \in \mathcal{M}(\mu, \Omega, \mathbb{R}^p)$. We say that f is a *simple rectangular function* on Ω if and only if $f \in \mathcal{E}(\mu, \Omega, \mathbb{R}^p)$ and

$$A = f^{-1}(\{\alpha\}) \in \mathfrak{R}(\Omega), \forall \alpha \in e^*(\Omega),$$

i.e. the inverse image $f^{-1}(\{\alpha\})$ of each non-null α from the image of f is a rectangle. The set formed by the rectangular simple functions on Ω is:

$$\mathcal{E}_R(\mu, \Omega, \mathbb{R}^p) = \{f \in \mathcal{E}(\mu, \Omega, \mathbb{R}^p) : \\ A = f^{-1}(\{\alpha\}) \in \mathfrak{R}(\Omega), \forall \alpha \in e^*(\Omega) \}.\ \blacksquare$$

As we will see in the following, $\mathcal{E}_R(\mu, \Omega, \mathbb{R}^p)$ has an essential rule in the construction of integrals, due to its properties of density, namely in the situation where $\Omega = \mathcal{R}(\boldsymbol{a}, \boldsymbol{b}) \subset \mathbb{R}^n$. When $\Omega = (a, b) \subset \mathbb{R}$ is an interval of real numbers, f is a rectangular simple function if and only if there are $n > 0$, $(f_1, ..., f_n) \in [\mathbb{R}^p]^n$ and n subintervals $A_i = (a_i, b_i) \subset (a, b)$, $1 \leq i \leq n$ such that:

$$f = \sum_{i=1}^{m} f_i 1_{A_i}.$$

In this case, we may consider only simple functions generated by partitions:

DEFINITION 1.17 (partition of an interval).– Let $\Omega = (a, b) \subset \mathbb{R}$. A n–partition of Ω is an element $t = (t_0, ..., t_n) \in \mathbb{R}^{n+1}$ such that:

$$t_0 = a, t_n = b, t_{i-1} < t_i \text{ for } 1 \leq i \leq n.$$

The diameter of the partition is:

$$\delta(t) = \max\{t_i - t_{i-1} : 1 \leq i \leq n\}.$$

The set formed by the n–partitions of Ω is:

$$\mathfrak{Part}_n(\Omega) = \{t = (t_0, ..., t_n) \in \mathbb{R}^{n+1} : t \text{ is a } n\text{-partition of } \Omega\}$$

and the set of the partitions of Ω is:

$$\mathfrak{Part}(\Omega) = \bigcup_{n=1}^{+\infty} \mathfrak{Part}_n(\Omega). \blacksquare$$

DEFINITION 1.18 (simple function defined by a partition).– Let $\Omega = (a, b) \subset \mathbb{R}$ and $f \in \mathcal{M}(\mu, \Omega, \mathbb{R}^p)$. We say that f is a *simple function defined by a partition* if and only if there exists an n–partition $t \in \mathfrak{Part}(\Omega)$ and $(f_1, ..., f_n) \in [\mathbb{R}^p]^n$ such that:

$$t = (t_0, ..., t_n) \in \mathbb{R}^{n+1}, \quad f = \sum_{i=1}^{n} f_i 1_{A_i} \text{ and } A_i = (t_{i-1}, t_i).$$

The set formed by the simple functions defined by partitions is denoted by $\mathcal{E}_P(\mu, \Omega, \mathbb{R}^p)$. $\blacksquare$

1.8.4. *Integrals*

The definition of measures aims generally the manipulation of integrals: the goal is to give a definition to expressions having the form $\int_\Omega f$, where $\Omega \subset \mathbb{R}^p$ is measurable and $f : \Omega \longrightarrow \mathbb{R}$ is a function. Naively, we may consider such an integral as the limit of a sequence of finite sums, which may be defined as:

$$\int_\Omega f \approx \sum_{i=1}^n f(x_i) \ell(A_i); \quad \bigcup_{i=1}^n A_i = \Omega; x_i \in A_i \text{ for } 1 \leq i \leq n$$

and the limit of these finite sums are taken for, on the one hand, $n \longrightarrow +\infty$ and, on the other hand, the maximal diameter of the subsets A_i going to zero.

More formally, we start by the definition of integrals over the set of the *simple functions*. For instance,

DEFINITION 1.19 (integral of a simple function).– For $e \in \mathcal{E}(\mu, \Omega, \mathbb{R})$ such that $e^*(\Omega)$ has exactly n elements and $e^*(\Omega) = \{\alpha_1, ..., \alpha_n\}$, we define:

$$\int_\Omega e\,\mu(dx) = \sum_{i=1}^n \alpha_i \mu(A_i); \ A_i = e^{-1}(\{\alpha_i\}) \text{ for } 1 \leq i \leq n. \ \blacksquare$$

The definition on $\mathcal{E}(\mu, \Omega, \mathbb{R})$ is extended to the set of positive measurable functions:

$$\mathcal{M}_+(\mu, \Omega, \mathbb{R}) = \{m \in \mathcal{M}(\mu, \Omega, \mathbb{R}) : m \geq 0\}$$

as follows: let $f \in \mathcal{M}_+(\mu, \Omega, \mathbb{R})$ and

$$\mathcal{E}_+(\mu, \Omega, f) = \{\text{ and } \in \mathcal{E}(\mu, \Omega, \mathbb{R}) : 0 \leq \text{ and } \leq f\}.$$

$\mathcal{E}_+(\mu, \Omega, f)$ is the set of the positive simple functions which are upper bounded by f. We set:

$$\int_\Omega f\,\mu(dx) = \sup\left\{\int_\Omega e\mu(dx) : e \in \mathcal{E}_+(\mu, \Omega, f)\right\}.$$

The next step consists of considering a function $f \in \mathcal{M}(\mu, \Omega, \mathbb{R})$ having an arbitrary sign. In this case, we use the decomposition

$$f = f^+ - f^-, f^+(x) = \max\{f(x), 0\}, f^-(x) = \max\{-f(x), 0\}$$

and we define

$$\int_\Omega f \, \mu(dx) = \int_\Omega f^+ \, \mu(dx) - \int_\Omega f^- \, \mu(dx).$$

We notice that the right member is well-defined, given that f^+ and f^- are elements of $\mathcal{M}_+(\mu, \Omega, \mathbb{R})$. Sometimes, we will use one of the notations $\int_\Omega f \, \mu \, (x \in dx)$, $\int_\Omega f \, d\mu$ or $\int_\Omega f \, d\mu \, (x)$ instead of $\int_\Omega f \, \mu(dx)$.

This definition leads to the usual properties of integrals. For instance, for $\alpha, \beta \in \mathbb{R}; f, g \in \mathcal{M}(\mu, \Omega, \mathbb{R})$:

$$\int_\Omega (\alpha f + \beta g) \, \mu(dx) = \alpha \int_\Omega f \mu(dx) + \beta \int_\Omega g \mu(dx);$$

$$\int_\Omega f \, \mu(dx) = \sum_{i=1}^n \int_{A_i} f \, \mu(dx),$$

if $\bigcup_{i=1}^n A_i = \Omega$ and $A_i \cap A_j = \varnothing$ for $i \neq j$;

$$f \leq g \; \mu\text{-a.e. on } \Omega \implies \int_\Omega f \, \mu(dx) \leq \int_\Omega g \, \mu(dx);$$

$$\left| \int_\Omega f \, \mu(dx) \right| \leq \int_\Omega |f| \, \mu(dx);$$

$$\left| \int_\Omega f \, \mu(dx) \right| \leq \int_\Omega |f| \, \mu(dx);$$

$$\mu(A) = \int_\Omega 1_A \mu(dx) = \int_A 1 \mu(dx) = \int_A \mu(dx), \; \forall A \in \mathfrak{M}(\mu, \Omega).$$

We have also *Jensen's inequality*: if μ is a probability, $f \in \mathcal{M}(\mu, \Omega, \mathbb{R})$ and $g : \mathbb{R} \longrightarrow \mathbb{R}$ is a continuous *convex* function such that $g(f)$ is measurable, we have:

$$g\left(\int_\Omega f\mu(dx)\right) \leq \int_\Omega g(f)\mu(dx). \qquad [1.10]$$

An immediate consequence of Jensen's inequality is the following:

PROPOSITION 1.18.– $|E(U)|_p \leq E\left(|U|_p\right)$, for all $1 \leq p \leq \infty$ such that $|U|_p$ is measurable. ∎

PROOF.– It results from the application of equation [1.10] to $g(\xi) = |\xi|_p$. ∎

We have:

PROPOSITION 1.19.– Let $f \in \mathcal{M}(\mu, \Omega)$ such that $\forall A \in \mathfrak{M}(\mu, \Omega)$: $\int_A f\,\mu(dx) = 0$. Then $f = 0$ μ–a.e. on Ω. ∎

PROOF.– Let $A = \{\omega \in \Omega : |f| > 0\}$: let us show that $\mu(A) = 0$.

Let $n, k > 0$. Let us consider:

$$A_k = \left\{x \in \Omega : |f(x)| > \frac{1}{k}\right\}$$

and

$$A_{n,k} = \left\{x \in \Omega : n > |f(x)| > \frac{1}{k}\right\}.$$

We have $A = \bigcup_{k \in N} A_k$ and $A_k = \bigcup_{n \in N} A_{n,k}$.

In addition, $\left(\frac{1}{k}, n\right) \subset \mathbb{R}$ is measurable, so that $A_{n,k} = f^{-1}\left(\left(\frac{1}{k}, n\right)\right)$ is measurable. We have:

$$0 = \int_{A_{n,k}} f\,\mu(dx) \geq \frac{1}{k}\int_{A_{n,k}} \mu(dx) = \frac{1}{k}\mu(A_{n,k}).$$

Thus, $\mu(A_{n,k}) \leq 0 \Longrightarrow \mu(A_{n,k}) = 0$. So, $A_k = \bigcup_{n \in N} A_{n,k}$ is a null set, what implies that $A = \bigcup_{k \in N} A_k$ is also a null set. ∎

COROLLARY 1.7.– Let $f \in \mathcal{M}(\mu, \Omega)$ be such that:

$$\forall g \text{ such that } fg \in \mathcal{M}(\mu, \Omega) : \int_\Omega fg\mu(dx) = 0.$$

Then $f = 0$ μ–a.e. on Ω. ∎

PROOF.– Let $A \in \mathfrak{M}(\mu, \Omega)$. We have:

$$\int_A f\mu(dx) = \int_\Omega f 1_A \mu(dx) = 0.$$

Since A is arbitrary, the result follows from the preceding proposition. ∎

COROLLARY 1.8.– Let $f \in \mathcal{M}(\mu, \Omega)$ be such that $f \geq 0$. If $\int_\Omega f\, \mu(dx) = 0$, then $f = 0$ μ–a.e. on Ω. ∎

PROOF.– Let $A \in \mathfrak{M}(\mu, \Omega)$. We have:

$$0 \leq \int_A f\mu(dx) = \int_\Omega f\mu(dx) - \int_{\Omega-A} f\mu(dx) \leq \int_\Omega f\mu(dx) = 0.$$

Since A is arbitrary, the result follows from the preceding proposition. ∎

Finally, we define:

$$\int_\Omega f = \int_\Omega f\, \ell(dx).$$

Thus, whenever the measure is not specified, the integral is with respect to the Lebesgue measure.

In the following, we use the usual notation $L^1(\Omega)$ to refer to the set of the *classes of equivalence* $[f]_\ell$ of the Lebesgue- measurable functions:

$$L^1(\Omega) = \{[f]_\ell : f \in \mathcal{M}(\ell, \Omega, \mathbb{R})\}.$$

Naively, $L^1(\Omega)$ may be considered as the set of the *functions* which are Lebesgue-measurable: we may identify $[f]_\ell$ to f. This identification is justified by the fact that the integral takes the same value for all the members of the class:

$$g \approx_\mu f \implies \int_\Omega g\,\mu(dx) = \int_\Omega f\,\mu(dx),$$

so that we may consider the common value as the value of the integral of the class:

$$\int_\Omega [f]_\mu \mu(dx) = \int_\Omega f\,\mu(dx).$$

In addition, the last result shows that:

$$g \approx_\mu f \iff \int_\Omega |f - g|\,\mu(dx) = 0.$$

In an analogous way, we will denote:

$$L^p(\Omega) = \{f : |f|^p \in L^1(\Omega)\}.$$

As previously observed, the measure on $\mathbb{R}^p$ may be considered as a product of p measures on $\mathbb{R}$. In fact, this is not an ordinary product, but the equality:

$$\int_\mathbb{R} f\,d\ell = \underbrace{\int_\mathbb{R} \int_\mathbb{R} \cdots \int_\mathbb{R}}_{p \text{ times}} f\,d\ell(x_1)\,d\ell(x_2)\,...d\ell(x_p)$$

This decomposition leads to *Fubini's theorem:*

$$\int_{\mathbb{R}^2} f\,d\ell = \int_\mathbb{R} \left[\int_\mathbb{R} f(x,y)\,d\ell(x)\right] d\ell(y) = \int_\mathbb{R} \left[\int_\mathbb{R} f(x,y)\,d\ell(y)\right] d\ell(x).$$

Finally, the definitions above extend to the situation where $f : \Omega \longrightarrow \mathbb{R}^p$ is a function taking its values on $\mathbb{R}^p$ through an evaluation component by component:

$$f = (f_1, ..., f_p) \implies \int f\,\mu(dx) = \left(\int f_1\,\mu(dx), ..., \int f_p\,\mu(dx)\right).$$

1.8.5. *Measures defined by densities*

When $\Omega \subset \mathbb{R}^p$, we may generate measures by using *densities*: naively, a measure μ is defined by a density m when $\mu(x \in dx) = m(x)\,dx$. Such a measure may be considered as a transformation of the Lebesgue measure which generates the same measurable sets. Into a more formal manner,

DEFINITION 1.20.– Let $\Omega \subset \mathbb{R}^p$ and μ be such that $\mathfrak{M}(\mu, \Omega) = \mathfrak{M}(\ell, \Omega)$ (in this case, $\mathcal{M}(\mu, \Omega) = \mathcal{M}(\ell, \Omega)$). We say that μ is defined by the density m if and only if m is a Lebesgue-measurable function and

$$\int_\Omega f\mu(dx) = \int_\Omega fm, \ \forall f \in \mathcal{M}(\mu, \Omega) = \mathcal{M}(\ell, \Omega). \blacksquare$$

The existence of densities is studied by the *Radon–Nikodym theorem*: ℓ must be dominant with respect to μ (i.e. $\ell(A) = 0 \implies \mu(A) = 0$). This theorem will not be studied in this text.

We have

LEMMA 1.4.– If μ is defined by the density m, then:

$$\mu(A) = \int_A m, \ \forall A \in \mathfrak{M}(\ell, \Omega). \blacksquare$$

PROOF.– Let us consider the characteristic function of A: $f = 1_A$. We have $1_A(\Omega) = \{0, 1\}$. We have $1_A \in \mathcal{M}(\ell, \Omega)$ (since $(1_A)^{-1}(\{0\}) = \Omega - A$ and $(1_A)^{-1}(\{1\}) = A$). Thus:

$$\mu(A) = \int_A \mu(dx) = \int_\Omega 1_A \mu(dx) = \int_\Omega 1_A m = \int_A m. \blacksquare$$

PROPOSITION 1.20.– If μ is defined by the density m then $m \geq 0$ μ-a.e. and Lebesgue-a.e. on Ω. $\blacksquare$

PROOF.– Let $A = \{\omega \in \Omega : m(\omega) < 0\}$: let us show that $\mu(A) = \ell(A) = 0$.

Let $n, k > 0$. Let us consider:

$$A_k = \left\{\omega \in \Omega : m(\omega) \leq -\frac{1}{k}\right\}$$

and

$$A_{n,k} = \left\{\omega \in \Omega : -n < m(\omega) < -\frac{1}{k}\right\}.$$

We have $A = \bigcup_{k \in N} A_k$ and $A_k = \bigcup_{n \in N} A_{n,k}$.

$\left(-n, -\frac{1}{k}\right) \subset \mathbb{R}$ is measurable, so that $A_{n,k} = m^{-1}\left(\left(-n, -\frac{1}{k}\right)\right)$ is measurable. We have:

$$0 \leq \mu(A_{n,k}) = \int_{A_{n,k}} m \leq -\frac{1}{k} \int_{A_{n,k}} 1 = -\frac{1}{k}\ell(A_{n,k}).$$

Thus, on the one hand, $\ell(A_{n,k}) \leq 0 \Longrightarrow \ell(A_{n,k}) = 0$ and, on the other hand, $\mu(A_{n,k}) = 0$. So, $A_k = \bigcup_{n \in N} A_{n,k}$ is a null set, which implies that $A = \bigcup_{k \in N} A_k$ is also a null set. ∎

1.9. Measures, probability and integrals in infinite dimensional spaces

Let $m \in \mathbb{R}_e^k$ and $M \in \mathbb{R}_e^k$. An open rectangle of $\mathbb{R}^k$ is denoted as $\mathcal{R}_k(m, M)$:

$$\mathcal{R}_k(m, M) = \left\{x \in \mathbb{R}^k : m_n < x_n < M_n, 1 \leq n \leq k\right\} = \prod_{n=1}^{k}(m_n, M_n).$$

The set formed by the open rectangles of $\mathbb{R}^k$ is:

$$\mathfrak{R}_k = \left\{\mathcal{R}_k(m, M) : m \in \mathbb{R}_e^k \text{ and } M \in \mathbb{R}_e^k\right\}.$$

A closed rectangle is denoted as $\overline{\mathcal{R}}_k(m, M)$ and the set of all the closed rectangles of $\mathbb{R}^k$ is $\overline{\mathfrak{R}}_k$. It may be useful to recall that $\mathbb{R}^k$ may be covered by a finite or countable number of subsets of $\mathfrak{R}_k$ or $\overline{\mathfrak{R}}_k$. For instance, for $\varepsilon > 0$, $\mathcal{R}_k(-\infty, \varepsilon) \cup \mathcal{R}_k(-\varepsilon, +\infty) = \overline{\mathcal{R}}_k(-\infty, 0) \cup \overline{\mathcal{R}}_k(0, +\infty) = \mathbb{R}^k$. More useful coverings may be generated by the translations of a fixed rectangle, in order to get a covering formed by identical rectangles. This property is often used in numerical applications.

We use the notation $\mathbb{R}^\infty$ to refer to the space of the sequences of real numbers:

$$\mathbb{R}^\infty = \{x = (x_1, x_2, x_3, ...) : x_n \in \mathbb{R}, \forall n \in \mathbb{N}^*\}.$$

It may be interesting to consider $\mathbb{R}^\infty$ as a countable Cartesian product:

$$\mathbb{R}^\infty = \mathbb{R} \times \mathbb{R} \times ... = \prod_{n=1}^{+\infty} \mathbb{R}.$$ We use an analogous notation for $\mathbb{R}_e^\infty$. An open rectangle of $\mathbb{R}^k$ is denoted as $\mathcal{R}_\infty(m, M)$, where $m \in \mathbb{R}_e^\infty$ and $M \in \mathbb{R}_e^\infty$:

$$\mathcal{R}_\infty(m, M) = \{x \in \mathbb{R}^\infty : m_n < x_n < M_n, \forall n \in \mathbb{N}^*\} = \prod_{n=1}^{+\infty}(m_n, M_n).$$

The set formed by the open rectangles of $\mathbb{R}^\infty$ is:

$$\mathfrak{R}_\infty = \{\mathcal{R}_\infty(m, M) : m \in \mathbb{R}_e^\infty \text{ and } M \in \mathbb{R}_e^\infty\}$$

As in the finite dimensional situation, $\mathbb{R}^\infty$ may be covered by a finite or countable reunion of elements from $\mathfrak{R}_\infty$. For instance, $\mathcal{R}_\infty(-\infty, \varepsilon) \cup \mathcal{R}_\infty(-\varepsilon, +\infty) = \overline{\mathcal{R}}_\infty(-\infty, 0) \cup \overline{\mathcal{R}}_\infty(0, +\infty) = \mathbb{R}^\infty$. This property is also used in numerical applications.

The notation $\mathbb{R}_0^\infty$ refers to the subset of $\mathbb{R}^\infty$ by the sequences having *only a finite number of non-null elements*, i.e. the sequences for which the set $e(x) = \{n \in \mathbb{N}^*, x_n \neq 0\}$ has a finite number of elements:

$$\mathbb{R}_0^\infty = \{x \in \mathbb{R}^\infty : card(e(x)) = k \in \mathbb{N}\}.$$

Here, $card(\bullet)$ is the cardinality (number of elements). $\mathbb{R}_0^\infty$ may also be identified as $\Pi = \bigcup_{k \in \mathbb{N}^*} \left((\mathbb{N}^*)^k \times \mathbb{R}^k\right)$: let $x \in \mathbb{R}_0^\infty$ and $e(x) = \{n_1, ..., n_k\}$. The application:

$$x = (x_1, x_2, x_3, ...) \xrightarrow{\pi} (n_1, ..., n_k, x_{n_1}, ..., x_{n_k}) \in \Pi$$

is a bijection between $\mathbb{R}_0^\infty$ and Π. Π does not possess open rectangles, given that $(\mathbb{N}^*)^k$ is a discrete space, but an analogous rule is performed by the set:

$$\mathfrak{P} = \bigcup_{k \in \mathbb{N}^*} \left((\mathbb{N}^*)^k \times \mathfrak{R}_k\right)$$

These properties are used in the following.

Let us consider a *real separable Hilbert space* (V, ($\bullet, \bullet$)), i.e. V is a linear space; ($\bullet, \bullet$) is a scalar product on V; V is complete for the norm $\| \bullet \|$ associated with the scalar product ($\|v\| = \sqrt{(v,v)}$); there exists a *countable* subset $\mathcal{S} \subset V$ which is *dense* on ($V, \| \bullet \|$). Since one of our aims is the application of the theory to optimization problems – thus, to the Calculus of Variations – our model for V is a Sobolev functional space: the elements of V are assumed to take their values on $\mathbb{R}^d$ for some $d \in \mathbb{N}^*$. The theory extends to Banach spaces, such as $[L^p(\Omega)]^d$, but this is not our objective.

In order to simplify the notations, $(\bullet,\bullet)$ and $\|\bullet\|$ are used in the following for scalar products and norms on other spaces. No confusion is induced by this simplification. If necessary, the space will be specified by writing $\|\bullet\|_V$.

Since V is separable, this space possesses a *Schauder basis*, i.e. a *countable orthonormal family* $\Phi = \{\varphi_n\}_{n \in \mathbb{N}^*} \subset V$ such that any $v \in V$ may be represented by a unique convergent series as $v = \sum_{n=1}^{+\infty} v_n \varphi_n$. If convenient, we may assume that this basis is *orthonormal*, i.e. a *Hilbertian basis*: every Schauder basis may be transformed into a Hilbertian basis by the Gram–Schmidt procedure of orthonormalization.

Thus, each $v \in V$ is entirely characterized by the sequence of real numbers $\boldsymbol{v} = (v_1, v_2, v_3, ...) \in \mathbb{R}^\infty$ which is uniquely determined and V may be identified to a subset $\mathcal{V}$ of $\mathbb{R}^\infty$, given by $\mathcal{V} = \{\boldsymbol{a} = (a_1, a_2, a_3, ...) : \|\boldsymbol{a}\| < \infty\}$; $\|\boldsymbol{a}\| = \left(\sum_{n=1}^{+\infty} a_n^2\right)^{1/2}$. Let:

$$\mathcal{V} = \{\boldsymbol{a} = (a_1, a_2, a_3, ...) : \|\boldsymbol{a}\| < \infty\}; \|\boldsymbol{a}\| = \left(\sum_{n=1}^{+\infty} a_n^2\right)^{1/2}.$$

Let us consider the application $\boldsymbol{I} : V \longrightarrow \mathcal{V}$ given by:

$$v = \sum_{n=1}^{+\infty} v_n \varphi_n \xrightarrow{I} \boldsymbol{v} = (v_1, v_2, v_3, ...)$$

and

$$(\boldsymbol{x}, \boldsymbol{y}) = \sum_{n=1}^{+\infty} x_n y_n.$$

Since

$$(\boldsymbol{I}(v), \boldsymbol{I}(w)) = (v, w),$$

$\boldsymbol{I}$ is an isometric bijection between V and $\mathcal{V}$; $(\mathcal{V}, (\bullet, \bullet))$, which is a Hilbert space.

This identification leads to the construction of topological basis and measures on V by using those defined on $\mathcal{V}$. For instance, we may use $\mathfrak{B}_\infty = \mathfrak{R}_\infty \cap \mathcal{V}$, formed by the open rectangles of $\mathcal{V}$:

$$\mathfrak{B}_\infty = \{\mathcal{R}_\infty(\boldsymbol{m}, \boldsymbol{M}) : \boldsymbol{m} \in \mathcal{V} \text{ and } \boldsymbol{M} \in \mathcal{V}\} \subset \mathfrak{R}_\infty.$$

The finite dimensional version $\mathfrak{B}_k$ of $\mathfrak{B}_\infty$ associated to $\mathbb{R}^k$ is simply $\mathfrak{B}_k = \mathfrak{R}_k$.

For practical purposes, such as numerical approximations, we also consider a *countable family* $\Psi = \{\psi_n\}_{n \in \mathbb{N}} \subset V$ such that its linear span,

$$\mathcal{D} = [\Psi] = \left\{ \sum_{i=1}^{k} a_{n_i} \psi_{n_i} : k \in \mathbb{N}^*, a_{n_i} \in \mathbb{R}, \text{ for } 1 \leq i \leq k \right\}$$

is *dense* in V, i.e.,

$$\forall v \in V : \forall \varepsilon > 0 : \exists v_\varepsilon \in \mathcal{D} \text{ such that } \|v - v_\varepsilon\| \leq \varepsilon.$$

$\mathcal{D}$ may be identified to a subset $\mathcal{V}_0$ of $\mathcal{V}$ formed by the sequences $\boldsymbol{a}$ containing only a finite number of non-null elements, i.e.,

$$\mathcal{V}_0 = \{\boldsymbol{a} \in \mathcal{V} : \boldsymbol{a} \in \mathbb{R}_0^\infty\}.$$

Thus, $\boldsymbol{I} \circ \pi$ is a bijection between $\mathcal{D}$ and $\mathfrak{V}_0 = \pi(\mathcal{V}_0)$. This property may be used in order to construct measures on $\mathcal{D}$. Moreover, the introduction of the family Ψ leads to an extension of the results to separable Banach spaces.

Let $(W, (\bullet, \bullet))$ be another Hilbert space. (W is also a functional space and the elements of W are assumed to take their values on $\mathbb{R}^d$ for some $d \in \mathbb{N}^*$, analogous to V).

We denote by $\mathcal{L}(V, W)$ the set of all the linear continuous maps $\ell : V \longrightarrow W$:

$$\mathcal{L}(V, E) = \{\ell : V \longrightarrow W : \ell \text{ is linear continuous}\}.$$

For $\ell \in \mathcal{L}(V, W)$,

$$\|\ell\| = \sup\{\|\ell(v)\| : \|v\| \leq 1\} < \infty.$$

Thus, there exists a real number $M \in \mathbb{R}$ such that:

$$\forall v \in V : \|\ell(v)\| \leq M \|v\|. \qquad [1.11]$$

When $W = \mathbb{R}^d$, there exists a real number $M_p \in \mathbb{R}$ such that:

$$\forall v \in V : |\ell(v)|_p \leq M_p \|v\|. \qquad [1.12]$$

For a subset $A \subset V$, we denote $\mathcal{P}(A)$, the power set of A, i.e., the set of all the subsets of A.

1.9.1. *Finite measures on infinite dimensional Hilbert spaces*

The most common way for the manipulation of finite measures on general Banach or Hilbert spaces is the use of *cylindrical measures* [SCH 69]. Naively, a *cylindrical set* of V is a subset of a finite dimensional subspace of V and a *cylindrical measure* is a measure defined on the cylindrical sets of V, i.e. an application $\mu : \mathfrak{C} \longrightarrow \mathbb{R}_e$, where $\mathfrak{C} \subset \mathcal{P}(V)$ is the set of all the cylindrical sets of V. More formally, we consider $k \in \mathbb{N}$ and we define $\mathfrak{C}$ as the reunion of all the inverse images of measurable parts of $\mathbb{R}^k$ for the elements of $\mathcal{L}(V, \mathbb{R}^k)$ and $k \in \mathbb{N}$:

$$\mathfrak{C} = \{C : C = \ell^{-1}(A), \ell \in \mathcal{L}(V, \mathbb{R}^k), k \in \mathbb{N}, \\ A \text{ measurable part of } \mathbb{R}^k\} \subset \mathcal{P}(V).$$

Then, a *cylindrical measure* is an application $\mu: \sigma(\mathfrak{C}) \longrightarrow \mathbb{R}_e$ such that $\mu_\ell = \mu \circ \ell^{-1}$ is a measure on $\mathbb{R}^k$, $\forall \ell \in \mathcal{L}(V, k)$, $k \in \mathbb{N}$(so, $\mu_\ell(A) = \mu(\ell^{-1}(A))$, A measurable subset of $\mathbb{R}^k$). The restriction of a Radon measure to $\mathfrak{C}$ defines a cylindrical measure: in the classical analysis, this procedure is the most usual manner for the manipulation of measures in infinite dimensional spaces.

Here, we will adopt a different approach: as previously observed, measures on $\mathbb{R}$ are defined by using Borel algebras and families of geometrically simple subsets: $\mathfrak{B}_1$: $\Sigma(\mathbb{R}) = \sigma(\mathfrak{B}_1)$. In practice, we may define a measure ν on $\mathbb{R}$ by using a density f, i.e. a function $f: \mathbb{R} \longrightarrow \mathbb{R}$ such that $f \geq 0$ on $\mathbb{R}$, $\int_\mathbb{R} f < \infty$ and $d\nu = f(x)\,dx$. In this case, we may consider:

$$\nu((a, b)) = \int_a^b f.$$

When considering probabilities, f is the probability density and $\int_\mathbb{R} f = 1$. In this case,

$$P(dx) = P(x \in dx) = d\nu = f(x)\,dx.$$

This procedure extends directly to $\mathbb{R}^k$ by using $\mathfrak{R}_k$: $\Sigma(\mathbb{R}^k) = \sigma(\mathfrak{R}_k)$. A simple way in order to generate measures on $\mathbb{R}^k$ consists of the use of a product of measures: let $(\nu_1, ..., \nu_k)$ be measures on $\mathbb{R}$. Then,

$$\nu(\mathcal{R}_k(\boldsymbol{m}, \boldsymbol{M})) = \prod_{n=1}^{k} \nu_n((m_n, M_n))$$

defines a measure on $\mathbb{R}^k$. Analogous to $\mathbb{R}$, ν may be defined by a density f, i.e. by a function $f: \mathbb{R}^k \longrightarrow \mathbb{R}$ such that $f \geq 0$ on $\mathbb{R}^k$, $\int_{\mathbb{R}^k} f < \infty$ and $d\nu = f(\boldsymbol{x})\,dx_1 ... dx_k$. Thus,

$$\nu(\mathcal{R}_k(\boldsymbol{m}, \boldsymbol{M})) = \int_{\mathcal{R}_k(\boldsymbol{m}, \boldsymbol{M})} f.$$

Analogous to the one-dimensional situation, when considering probabilities, f is the probability density and $\int_{\mathbb{R}^k} f = 1$. Thus,

$$P(d\boldsymbol{x}) = P(\boldsymbol{x} \in d\boldsymbol{x}) = d\nu = f(\boldsymbol{x})\, dx_1...dx_k.$$

Densities may also be generated by using products: if, for $1 \leq n \leq k$, $d\nu_n = f_n(t)\, dt$, then $f = \prod_{n=1}^{k} f_n$ and

$$\nu(\mathcal{R}_k(\boldsymbol{m}, \boldsymbol{M})) = \prod_{n=1}^{k} \left(\int_{m_n}^{M_n} f_n(x_n)\, dx_n \right) = \int_{\mathcal{R}_k(\boldsymbol{m}, \boldsymbol{M})} f.$$

This procedure extends also to $\mathbb{R}^\infty$: "naively", a sequence $\{\nu_n\}_{n \in \mathbb{N}}$ of measures on $\mathbb{R}$ may be used in order to generate a measure ν on $\mathbb{R}^\infty$ by using $\nu(A_1 \times A_2 \times ...) = \nu_1(A_1)\nu_2(A_2)\,...$. More formally, we use $\mathfrak{R}_\infty$:

$$\nu(\mathcal{R}_\infty(\boldsymbol{m}, \boldsymbol{M})) = \prod_{n=1}^{+\infty} \nu_n(\,(m_n, M_n)\,).$$

defines a measure on $\mathcal{V}$. As previously observed, the use of densities may be more convenient for practical purposes: if $d\nu_n = f_n(t)\, dt$, $\forall\, n \in \mathbb{N}$, then:

$$\nu_n(\,(m_n, M_n)\,) = \int_{m_n}^{M_n} f_n(x_n)\, dx_n$$

and

$$\nu(\mathcal{R}_\infty(\boldsymbol{m}, \boldsymbol{M})) = \prod_{n=1}^{+\infty} \left(\int_{m_n}^{M_n} f_n(x_n)\, dx_n \right).$$

"Naively", this procedure corresponds to the construction of a density f: $\mathbb{R}^\infty \longrightarrow \mathbb{R}$ such that $d\nu = f(\boldsymbol{x})\, dx_1 dx_2\, ...$, i.e. to a density

$$f(\boldsymbol{x}) = \prod_{n=1}^{+\infty} f_n(x_n).$$

An analogous approach may be used in order to construct finite measures on Π: for each $k \in \mathbb{N}^*$, let us consider a finite measure ν_k on $\mathbb{R}^k$ and k summable sequences of strictly positive real numbers having a sum inferior to 1:

$$\mathbf{q}_k = (\mathbf{q}_{k,1}, ..., \mathbf{q}_{k,k}) \in (\mathbb{R}^\infty)^k, \ \mathbf{q}_{k,i} = (q_{k,i,1}, q_{k,i,2}, q_{k,i,3}, ...),$$
$$q_{k,i,n} \geq 0, \text{ for } k, i, n \in \mathbb{N}^*; \ \sum_{n=1}^{+\infty} q_{k,i,n} = Q_{k,i} \leq 1.$$
[1.13]

Let us assume that ν_k is a finite measure:

$$\nu_k\left(\mathbb{R}^k\right) = N_k \leq 1$$

Let $(n_1, ..., n_k) \times \mathcal{R}_k(\mathbf{m}, \mathbf{M}) \in (\mathbb{N}^*)^k \times \mathfrak{R}_k$ be an element of $\mathfrak{P}$ and a summable sequence of strictly positive real numbers:

$$\mathbf{p} = (p_1, p_2, p_3, ...) \in \mathbb{R}^\infty; \ p_n \geq 0, \forall n \in \mathbb{N}^*; \ \sum_{n=0}^{+\infty} p_n = p \geq 0 \qquad [1.14]$$

We define:

$$\eta\left((n_1, ..., n_k) \times \mathcal{R}_k(\mathbf{m}, \mathbf{M})\right) = p_k \, \nu_k\left(\mathcal{R}_k(\mathbf{m}, \mathbf{M})\right) \prod_{i=1}^{k} q_{k,i,n_i}.$$

Then η is a finite measure on $\mathfrak{P}$. As usual, we may use densities instead of measures:

$$\eta\left((n_1, ..., n_k) \times \mathcal{R}_k(\mathbf{m}, \mathbf{M})\right) = p_k \prod_{i=1}^{k} q_{k,i,n_i} \left(\int_{m_i}^{M_i} f_{k,i,n_i}(x_{n_i}) \, dx_{n_i} \right),$$

where:

$$f_{k,i,n} \geq 0 \text{ on } \mathbb{R} \text{ and } \int_{\mathbb{R}} f_{k,i,n} \leq 1, \text{ for } k, i, n \in \mathbb{N}^*. \qquad [1.15]$$

This procedure generates a finite measure on $\mathbb{R}_0^\infty$: $\nu = \eta \circ \pi$ (i.e. $\nu(A) = \eta(\pi(A))$) is a finite measure on $\mathbb{R}_0^\infty$.

When V has a basis, the identification between V and $\mathcal{V} \subset \mathbb{R}^\infty$ may be combined with this procedure: if ν is a finite measure on $\mathbb{R}^\infty$, then $\mu = \nu \circ I$ (i.e. $\mu(A) = \nu(I(A))$) is a finite measure on V. Analogously, we may consider $\mu = \eta \circ \pi \circ I$ (i.e. $\mu(A) = \eta(\pi(I(A)))$): μ is a finite measure on $[\mathcal{D}]$.

An alternative measure may be defined as follows: let us consider the situation where the space is $\mathbb{R}^k$. Let $p = (p_1, ..., p_k)$ be a vector of positive real numbers ($p_n \geq 0$, for $1 \leq n \leq k$) and $f = (f_1, ..., f_k)$ be a bounded set of positive functions which are integrable on $\mathbb{R}$ (i.e. $f_n \geq 0$ on $\mathbb{R}$ for $1 \leq n \leq k$ and there exists $\Lambda \in \mathbb{R}$ such that $\int_\mathbb{R} f_n \leq \Lambda$, for $1 \leq n \leq k$). We may use p and f in order to define a finite measure on $\mathbb{R}^k$:

$$\nu(\mathcal{R}_k(m, M)) = \sum_{n=1}^{k} p_n \left(\int_{m_n}^{M_n} f_n(x_n) \, dx_n \right).$$

Indeed, ν is a finite measure on $\mathbb{R}^k$: $\nu \geq 0$, $\nu(\varnothing) = 0$, ν is countably additive, $\nu(\mathbb{R}^k) \leq \Lambda \sum_{n=1}^{k} p_n < \infty$.

This alternative approach may be extended to $\mathbb{R}^\infty$ by using a sequence p verifying [1.14] and a bounded sequence $f = (f_1, f_2, f_3, ...)$ of positive integrable functions on $\mathbb{R}$ (i.e. $\forall n \in \mathbb{N}: f_n \geq 0$ on $\mathbb{R}$ and there exists $\Lambda \in \mathbb{R}$ such that $\int_\mathbb{R} f_n \leq \Lambda, \forall n \in \mathbb{N}$). In this case, we set:

$$\nu(\mathcal{R}_\infty(m, M)) = \sum_{n=1}^{+\infty} p_n \left(\int_{m_n}^{M_n} f_n(x_n) \, dx_n \right).$$

This series converges when $\mathcal{R}_\infty(m, M) \in \mathfrak{R}_\infty$: for instance, we have:

$$\left| \sum_{n=k}^{+\infty} p_n \int_{m_n}^{M_n} f_n(x_n) \, dx_n \right| \leq \sum_{n=k}^{+\infty} p_n \left| \int_{m_n}^{M_n} f_n(x_n) \, dx_n \right| \leq$$

$$\Lambda \sum_{n=k}^{+\infty} p_n \longrightarrow 0, \text{ for } k \longrightarrow +\infty.$$

Analogous to the finite dimensional situation, where the space is $\mathbb{R}^k$, ν is a finite measure on $\mathbb{R}^\infty$: $\nu \geq 0$, $\nu(\varnothing) = 0$, ν is countably additive, $\nu(\mathbb{R}^\infty) \leq p\Lambda < \infty$.

This procedure may be applied to $\mathbb{R}_0^\infty$ into an analogous way: let us assume that p satisfies [1.14], q_k verifies [1.13], $f_{k,i,n}$ satisfies [1.15] and

$$\sum_{n=1}^{+\infty} q_{k,i,n} = Q_{k,i}, \quad \sum_{i=1}^{k} Q_{k,i} = Q_k \leq \Lambda \in \mathbb{R},$$

where Λ is independent of k. For $(n_1, ..., n_k) \times \mathcal{R}_k(\boldsymbol{m}, \boldsymbol{M}) \in (\mathbb{N}^*)^k \times \mathfrak{R}_k \subset \mathfrak{P}$, we define:

$$\eta((n_1, ..., n_k) \times \mathcal{R}_k(\boldsymbol{m}, \boldsymbol{M})) = p_k \sum_{i=1}^{k} q_{k,i,n_i} \left(\int_{m_i}^{M_i} f_{k,i,n_i}(x_{n_i}) dx_{n_i} \right).$$

Then η is a finite measure on $\mathfrak{P}$: $\eta \geq 0$, $\eta(\varnothing) = 0$, η is countably additive, $\eta(\mathfrak{P}) \leq p\Lambda < \infty$. Thus, $\nu = \eta \circ \pi$ is a finite measure on $\mathbb{R}_0^\infty$.

Analogous to the preceding situation, $\mu = \nu \circ \boldsymbol{I}$ is a finite measure on V – when V has a basis – and $\mu = \eta \circ \pi \circ \boldsymbol{I}$ is a finite measure on $[\mathcal{D}]$.

1.9.2. *Integration involving a measure on an infinite dimensional Hilbert space*

As previously remarked, our objective is the application of the theory to problems of calculus of variations: so, V and W are real functional spaces such as, for instance, $[L^p(\Omega)]^n$ or the standard Sobolev spaces, and their elements take their values on $\mathbb{R}^k$ for some $k > 0$. In this case, the extension of the preceding ideas, namely of the integrals defined for finite dimensional situations are quite natural. For instance, we are interested in the evaluation of the mean value $E(U)$ on $S \subset V$ for a random variable U defined on V, i.e. the mean value on S of a function $U : V \longrightarrow W$, associating with each $v \in V$ an element $U(v) \in W$. We have:

$$E(U) = \int_S U d\mu \Big/ \mu(S)$$

The evaluation of $\int_S U(v)\,d\mu$ involves a non-standard integration, since μ is defined on an infinite dimensional space. Nevertheless, the formal definition is entirely analogous to the definition introduced for the finite dimensional situation: for instance, we initially restrict ourselves to the situation where the elements of W are either real numbers or real functions (i.e. $W = \mathbb{R}$ or $W = L^p(\Omega)$).

Analogous to the finite dimensional case, let $\mathcal{M}$ be the set of the *measurable functions*, which are the functions which transform measurable sets into measurable sets. We consider the set of *measurable finite range functions*: $\mathcal{F} = \{f \in \mathcal{M} : f(V) \text{ is a finite set }\}$. For $F \in \mathcal{F}$, we define $R^*(f, S) = \{\alpha \in \mathbb{R} : \alpha \neq 0 \text{ and } \alpha \in f(S)\}$. The set of *elementary functions* is:

$$\mathcal{E}(\mu, S) = \{e \in \mathcal{F} : \mu\left(e^{-1}(\{y\})\right) < \infty, \forall y \in R^*(e, S)\}$$

If μ is a *finite* measure, we have $\mathcal{E}(\mu, S) = \mathcal{F}$. For $e \in \mathcal{E}(\mu, S)$ such that $R^*(e, S) = \{\alpha_1, ..., \alpha_n\}$, the integral is defined as:

$$\int_S e\,d\mu = \sum_{i=1}^n \alpha_i \mu(A_i)\,;\, A_i = e^{-1}(\{\alpha_i\}) \text{ for } 1 \leq i \leq n.$$

This definition is extended to the set of positive measurable functions $\mathcal{M}_+ = \{m \in \mathcal{M} : m \geq 0\}$ as follows: let $U \in \mathcal{M}_+$. We consider the positive elementary functions upper bounded by U: $\mathcal{E}_+(\mu, S, U) = \{e \in \mathcal{E}(\mu, S) : 0 \leq e \leq U\}$ and we define:

$$\int_S U\,d\mu = \sup\left\{\int_S e : e \in \mathcal{E}_+(\mu, S, U)\right\}.$$

Finally, an arbitrary $U \in \mathcal{M}$ is decomposed as $U = U^+ - U^-$, $U^+(x) = \max\{U(x), 0\}$, $U^-(x) = \max\{-U(x), 0\}$ and we define:

$$\int_S U\,d\mu = \int_S U^+\,d\mu - \int_S U^-\,d\mu.$$

The right member is well-defined, since both U^+ and U^- are elements of $\mathcal{M}_+$. The definition yields the classical inequality: $\left|\int_S U\,d\mu\right| \leq \int_S |U|\,d\mu$,, which corresponds to a particular form of Jensen's inequality: for a *convex* function $g : \mathbb{R} \longrightarrow \mathbb{R}$ such that $g(U)$ is measurable, we have $g(E(U)) \leq E(g(U))$.

The extension to the situation where the elements of W are either real vectors or real vector-valued functions (i.e. $W = \mathbb{R}^d$ or $W = [L^p(\Omega)]^d$) is performed by considering the components of U: we have $U = (U_1, ..., U_d)$ and we define:

$$\int_S U d\mu = \left(\int_S U_1 d\mu, ..., \int_S U_d d\mu \right).$$

Let $g : \mathbb{R}^d \longrightarrow \mathbb{R}$ be a continuous *convex* function such that $g(U)$ is measurable. Since g is the pointwise supremum of all its affine minorants [TIE 84], we have Jensen's inequality:

$$g\left(\int_S U d\mu \right) \leq \int_S g(U) d\mu.$$

and also:

$$|E(U)|_p \leq E\left(|U|_p \right), \text{ for all } 1 \leq p \leq \infty \text{ such that } |U|_p \text{ is measurable.}$$

As usual, the integral of $U : V \longrightarrow W$ on $S \subset V$ may be approximated by:

$$\int_S U d\mu \approx \sum_{i=1}^{n} U(x_i) \mu(A_i) ; \bigcup_{i=1}^{n} A_i = S; x_i \in A_i \text{ for } 1 \leq i \leq n \quad [1.16]$$

and corresponds to a limit for $n \longrightarrow +\infty$, with the maximal diameter of the subsets A_i going to zero. In our numerical calculations, we *do not* use the approximation [1.16], but a *Monte Carlo approximation*: we generate a sample $U_1, ..., U_{nr}$ of nr variates from the distribution μ and we approximate:

$$E(U) \approx \frac{1}{nr} \sum_{i=1}^{nr} U_i. \qquad [1.17]$$

This approach is more convenient for the situation under consideration: when using standard distributions, such as, for instance, Gaussian or Poissonian distributions, efficient methods of generation exist in the literature.

1.10. Random variables

Let us consider Ω non-empty and a probability P on Ω (i.e. a *probability space* (Ω, P)) (since the σ-algebra is always the Borel algebra we do not mention it explicitly, but of course the probability is defined for the Borel sets).

A *random variable* X on Ω is a *measurable* function $X : \Omega \longrightarrow \mathbb{R}$. It may be seen as a generalization of the concept of numerical characteristic. A particular value of X is called a *realization of X* or a *variate from the distribution of X* (or simply a *variate from X*). A variate may be considered as the value $X(\omega)$ for a particular ω.

The image of X is $I = X(\Omega) \subset \mathbb{R}$. For any part $J \subset \mathbb{R}$, we set:

$$P(X \in J) = P(X^{-1}(J)).$$

Since X is measurable, the probability above is defined for every measurable part of $\mathbb{R}$ – namely, for the intervals of $\mathbb{R}$. Thus, we may characterize the global behavior of X by using its *cumulative function*, i.e.,

$$F(x) = P(X < x) = P(X \in (-\infty, x)).$$

In this case, we say that X *follows the distribution F* or X *has the distribution F*. We have

PROPOSITION 1.21.– Let F be the cumulative function of X. Then:

i) $0 \leq F(x) \leq 1, \forall\, x \in \mathbb{R}$;

ii) F is monotonically increasing: $a \leq b \Longrightarrow F(a) \leq F(b)$;

iii) $P(X \geq x) = 1 - F(x)$ and $P(a \leq X < b) = F(b) - F(a)$;

iv) $\lim\limits_{x \longrightarrow +\infty} F(x) = 1$ and $\lim\limits_{x \longrightarrow -\infty} F(x) = 0$;

v) F is left-continous: $F(x-) = F(x)$ $\left(F(x-) = \lim\limits_{h \longrightarrow 0+} F(x-h)\right)$;

vi) F is a bounded variation function and its total variation is 1;

vii $P(X = x) = F(x+) - F(x-)$ $\left(F(x+) = \lim\limits_{h \longrightarrow 0+} F(x+h)\right)$. ∎

PROOF.–

i) Since P is a probability and $F(x) = P(X < x)$, we have $0 \leq F(x) \leq 1$.

ii) Let $a, b \in \mathbb{R}$ such that $a \leq b$. Then:

$$F(b) = P(X \in (-\infty, b)) = P(X \in (-\infty, a) \cup [a, b)),$$

so that:

$$F(b) = \underbrace{P(X \in (-\infty, a))}_{F(a)} + P(X \in [a, b)) = F(a) + \underbrace{P(X \in [a, b))}_{\geq 0}$$

and we have $F(b) \geq F(a)$.

iii) Since,

$$F(b) = F(a) + P(X \in [a, b)),$$

we have:

$$P(X \in [a, b)) = F(b) - F(a).$$

In addition,

$$P(X \in (-\infty, +\infty)) = 1,$$

so that:

$$\underbrace{P(X \in (-\infty, x))}_{F(x)} + P(X \in [x, +\infty)) = 1$$

we have:

$$P(X \geq x) = P(X \in [x, +\infty)) = 1 - F(x).$$

iv) Since F is increasing and $0 \leq F(x) \leq 1, \forall x \in \mathbb{R}$, there exists $a \in \mathbb{R}$ such that:

$$\forall n \in \mathbb{N} : 0 \leq \lim_{x \longrightarrow +\infty} F(x) = a \leq 1.$$

Let us assume that $a < 1$: we consider $a_n = F(n) = P(X < n)$. Then:

$a_n \longrightarrow a$ when $n \longrightarrow +\infty$ and $\forall n \in \mathbb{N}: 0 \leq a_n \leq 1$.

Let $b_n = F(n+1) - F(n) = P(n \leq X < n+1)$. Then:

$\forall n \in \mathbb{N}: 0 \leq b_n \leq 1$

and

$$\forall n \in \mathbb{N}: 1 - a_n = P(X \geq n) = \sum_{i=n}^{+\infty} P(n \leq X < n+1) = \sum_{i=n}^{+\infty} b_i.$$

In this case,

$$\forall n \in \mathbb{N}: 0 < 1 - a \leq 1 - a_n = \sum_{i=n}^{+\infty} b_i.$$

This inequality yields that $\sum_{i=0}^{+\infty} b_i = +\infty$, since all the terms of the series are non-negative and the complements of its partial sums do not converge to zero. Thus:

$$1 - a_0 = \sum_{i=0}^{+\infty} b_i = +\infty,$$

what is a contradiction, since $0 \leq a_0 \leq 1$. So, $a = 1$.

Let us consider $Y = -X$. Let G be the cumulative function of Y: as previously established:

$$\lim_{y \to +\infty} G(y) = 1.$$

or: $P(Y < y) = P(X > -y) \leq P(X \geq -y) = 1 - F(-y)$, so that

$F(-y) = 1 - G(y)$.

Thus,

$$\lim_{x \to +\infty} F(x) = \lim_{y \to +\infty} F(-y) = 1 - \lim_{y \to +\infty} G(y) = 1 - 1 = 0.$$

v) Let $f(h) = F(x) - F(x-h) = P(x-h \leq X < x)$. It is enough to show that

$$\lim_{h \longrightarrow 0+} f(h) = 0.$$

f is increasing and bounded on the positive reals: if $0 \leq h_1 \leq h_2$ then $0 \leq f(h_1) \leq f(h_2) \leq 1$. Thus, there exists a real number $a \geq 0$ such that $a = \lim_{h \longrightarrow 0+} f(h)$.

Assume that $a > 0$: let $a_p = f\left(\frac{1}{p+1}\right)$. Then:

$$a_p \longrightarrow a \text{ when } p \longrightarrow +\infty \text{ and } \forall p \in \mathbb{N} : 0 \leq a \leq a_p \leq 1.$$

Let us consider the event:

$$A_n = \left\{ \omega \in \Omega : x - \frac{1}{n+1} \leq X(\omega) < x - \frac{1}{n+2} \right\}.$$

We have $A_i \cap A_j = \emptyset$ when $i \neq j$, so that $A = \bigcup_{n \in \mathbb{N}} A_n$ satisfies:

$$P(A) = \sum_{n=0}^{+\infty} P(A_n).$$

or,

$$A = \{\omega \in \Omega : x - 1 \leq X(\omega) < x\},$$

and we have:

$$F(x) - F(x-1) = P(x - 1 \leq X < x) = \sum_{n=0}^{+\infty} P(A_n).$$

In addition, let us consider $p > 0$ and $B_n = A_{n+p}$. We have yet $B_i \cap B_j = \emptyset$ when $i \neq j$, so that $B = \bigcup_{n \in \mathbb{N}} B_n$ satisfies:

$$P(B) = \sum_{n=0}^{+\infty} P(B_n) = \sum_{n=0}^{+\infty} P(A_{n+p}) = \sum_{n=p}^{+\infty} P(A_n).$$

or,

$$B = \left\{\omega \in \Omega : x - \frac{1}{n+p} \leq X(\omega) < x\right\},$$

and we have $P(B) = a_p$. Thus,

$$\sum_{n=p}^{+\infty} P(A_n) = a_p \geq a > 0.$$

This inequality yields that $\sum_{i=0}^{+\infty} P(A_n) = +\infty$, since all the terms of the series are non-negative and the complements of its partial sums do not converge to zero. Thus:

$$F(x) - F(x-1) = +\infty,$$

which is a contradiction, since $F(x) - F(x-1) = P(x-1 \leq X < x) \leq 1$. So, $a = 0$ and we obtain the result claimed.

vi) is immediate.

vii) We have $F(x+h) - F(x) = P(x < X < x+h) + P(X = x)$. So, it is enough to show that $f(h) = P(x < X < x+h)$ satisfies:

$$\lim_{h \to 0+} f(h) = 0.$$

f is increasing and bounded on the positive reals: if $0 \leq h_1 \leq h_2$ then $0 \leq f(h_1) \leq f(h_2) \leq 1$. Thus, there exists a real number $a \geq 0$ such that $a = \lim_{h \to 0+} f(h)$.

Assume that $a > 0$: let $a_p = f\left(\frac{1}{p+1}\right)$. Then:

$$a_p \longrightarrow a \text{ when } p \longrightarrow +\infty \text{ and } \forall p \in \mathbb{N} : 0 \leq a \leq a_p \leq 1.$$

Let:

$$A_n = \left\{\omega \in \Omega : x + \frac{1}{n+2} \leq X(\omega) < x + \frac{1}{n+1}\right\}.$$

We have $A_i \cap A_j = \emptyset$ when $i \neq j$, so that $A = \bigcup_{n \in \mathbb{N}} A_n$ satisfies:

$$P(A) = \sum_{n=0}^{+\infty} P(A_n).$$

or,

$$A = \{\omega \in \Omega : x < X(\omega) < x+1\},$$

and we have:

$$F(x+1) - F(x) = P(x \leq X < x+1) \geq \sum_{n=0}^{+\infty} P(A_n).$$

Thus, $\sum_{n=0}^{+\infty} P(A_n)$ converges and the complements of its partial sums satisfy:

$$\sum_{n=p}^{+\infty} P(A_n) \longrightarrow 0 \text{ when } p \longrightarrow +\infty.$$

or,

$$\sum_{n=p}^{+\infty} P(A_n) = P\left(x < X < x + \frac{1}{p+1}\right) = f\left(\frac{1}{p+1}\right) = a_p,$$

so that $a_p \longrightarrow 0$ and we have $a = 0$.

The cumulative function may be used in order to define a measure of probability on $\mathbb{R}$: for all rectangle (a, b),

$$\mu_F((a,b)) = F(b) - F(a).$$

μ_F extends to the measurable sets of $\mathbb{R}$ by the procedure described in the preceding section. When μ_F is defined by the density f, we say that f is the *probability density* of X. In this case, we have:

$$P(X \in J) = \mu_F(J) = \int_J f \text{ for all } J \subset \mathbb{R}$$

Thus,
$$\forall\, a, x \in \mathbb{R} : F(x) - F(a) = \int_a^x f \iff F' = f,$$

and the probability density is the derivative of the cumulative function.

Let $g : \mathbb{R} \longrightarrow \mathbb{R}$ be a function. The mean of $Y = g(X)$ is:
$$E(Y) = E(g(X)) = \int_\mathbb{R} g\, \mu_F(dx).$$

Sometimes, the notations $\int_\mathbb{R} g\, P(dx)$ or $\int_\mathbb{R} g\, P(X \in dx)$ are used. When μ_F is defined by the density f, we have:
$$E(Y) = E(g(X)) = \int_\mathbb{R} g\, f.$$

Two particular cases corresponding to these expression are the *mean*, or *expectation*, of X, given by:
$$E(X) = \int_\mathbb{R} x\, \mu_F(dx)$$

and the *moment of order p* of X (or p-moment of X), given by:
$$M_p(X) = E(X^p) = \int_\mathbb{R} x^p\, \mu_F(dx).$$

When μ_F is defined by the density f, we have:
$$E(X) = \int_\mathbb{R} x\, f; \quad M_p(X) = \int_\mathbb{R} x^p\, f.$$

The *variance* of X is:
$$V(X) = E\left((X - E(X))^2\right) = E(X^2) - [E(X)]^2.$$

The *standard deviation* $\sigma(X)$ of X is the square root of $V(X)$:
$$\sigma(X) = \sqrt{V(X)}.$$

The properties established for finite populations are still valid in the general case of a random variable: for instance, $E(\alpha X + \beta Y) = \alpha E(X) + \beta E(Y)$. We also have:

PROPOSITION 1.22.– $V(X) \geq 0$, $\forall\, X$. In addition, $V(X) = 0$ if and only if $X = E(X)$ P–a.s.. ∎

PROOF.– Let $g(X) = (X - E(X))^2$. Since $g(X) \geq 0$, we have $\int_\mathbb{R} g\, \mu_F(dx) \geq 0$.

Assume that $\int_\mathbb{R} g\, \mu_F(dx) = 0$: let $A = \{\omega \in \Omega : g(X(\omega)) > 0\}$: we show that $P(A) = 0$.

Let $n, k > 0$. Let us consider:

$$A_k = \left\{ \omega \in \Omega : g(X(\omega)) > \frac{1}{k} \right\}$$

and

$$A_{n,k} = \left\{ \omega \in \Omega : n > g(X(\omega)) > \frac{1}{k} \right\}.$$

$X(A_{n,k})$ is a finite reunion of intervals, so that $X(A_{n,k})$ is μ_F–measurable and $P(A_{n,k}) = \mu_F(X(A_{n,k}))$.

Or, on the one hand,

$$0 \leq \int_{X(A_{n,k})} g\mu_F(dx) = \int_\mathbb{R} g\mu_F(dx) - \underbrace{\int_{\mathbb{R}-X(A_{n,k})} g\mu_F(dx)}_{\geq 0}$$

$$\leq \int_\mathbb{R} g\mu_F(dx) = 0.$$

and, on the other hand,

$$\int_{X(A_{n,k})} g\mu_F(dx) \geq \frac{1}{k} \int_{X(A_{n,k})} \mu_F(dx) = \frac{1}{k}\mu(X(A_{n,k})).$$

So, $\mu(X(A_{n,k})) = 0$. In addition, we have $A = \bigcup_{k \in N} A_k$ and $A_k = \bigcup_{n \in N} A_{n,k}$. As a consequence, $A_k = \bigcup_{n \in N} A_{n,k}$ is negligible, which implies that $A = \bigcup_{k \in N} A_k$ is negligible too. ∎

COROLLARY 1.9.– $E(X^2) \geq 0, \forall\, X$. In addition, $E(X^2)$ if and only if $X = 0$ P–a.s. ∎

PROOF.– We have $E(X^2) = V(X) + [E(X)]^2$. Thus, on the one hand, $E(X^2) \geq 0$ and, on the other hand, $E(X^2) = 0$ if and only if $V(X) = 0$ e $E(X) = 0$. ∎

The *characteristic function* of X is:

$$\varphi(t) = E\left(e^{itX}\right).$$

φ has some useful properties. For instance:

– φ is uniformly continuous on $\mathbb{R}$, $\varphi(0) = 1$ and $|\varphi(t)| \leq 1, \forall\, t \in \mathbb{R}$;

– if $M_p(X) < \infty$ then $\varphi^{(p)}$ (derivative of order p of φ) satisfies $\varphi^{(p)}(t) = i^p E\left(X^p e^{itX}\right)$. Namely, $\varphi^{(p)}(0) = i^p M_p(X)$;

– if $M_p(X) < \infty, \forall\, p \in \mathbb{N}$ and the series $S(t) = \sum_{n \in \mathbb{N}} \frac{(it)^n}{n!} M_n(X)$ has a strictly positive radius of convergenge then $\varphi(t) = S(t)$;

– if X and Y are independent and $Z = X + Y$, then $\varphi_Z(t) = \varphi_X(t)\varphi_Y(t)$;

– if $Y = aX + b$, $a \in \mathbb{R}$ and $b \in \mathbb{R}$, then $\varphi_Y(t) = e^{itb}\varphi_X(at)$;

– if there exists $t \neq 0$ such that $|\varphi(t)| = 1$, then there are $a \in \mathbb{R}$ and $b \in \mathbb{R}$ such that $J = \{an + b : n \in \mathbb{Z}\}$ verifies $P(X \in J) = 1$;

– if f is the *probability density* of X then $\varphi(t) = \int_\mathbb{R} e^{itx} f = TF(f)$, i.e. φ is the Fourier transform of f. Reciprocally, if $\varphi \in L^1(\mathbb{R})$, then $f = TF^{-1}(f) = \frac{1}{2\pi} \int_\mathbb{R} e^{-itx} \varphi$.

1.10.1. *Matlab implementation*

Analogous to the case of finite populations, Matlab may be used in order to evaluate statistics of random variables.

Let us start by considering the situation where the probability density f of the scalar random variable X is known. In this case, we may use the numerical integration `quad` (or, for recent versions, `integral`) in order to evaluate its mean, variance or moments. For instance, by assuming that X is a scalar random variable having density f, supported by (a, b), we may use the following code:

Listing 1.17. *Evaluation of statistics*

```
function [mX,vX,sdX] = statsf(f,a,b)
%
% evaluation of the mean
%
f1 = @(x) x*f(x);
mX = numint(f1,a,b);
%
% evaluation of the second moment
%
f2 = @(x) x^2*f(x);
m2X = numint(f2,a,b);
%
% evaluation of the variance
%
vX = m2X - mX^2;
%
% evaluation of the standard deviation
%
sdX = sqrt(vX);
end

function v = map(f,x, dimf)
%
% maps function f on a set of points x
% each column of x is a point
%
% IN:
% f : the function - type anonymous function
% x : table dimx x np - type array of double
% dimf : dimension of f (number of lines) - type integer
%
% OUT:
% v : table dimf x np - type array of double
%
np = size(x,2);
v = zeros(dimf, np);
for i = 1: np
    xx = x(:,i);
    v(:,i) = f(xx);
end;
return;
end
```

```
function v = numint(f,a,b)
%
% evaluates integral(f,a,b)
% f is scalar
%
% IN:
% f : the function - type anonymous function
% a : lower bound of integration - type double
% b : upper bound of integration - type double
%
% OUT:
% v : value of the integral - type array of double
%
integrand = @(x) map(f,x, dimf);
v = quad(integrand,a,b); % old versions of Matlab
% v = integral(integrand,a,b); % new versions of Matlab
return;
end
```

A second situation of interest is the one where the map $X = X(\omega)$ is given and the density φ of ω is known and supported by (a, b). In this case, we just need to modify the first function:

Listing 1.18. *Evaluation of statistics*

```
function [mX,vX,sdX] = statsphi(phi,X,a,b)
%
% evaluation of the mean
%
f1 = @(om) X(om)*phi(om);
mX = numint(f1,a,b);
%
% evaluation of the second moment
%
f2 = @(om) X(om)^2*phi(om);
m2X = numint(f2,a,b);
%
% evaluation of the variance
%
vX = m2X - mX^2;
%
% evaluation of the standard deviation
%
sdX = sqrt(vX);
end
```

Finally, in the situation where only a sample of X is available, the statistics are obtained by considering the sample as a data vector of a finite population (see section 1.3).

1.10.2. *Couples of random variables*

Couples of random variables generalize couples of numerical characteristics.

Let us consider $X, Y : \Omega \longrightarrow \mathbb{R}$. The cumulative function of (X, Y) is:

$$F(x, y) = P(X < x, Y < y).$$

This function has the following properties:

– $F \geq 0$;
– $F(-\infty, y) = 0, \forall\, y \in \mathbb{R}$;
– $F(x, -\infty) = 0, \forall\, x \in \mathbb{R}$;
– $F(+\infty, +\infty) = 1$;
– If $x_1 \leq x_2$ and $y_1 \leq y_2$ then $0 \leq F(x_2, y_2) - F(x_1, y_2) - F(x_2, y_1) + F(x_1, y_1) \leq 1$.

The distribution of the single variable X is called *marginal distribution* of X. It is given by $F_X(x) = P(X < x, Y < +\infty) = F(x, +\infty)$. In an analogous way, the distribution of Y is called *marginal distribution* of Y. It is given by $F_Y(x) = F(+\infty, y)$.

Analogous to the one-dimensional situation, the cumulative function F may be used in order to define a measure on $\mathbb{R}^2$: let us consider a rectangle:

$$\mathcal{R} = \left\{(x, y) \in \mathbb{R}^2 : x_1 \leq x \leq x_2 \text{ and } y_1 \leq y \leq y_2\right\}.$$

We set

$$\mu_F(\mathcal{R}) = F(x_2, y_2) - F(x_1, y_2) - F(x_2, y_1) + F(x_1, y_1).$$

μ_F is extended to the measurable sets of $\mathbb{R}^2$ by the same way as in the case of a single random variable (procedure previously exposed). Similarly to the preceding situation, we say that f is the *probability density* of the couple (X, Y) when μ_F is defined by the density f. In this case,

$$P((X, Y) \in J) = \mu_F(J) = \int_J f \text{ for all } J \subset \mathbb{R}^2,$$

and the probability density of X is $f_X(x) = \int_\mathbb{R} f(x, y) \, dy$, while the density of probability of Y is $f_Y(y) = \int_\mathbb{R} f(x, y) \, dx$.

Let $g : \mathbb{R} \longrightarrow \mathbb{R}$ a function. The expectation of $Z = g(X, Y)$ is:

$$E(Z) = E(g(X, Y)) = \int_{\mathbb{R}^2} g \, d\mu_F.$$

When μ_F is defined by the density f, we have:

$$E(Z) = E(g(X, Y)) = \int_{\mathbb{R}^2} g \, f.$$

Two particular cases corresponding to this general expression are, on the one hand, the *mean* or *expectation* of X and Y, given by:

$$E(X) = \int_{\mathbb{R}^2} x \, d\mu_F; E(Y) = \int_{\mathbb{R}^2} y \, d\mu_F$$

and the *moments of order p* (or p-moments) of X and Y, given by:

$$M_p(X) = E(X^p) = \int_{\mathbb{R}^2} x^p \, d\mu_F; \; M_p(Y) = E(Y^p) = \int_{\mathbb{R}^2} y^p \, d\mu_F.$$

When μ_F is defined by the density f, we have:

$E(X) = \int_{\mathbb{R}^2} x \, f; \; M_p(X) = \int_{\mathbb{R}^2} x^p \, f;$
$E(Y) = \int_{\mathbb{R}^2} y \, f; \; M_p(Y) = \int_{\mathbb{R}^2} y^p \, f.$

The *variance* of X is:

$$V(X) = E\left((X - E(X))^2\right) = E(X^2) - [E(X)]^2$$

and *variance* of Y is:

$$V(Y) = E\left((Y - E(Y))^2\right) = E(Y^2) - [E(Y)]^2.$$

The *covariance* of X and Y is:

$$Cov(X,Y) = E((X - E(X))(Y - E(Y))) = E(XY) - E(X)E(Y).$$

The properties established for finite populations remain valid: for instance,

$$V(\alpha X + \beta Y) = \alpha^2 V(X) + \beta^2 V(Y) + 2\alpha\beta Cov(X,Y).$$

and

$$|Cov(X,Y)| \leq \sqrt{V(X)}\sqrt{V(Y)}.$$

The Matlab implementation is an extension of the situation for scalar variables. If the probability density f of the pair (X,Y) is known and supported by $(a,b) \times (c,d)$, we may use 'quad2d' (or, for recent versions, integral2):

Listing 1.19. *Evaluation of statistics*

```
function [mX,mY, C] = statsfpair(f,a,b,c,d)
%
% evaluation of the mean of X
%
f1  = @(x,y) x*f(x,y);
mX = numint2d(f1,a,b,c,d);
%
% evaluation of the mean of Y
%
f1  = @(x,y) y*f(x,y);
mY = numint2d(f1,a,b,c,d);
%
% evaluation of the covariance matrix
%
C = zeros(2,2);
%
f2 = @(x,y) (x-mX)^2*f(x,y);
C(1,1) = numint2d(f2,a,b,c,d);
%
f2 = @(x,y) (y-mY)^2*f(x,y);
C(2,2) = numint2d(f2,a,b,c,d);
%
f2 = @(x,y) (x-mX)*(y-mY)*f(x,y);
C(1,2) = numint2d(f2,a,b,c,d);
C(2,1) = C(1,2);
end
```

```
function v = map2d(f,x,y)
%
% maps scalar function f on a set of points (x,y)
%
% IN:
% f : the function - type anonymous function
% x : table nx x ny - type array of double
% y : table nx x ny - type array of double
%
% OUT:
% v : table dimf x np - type array of double
%
nx = size(x,1);
ny = size(x,2);
v = zeros(nx,ny);
for i = 1: nx
    for j = 1: ny
        xx = x(i,j);
        yy = y(i,j);
        v(i,j) = f(xx, yy);
    end;
end;
return;
end

function v = numint2d(f,a,b,c,d)
%
% evaluates integral(f,a,b)
% f is scalar
%
% IN:
% f : the function - type anonymous function
% a,b,c,d : bounds of integration - type double
%
% OUT:
% v : value of the integral - type array of double
%
integrand = @(x,y) map2d(f,x,y);
v = quad2d(integrand,a,b,c,d); % old versions of Matlab
% v = integral2(integrand,a,b,c,d); % new versions of Matlab
return;
end
```

In an analogous way, we may consider the situation where $X = X(\omega)$ and $Y = Y(\omega)$. For instance, ω is one-dimensional and its density is φ, supported by (a, b), we may use the code:

Listing 1.20. *Evaluation of statistics*

```
function [mX,mY,C] = statsphipair(phi,X,Y,a,b)
%
% evaluation of the mean of X
%
f1 = @(om) X(om)*phi(om);
mX = numint(f1,a,b);
%
% evaluation of the mean of Y
%
f1 = @(om) Y(om)*phi(om);
mY = numint(f1,a,b);
%
% evaluation of the covariance matrix
%
C = zeros(2,2);
%
f2 = @(om) (X(om)-mX)^2*phi(om);
C(1,1) = numint(f2,a,b);
%
f2 = @(om) (Y(om)-mY)^2*phi(om);
C(2,2) = numint(f2,a,b);
%
f2 = @(om) (X(om)-mX)*(Y(om)-mY)*phi(om);
C(1,2) = numint(f2,a,b);
C(2,1) = C(1,2);
end
```

The extension to the situation where ω is obtained by modifying this code in order to perform multidimensional integration and will not be stated here.

1.11. Hilbertian properties of random variables

In the same way as numerical characteristics, random variables also possess a Hilbertian structure. The main difference between finite populations and the general situation actually under study is the use of a probability measure P, defined on Ω, which is the fundamental stone in the construction: P defines null sets, almost sure and negligible events. P is the basis of the whole construction, but it remains an abstract probability, which intervenes only in the initial definitions and not in the practical calculations or in the most advanced results of the theory under construction. The fundamental rule of P is the definition of the equivalence between two random variables – analogous to the section 1.7), we have:

$$X \approx Y \iff A = \{\omega \in \Omega : X(\omega) \neq Y(\omega)\} \text{ is } P\text{-negligible.}$$

Let us consider the set formed by the random variables on Ω:

$$\mathcal{C}(\Omega) = \{X : \Omega \longrightarrow \mathbb{R}\}.$$

and the linear subspace formed by the *simple random variables* on Ω:

$$\mathcal{V}(\Omega) = \{X \in \mathcal{C}(\Omega) : X(\Omega) \text{ is finite}\}.$$

The relation $\approx$ is a relation of equivalence on $\mathcal{C}(\Omega)$:

$$X \approx X;\, X \approx Y \implies Y \approx X\,;\, X \approx Y \text{ and } Y \approx Z \implies X \approx Z.$$

The class of equivalence of X is:

$$\widetilde{X} = \{Y \in \mathcal{C}(\Omega) : Y \approx X\}$$

Let us denote by $\mathcal{C}(\Omega, P)$, the set of all the classes of equivalence:

$$\mathcal{C}(\Omega, P) = \left\{\widetilde{X} : X \in \mathcal{C}(\Omega)\right\}$$

and by $\mathcal{V}(\Omega, P)$, the set of the classes of equivalence of the simple random variables:

$$\mathcal{V}(\Omega, P) = \left\{\widetilde{X} \in \mathcal{C}(\Omega, P) : X \in \mathcal{V}(\Omega)\right\}$$

For $\widetilde{X}, \widetilde{Y} \in \mathcal{V}(\Omega, P)$, we consider:

$$\left(\widetilde{X}, \widetilde{Y}\right) = E(XY). \qquad [1.18]$$

Then:

PROPOSITION 1.23.– $(\bullet, \bullet)$ is a scalar product on $\mathcal{V}(\Omega, P)$. ∎

PROOF.– The definition corresponds to a bilinear symmetric form. In addition, $\left(\widetilde{X}, \widetilde{X}\right) = E\left(X^2\right) \geq 0$ for all $\widetilde{X} \in \mathcal{V}(\Omega, P)$. Finally, if $\left(\widetilde{X}, \widetilde{X}\right) = E\left(X^2\right) = 0$, then $\widetilde{X} = \widetilde{0}$ and the form is a positive definite one.
∎

Let $L^2(\Omega, P)$ be the completion of $\mathcal{V}(\Omega, P)$ for the scalar product defined in equation [1.3]: $L^2(\Omega, P)$ is a Hilbert space for the scalar product [1.18]. The norm of an element $X \in L^2(\Omega, P)$ is:

$$\|X\| = \sqrt{E(X^2)}. \qquad [1.19]$$

If we are interested in *vectors of random variables*, the elements above may be extended by considering product spaces. For instance, if we are interested in vectors $\mathbf{X} = (X_1, \ldots, X_k)$, we consider $[\mathcal{V}(\Omega)]^p$ and

$$(\mathbf{X}, \mathbf{Y}) = E(\mathbf{X}.\mathbf{Y}) = \sum_{i=1}^{k} E(X_i Y_i). \qquad [1.20]$$

In this case,

$$\|\mathbf{X}\| = \sqrt{E\left(|\mathbf{X}|^2\right)} = \sqrt{\sum_{i=1}^{k} E(X_i^2)}. \qquad [1.21]$$

The completion of $[\mathcal{V}(\Omega, P)]^p$ is $[L^2(\Omega, P)]^p$. In a manner analogous to that of a finite population, all the approaches below extend to vectors of numerical characteristics by this way.

1.11.1. *Approximation*

As previously observed, the Hilbertian structure of $L^2(\Omega)$ makes the application of techniques and results issued from the theory of Hilbert spaces possible, namely those connected to *orthogonal projections*. Here also, the results established for finite populations extend to the general situation.

1.11.1.1. *Conditional probability and approximation*

One of the remarkable properties connected to orthogonal projections is the notion of *conditional probability*. Let us consider two events $A, B \subset \Omega$. As previously observed, an event may be represented by its *characteristic function*: 1_A and 1_B represent A and B, respectively. We may consider the approximation of 1_B in terms of 1_A: for instance, we may look for the coefficient $\lambda \in \mathbb{R}$ such that $\lambda 1_A$ is the closest possible to 1_B. Thus, we may look for the orthogonal projection of 1_B onto a linear subspace having dimension 1 and given by:

$$S = \{Z \in L^2(\Omega, P) : Z \text{ is constant: } Z(\omega) = s1_A(\omega) \in \mathbb{R}, \forall \omega \in \Omega\}$$

Analogous to the result for finite populations, we have $\lambda = P(B|A)$, where

$$P(B|A) = \frac{P(A \cap B)}{P(A)} \text{ if } P(A) \neq 0; P(B|A) = 0 \text{ otherwise.} \quad [1.22]$$

Symmetrically, we have:

$$P(A|B) = \frac{P(A \cap B)}{P(B)} \text{ if } P(B) \neq 0; P(A|B) = 0 \text{ otherwise.}$$

Here also, we have:

$$P(A \cap B) = P(B|A).P(A) = P(A|B).P(B). \quad [1.23]$$

As for finite populations, A and B are said to be *independent* if and only if $P(A|B) = P(A)$ or $P(B|A) = P(B)$. If A and B are independent, then $P(A \cap B) = P(A) P(B)$.

1.11.1.2. *Approximation of a random variable by a constant*

When considering the best approximation of a random variable X by a constant, we may use the same approach as for finite populations and look for the orthogonal projection of X onto a linear subspace having dimension 1 and given by:

$$S = \{Z \in L^2(\Omega, P) : Z \text{ is constant: } Z(\omega) = s \in \mathbb{R}, \forall \omega \in \Omega\}$$

The solution is – as in the finite population case – $PX = E(X)$ and the error in the approximation is $\|X - PX\| = \sqrt{V(X)}$.

As in the finite population case, when considering *vectors of random variables*, the best approximation of $\mathbf{X} = (X_1, \ldots, X_k)$ by a *constant vector* is $\mathbf{m} = E(\mathbf{X}) = (m_1, \ldots, m_k)$, where $m_i = E(X_i)$. The error in the approximation is $\|\mathbf{X} - \mathbf{m}\| = \sqrt{\sum_{i=1}^{k} E\left((X_i - m_i)^2\right)}$.

As previously observed, Matlab implementation may be performed (see section 1.10.1)

1.11.1.3. *Approximation of a random variable by an affine function of another random variable*

In the situation where we look for the best approximation of the random variable Y by an affine function of X, the approach is also entirely analogous to the approach introduced for finite populations: we look for the orthogonal projection of Y onto the linear subspace having dimension 2, given by:

$$S = \{s \in L^2(\Omega, P) : s = \alpha X + \beta; \alpha, \beta \in \mathbb{R}\},$$

The solution is $PX = aX + b$, where:

$$a = \frac{Cov(X, Y)}{V(X)}; \quad b = E(Y) - aE(X).$$

we have also:

$$\|PX - aX - b\| = \sqrt{V(Y)\left(1 - [\rho(X, Y)]^2\right)},$$

where:

$$\rho(X, Y) = \frac{Cov(X, Y)}{V(X)V(Y)}$$

is the *linear correlation coefficient* between X and Y. We have $|\rho(X, Y)| \leq 1$ and the error is null if and only if $|\rho(X, Y)| = 1$.

Analogous to the finite population case, when considering *vectors of random variables*, the best approximation of $\mathbf{Y} = (Y_1, \ldots, Y_k)$ by an affine function of $\mathbf{X} = (X_1, \ldots, X_m)$ is $\mathbf{Y} = \mathbf{AX} + \mathbf{B}$, where $\mathbf{A}$ is determined by solving, for $i = 1, \ldots, k$,

$$\mathbf{C}.\mathbf{a} = \mathbf{b},$$

with $\mathbf{a} = (a_1, \ldots, a_m)^t \in \mathcal{M}(m, 1)$ such that $a_j = A_{ij}$, $\mathbf{C} \in \mathcal{M}(m, m)$ such that $C_{rj} = Cov(X_j, X_r)$ and $\mathbf{b} = (b_1, \ldots, b_m)^t \in \mathcal{M}(m, 1)$ such that $b_r = Cov(Y_i, X_r)$. After the determination of $\mathbf{A}$, $\mathbf{B}$ is given by:

$$\mathbf{B} = \mathbf{Y} - \mathbf{A}.E(\mathbf{X}).$$

In the scalar situation (both X and Y are real-valued), Matlab implementation may be performed by using the programs introduced in section 1.10.2.

1.11.1.4. *Approximation of a random variable by a non-parametric family of functions of another one*

When we are interested in the best approximation of Y by a family of non-parametric functions of X, we may also use the same approach as for finite populations and look for the orthogonal projection of Y onto the linear subspace given by:

$$S = \{s \in L^2(\Omega, P) : s = \varphi(X); \varphi : \mathbb{R} \longrightarrow \mathbb{R}\}.$$

In this case, we have:

$$(Y - g(X), \varphi(X)) = 0, \forall \varphi : \mathbb{R} \longrightarrow \mathbb{R},$$

i.e.,

$$\int_{\mathbb{R}^2} (y - g(x)) \varphi(x) d\mu_F = 0, \forall \varphi : \mathbb{R} \longrightarrow \mathbb{R},$$

where μ_F is the cumulative function of the pair (X, Y). When μ_F is defined by a density f, we have:

$$\int_{\mathbb{R}^2} (y - g(x)) \varphi(x) f(x, y) = 0, \forall \varphi : \mathbb{R} \longrightarrow \mathbb{R}, \qquad [1.24]$$

so that:

$$\int_{\mathbb{R}^2} yf(x,y)\varphi(x) = \int_{\mathbb{R}^2} g(x) f(x,y)\varphi(x), \forall \varphi : \mathbb{R} \longrightarrow \mathbb{R},$$

and, $\forall \varphi : \mathbb{R} \longrightarrow \mathbb{R}$

$$\int_{\mathbb{R}} \left\{ \int_{\mathbb{R}} yf(x,y)\, d\ell(y) \right\} \varphi(x)\, d\ell(x) = \\ \int_{\mathbb{R}} \left\{ \left[\int_{\mathbb{R}} f(x,y)\, d\ell(y) \right] g(x) \right\} \varphi(x)\, d\ell(x).$$

Thus,

$$\int_{\mathbb{R}} yf(x,y)\, d\ell(y) = \left[\int_{\mathbb{R}} f(x,y)\, d\ell(y) \right] g(x) \quad \text{a.s.}$$

By assuming that $\int_{\mathbb{R}} f(x,y)\, d\ell(y) \neq 0$ for all x, the solution is:

$$g(x) = \int_{\mathbb{R}} yf(y \mid X = x)\, d\ell(y)$$

where:

$$f(y \mid X = x) = f(x,y) \,/\, \int_{\mathbb{R}} f(x,y)\, d\ell(y) = f(x,y) \,/\, f_X(x).$$

The function g thus defined is called *conditional mean of Y with respect to X* or *conditional expectation of Y with respect to X*, and is denoted by $E(Y \mid X)$ (i.e. $g(x) = E(Y \mid X = x)$).

$f(y \mid X = x)$ is the *conditional distribution* of Y with respect to X. Sometimes, the expression *distribution of Y conditional to X* is also used. In an analogous way, the *probability density of X conditional to Y* is:

$$f(x \mid Y = y) = f(x,y) \,/\, \int_{\mathbb{R}} f(x,y)\, d\ell(x) = f(x,y) \,/\, f_Y(y).$$

and we have:

$$E(X \mid Y = y) = \int_{\mathbb{R}} xf(x \mid Y = y)\, d\ell(x).$$

The random variables forming the couple are *independent* if and only if

$$\forall\, x, y : f(x \mid Y = y) = f_X(x) \quad \text{or} \quad f(y \mid X = x) = f_Y(y),$$

i.e.,

$$\forall\, x, y : f(x, y) = f_X(x)\, f_Y(y).$$

PROPOSITION 1.24.– Let X and Y be two independent random variables such that the density of the pair (X, Y) is f. Then, for $g, h : \mathbb{R} \longrightarrow \mathbb{R}$:

$$E(g(X) h(Y)) = E(g(X)) E(h(Y)).$$

PROOF.– We have $f(x, y) = f_X(x) f_Y(y)$, so that:

$$E(g(X) h(Y)) = \int_{\mathbb{R}^2} g(x) h(y) f(x, y) = \int_{\mathbb{R}^2} g(x) h(y) f_X(x) f_Y(y),$$

and Fubini's theorem shows that:

$$E(g(X) h(Y)) = \int_{\mathbb{R}} \left\{ \left[\int_{\mathbb{R}} h(y) f_Y(y)\, d\ell(y) \right] \right\} g(x) f_X(x)\, d\ell(x),$$

i.e.,

$$E(g(X) h(Y)) = \left[\int_{\mathbb{R}} h(y) f_Y(y)\, d\ell(y) \right] \left[\int_{\mathbb{R}} g(x) f_X(x)\, d\ell(x) \right]$$
$$= E(g(X)) E(h(Y)).\ \blacksquare$$

COROLLARY 1.10.– Let X and Y be two independent random variables such that the pair probability density (X, Y) is f and the characteristic function of the pair is φ. Then:

i) $E(XY) = E(X) E(Y)$ and $Cov(X, Y) = 0$;

ii) $E(Y \mid X = x) = E(Y)$ and $E(X \mid Y = y) = E(X)$ for all (x, y);

iii) $\varphi(t) = \varphi_X(t_1) \varphi_Y(t_2)$, $\forall\, t = (t_1, t_2) \in \mathbb{R}^2$. $\blacksquare$

PROOF.– The result is immediate. $\blacksquare$

PROPOSITION 1.25.– Let X and Y be two random variables such that the density of the pair (X, Y) is f and the pair characteristic function is φ. Then X and Y are independent if and only if $\varphi(t) = \varphi_X(t_1) \varphi_Y(t_2)$, $\forall\, t = (t_1, t_2) \in \mathbb{R}^2$.

PROOF.– $\Longrightarrow$: this implication follows from the preceding result.

$\Longleftarrow$: We have:

$$\varphi_X(t_1) \varphi_Y(t_2) = \left[\int_{\mathbb{R}} e^{it_2 y} f_Y(y)\, d\ell(y) \right] \left[\int_{\mathbb{R}} e^{it_1 x} f_X(x)\, d\ell(x) \right],$$

so that:

$$\varphi_X(t_1) \varphi_Y(t_2) = \int_{\mathbb{R}} \left\{ \left[\int_{\mathbb{R}} e^{it_2 y} f_Y(y)\, d\ell(y) \right] \right\} e^{it_1 x} f_X(x)\, d\ell(x)$$

and Fubini's theorem shows that:

$$\varphi_X(t_1) \varphi_Y(t_2) = \int_{\mathbb{R}^2} e^{i(t_1 x + t_2 y)} f_X(x)\, f_Y(y).$$

Since $\varphi_X(t_1) \varphi_Y(t_2) = \varphi(t)$, $\forall\, t \in \mathbb{R}^2$, we have:

$$\int_{\mathbb{R}^2} e^{i(t_1 x + t_2 y)} f_X(x)\, f_Y(y) = \int_{\mathbb{R}^2} e^{i(t_1 x + t_2 y)} f(x, y),$$

so that the injectivity of the Fourier Transform implies that:

$$f(x, y) = f_X(x)\, f_Y(y) \text{ on } \mathbb{R}^2. \blacksquare$$

1.12. Sequences of random variables

1.12.1. *Quadratic mean convergence*

The Hilbertian structure of the set $L^2(\Omega, P)$ induces a first notion of convergence, connected to the norm and the scalar product in this space: we say that the sequence $\{U_n\}_{n \in \mathbb{N}}$ converges to U *in quadratic mean* if and only if

$$\|U_n - U\| = \sqrt{E\left((U_n - U)^2 \right)} \longrightarrow 0 \text{ when } n \longrightarrow +\infty.$$

This convergence is denoted by $U_n \longrightarrow U$ q.m.. It is the equivalent for random variables of the *strong convergence* in the space of the square summable functions (L^2) and has analogous properties. For instance:

$$U_n \longrightarrow U \ q.m. \implies \|U_n\| = \sqrt{E(U_n^2)} \longrightarrow \|U\| = \sqrt{E(U^2)};$$

$$U_n \longrightarrow U \ q.m. \implies E(U_n^2) \longrightarrow E(U^2) \text{ and } E(U_n) \longrightarrow E(U).$$

1.12.2. Convergence in the mean

A second notion of strong convergence is the following: we say that the sequence $\{U_n\}_{n \in \mathbb{N}}$ converges to U *in the mean* if and only if:

$$E(|U_n - U|) \longrightarrow 0 \ \text{ when } n \longrightarrow +\infty.$$

This convergence is denoted as $U_n \longrightarrow U$ m. It is the equivalent for random variables of the *strong convergence* of summable functions (space L^1) and has analogous properties. For instance:

$$U_n \longrightarrow U \ m. \implies E(U_n) \longrightarrow E(U).$$

1.12.3. Convergence in probability

We say that the sequence $\{U_n\}_{n \in \mathbb{N}}$ converges to U *in probability* if and only if:

$$\forall \varepsilon > 0 : P(|U_n - U| \geq \varepsilon) \longrightarrow 0$$

This convergence is denoted by $U_n \longrightarrow U$ p. It is the equivalent for random variables of the *convergence in measure* usually introduced when considering measure spaces and has analogous properties. We observe that this convergence may be interpreted in terms of events, let:

$$E_n(\varepsilon) = "|U_n - U| \geq \varepsilon" = \{\omega \in \Omega : |U_n(\omega) - U(\omega)| \geq \varepsilon\}.$$

Then,

$$U_n \longrightarrow U \ p. \iff \forall \varepsilon > 0 : P(E_n(\varepsilon)) \longrightarrow 0 \ \text{ when } n \longrightarrow +\infty.$$

1.12.4. *Almost sure convergence*

We say that the sequence $\{U_n\}_{n \in \mathbb{N}}$ converges to U *almost surely* if and only if:

$$P(U_n \longrightarrow U) = 1.$$

This convergence is denoted as $U_n \longrightarrow U$ *a.s.* It is the equivalent for random variables of the *almost sure convergence* usually introduced when considering measure spaces and has analogous properties. This convergence has an interpretation in terms of events, let:

$$A = "U_n \longrightarrow U" = \{\omega \in \Omega : X_n(\omega) \longrightarrow X(\omega)\}.$$

Then,

$$U_n \longrightarrow U \; a.s. \iff P(A) = 1.$$

i.e. E is *almost sure*.

1.12.5. *Convergence in distribution*

Let $C_l(\mathbb{R}) = \{\varphi : \mathbb{R} \longrightarrow \mathbb{R} : \varphi \text{ is continuous and bounded}\}$. We say that the sequence $\{U_n\}_{n \in \mathbb{N}}$ converges to U *in distribution* if and only if:

$$\forall \varphi \in C_l(\mathbb{R}) : E(\varphi(U_n)) \longrightarrow E(\varphi(U)) \quad \text{when } n \longrightarrow +\infty.$$

This convergence is denoted as $U_n \longrightarrow U$ D. It is the equivalent for random variables of a particular type of weak convergence (i.e. convergence in the dual of a particular space). We have the following theorem:

THEOREM 1.1.– Let F_n be the cumulative function of U_n and F the cumulative function of U. Then,

$$U_n \longrightarrow U \; L. \iff F_n(x) \longrightarrow F(x) \quad \text{when } n \longrightarrow +\infty$$
$$\text{at any } x \text{ where } F \text{ is continuous.} \; \blacksquare$$

We also have the following theorem:

THEOREM 1.2.– Let F_n be the cumulative function of U_n and F the cumulative function of U. Then $U_n \longrightarrow U$ L if and only if both the two following conditions are satisfied:

i) $< 1_{\mathbb{R}}, F_n > \longrightarrow < 1_{\mathbb{R}}, F >$ when $n \longrightarrow +\infty$;

ii) $\forall \varphi \in C_s(\mathbb{R}) : < \varphi, F_n > \longrightarrow < \varphi, F >$ when $n \longrightarrow +\infty$ ($C_s(\mathbb{R}) = \{\varphi : \mathbb{R} \longrightarrow \mathbb{R}$ is continuous and has a bounded support, $1_{\mathbb{R}}$ is a constant function taking the value 1 at any point).

THEOREM 1.3 (Lévy's theorem).– Let φ_n be the characteristic function of U_n and φ the characteristic function of U. $U_n \longrightarrow U$ L if and only if $\varphi_n(t) \longrightarrow \varphi(t)$ a.e. on $\mathbb{R}$. ∎

1.12.6. *Relations among different types of convergence*

The connections between the convergences are the following:
THEOREM 1.4.–

1) $U_n \longrightarrow U$ q.m. $\Longrightarrow U_n \longrightarrow U$ m.

2) $U_n \longrightarrow U$ m. $\Longrightarrow U_n \longrightarrow U$ p.

3) $U_n \longrightarrow U$ p. $\Longrightarrow U_{n(k)} \longrightarrow U$ a.s. for a subsequence.

4) $U_n \longrightarrow U$ m. $\Longrightarrow U_{n(k)} \longrightarrow U$ a.s. for a subsequence.

5) $U_n \longrightarrow U$ a.s. $\Longrightarrow U_n \longrightarrow U$ p.

6) $U_n \longrightarrow U$ p. $\Longrightarrow U_n \longrightarrow U$ D. ∎

1.13. Some usual distributions

The reader may find in the literature the most used probability distributions. Here, we focus on the probability distributions which are used in the following.

1.13.1. *Poisson distribution*

DEFINITION 1.21 (Poisson distribution).– Let $\lambda \in \mathbb{R}$ be such that $\lambda > 0$. We say that a random variable X *has the Poisson distribution* $\mathcal{P}(\lambda)$ or that X is *Poissonian of parameter* λ if and only if, on the one hand, X takes the values $0, 1, ...$ (i.e. its image is $\mathbb{N}$) and, on the other hand,

$$\forall n \in \mathbb{N} \;:\; P(X = n) = \frac{\lambda^n}{n!} e^{-\lambda}.$$

In this case, $E(X) = V(X) = \lambda$. The characteristic function of X is $\psi(t) = \exp(\lambda(\exp(it) - 1))$.

1.13.2. *Uniform distribution*

DEFINITION 1.22 (uniform distribution).– Let $a, b \in \mathbb{R}$ be such that $a < b$. We say that a random variable X *has a uniform distribution* $U(a, b)$ or that X *is uniformly distributed on* (a, b) if and only if its probability density is:

$$f(x) = \begin{cases} \frac{1}{b-a}, & \text{if } x \in (a, b) \\ 0, & \text{otherwise.} \end{cases}$$

We have $E(X) = (a+b)/2$ and $V(X) = (b-a)^2/12$. The characteristic function associated to $U(a, b)$ is:

$$\varphi(t) = \frac{\exp(ibt) - \exp(iat)}{i(b-a)t} \frac{2}{(b-a)} \left[\frac{\sin((b-a)t)}{t} \right] \exp\left(i\left(\frac{b+a}{2}\right)t\right). \blacksquare$$

In an abbreviated way, we say simply that X is $U(a, b)$.

1.13.3. *Normal or Gaussian distribution*

DEFINITION 1.23 (Normal or Gaussian distribution).– Let $m \in \mathbb{R}, \sigma \in \mathbb{R}$ be such that $\sigma \geq 0$. We say that a random variable X *is normally distributed* $N(m, \sigma)$ or X *is Gaussian* $N(m, \sigma)$ if and only if its characteristic function is:

$$\varphi(t) = \exp\left(imt - \frac{\sigma^2}{2}t^2\right). \blacksquare$$

In an abbreviated way, we say simply that X is $N(m, \sigma)$.

$N(m, \sigma)$ admits moments of an arbitrary order and we have $E(X) = m$, $V(X) = \sigma^2$.

For $\sigma = 0$, the Gaussian distribution is *degenerated*: its probability density f is a Dirac measure $f(x) = \delta(x - m)$, so that $P(X = m) = 1$ and the value m is a.s., while any other value is negligible.

For $\sigma > 0$, the probability density of $N(m, \sigma)$ is:

$$f(x) = \frac{1}{\sigma\sqrt{2\pi}} \exp\left(-\frac{1}{2}\left(\frac{x-m}{\sigma}\right)^2\right).$$

The interest of Gaussian distributions is connected to the *central limit theorem*: the empirical distribution of the sequence of partial sums of independent and identically distributed random variables is approximately Gaussian.

When $m = 0$ and $\sigma = 1$, we say that the distribution is a *reduced* one, i.e. X is a *standard Gaussian variable* or simply that X is *standard Gaussian*. The standard Gaussian distribution is denoted as $N(0, 1)$ (for instance, we say that X has the distribution $N(0, 1)$ or that X is $N(0, 1)$) and its probability density is:

$$f(x) = \frac{1}{\sqrt{2\pi}} \exp\left(-\frac{x^2}{2}\right)$$

and its characteristic function is:

$$\varphi(t) = \exp\left(-\frac{t^2}{2}\right).$$

We have $E(X) = 0$ and $V(X) = 1$ for the standard distribution. The moments of the standard distribution are given by:

$$M_p(X) = 0, \text{ if } p = 2n + 1 \text{ (odd)}; \quad M_p(X) = \frac{(2n)!}{n!2^n}, \text{ if } p = 2n \text{ (even)}.$$

1.13.4. *Gaussian vectors*

DEFINITION 1.24 (Gaussian random vectors).– Let $\mathbf{X} = (X_1, ..., X_k)$ a random vector. We say that $\mathbf{X}$ is a *Gaussian random vector* (or simply *Gaussian vector*) having distribution $N(\mathbf{m}, \mathbf{C})$ if and only if its characteristic function is:

$$\varphi(\mathbf{t}) = \exp\left(i(\mathbf{t}, \mathbf{m})_k - \frac{1}{2}(\mathbf{t}, \mathbf{Ct})_k\right),$$

where $m_i = E(X_i)$ and $\mathbf{C} = (C_{ij})_{1 \leq i,j \leq k}$ is the covariance matrix of $\mathbf{X}$, defined as:

$$C_{ij} = Cov(X_i, X_j) = E(X_i X_j) - E(X_i) E(X_j). \ \blacksquare$$

The mean of $\mathbf{X}$ is $E(X) = \mathbf{m} = (m_1, ..., m_k)$. We have $V(X_i) = C_{ii}$.

Gaussian vectors may also be characterized as follows:

PROPOSITION 1.26.– $\mathbf{X} = (X_1, ..., X_k)$ is a Gaussian vector if and only if all linear combination of its components is Gaussian, i.e.,

$$\forall \mathbf{t} \in \mathbb{R}^k : (\mathbf{t}, \mathbf{X})_k \text{ is Gaussian. } \blacksquare$$

PROOF.– Let $\mathbf{t} \in \mathbb{R}^k$, $Z(\mathbf{t}) = (\mathbf{t}, \mathbf{X})_k$. The characteristic function of $Z(\mathbf{t})$ is:

$$\varphi_Z(u) = \exp(iuZ) = \exp\left(iu \sum_{i=1}^k t_i X_i\right) = \varphi_\mathbf{X}(u\mathbf{t}).$$

Let $\mathbf{C} = (C_{ij})_{1 \leq i,j \leq k}$ be the covariance matrix of $\mathbf{X}$: $C_{ij} = Cov(X_i, X_j)$. $\Longrightarrow$: Assume that $\mathbf{X}$ is Gaussian and has the distribution $N(\mathbf{m}, \mathbf{C})$. Then:

$$\varphi_\mathbf{X}(u\mathbf{t}) = \exp\left(iu(\mathbf{t}, m)_k - \frac{1}{2}u^2(\mathbf{t}, \mathbf{Ct})_k\right),$$

so that,

$$\varphi_Z(u) = \exp\left(im_z u - \frac{1}{2}\sigma_Z^2 u^2\right),$$

where $m_Z \in \mathbb{R}$, $\sigma_Z \in \mathbb{R}$ such that $\sigma_Z \geq 0$ and

$$m_z = (\mathbf{t}, \mathbf{m})_k, \quad \sigma_Z^2 = (\mathbf{t}, \mathbf{Ct})_k = \sum_{i,j=1}^{k} C_{ij} t_i t_j \geq 0,$$

since $\mathbf{C}$ is positive. Thus $Z(\mathbf{t})$ is Gaussian of distribution $N(m_z, \sigma_Z)$. $\Longleftarrow$:
Assume that $Z(t)$ is Gaussian for any $\mathbf{t} \in \mathbb{R}^k$. Then:

$$\varphi_Z(u) = \exp\left(im_z u - \frac{1}{2}\sigma_Z^2 u^2\right)$$

Taking $\mathbf{t} = \mathbf{e}_i$, where $\{\mathbf{e}_1, ..., \mathbf{e}_k\}$ is a canonical basis of $\mathbb{R}^k$, we have $Z(\mathbf{e}_i) = X_i$, so that each X_i is Gaussian. Let $N(m_i, \sigma_i)$ be the distribution of X_i. Then, on the one hand,

$$m_z = E(Z(\mathbf{t})) = E\left(\sum_{i=1}^{k} t_i X_i\right) = \sum_{i=1}^{k} t_i E(X_i) = \sum_{i=1}^{k} t_i m_i = (t, m)_k;$$

and, on the other hand,

$$\sigma_Z^2 = V(Z(t)) = \sum_{i,j=1}^{k} C_{ij} t_i t_j = (t, Ct)_k,$$

since:

$$E\left(\left(\sum_{i=1}^{k} t_i X_i\right)^2\right) = E\left(\sum_{i=1}^{k} t_i X_i \cdot \sum_{j=1}^{k} t_j X_j\right) =$$

$$E\left(\sum_{i,j=1}^{k} t_i t_j X_i X_j\right) = \sum_{i,j=1}^{k} t_i t_j E(X_i X_j)$$

and

$$\left[E\left(\sum_{i=1}^{k} t_i X_i\right)\right]^2 = \left(\sum_{i=1}^{k} t_i m_i\right)^2 = \sum_{i,j=1}^{k} t_i t_j m_i m_j = \sum_{i,j=1}^{k} t_i t_j E(X_i) E(X_j),$$

so that:

$$\varphi_Z(u) = \exp\left(i(\mathbf{t}, \mathbf{m})_k u - \frac{1}{2}(\mathbf{t}, \mathbf{C}\mathbf{t})_k u^2\right).$$

Consequently,

$$\varphi_{\mathbf{X}}(\mathbf{t}) = \varphi_Z(1) = \exp\left(i(\mathbf{t}, \mathbf{m})_k - \frac{1}{2}(\mathbf{t}, \mathbf{C}\mathbf{t})_k\right)$$

and $\mathbf{X}$ is Gaussian and has the distribution $N(\mathbf{m}, \mathbf{C})$. ∎

We may generate a Gaussian vector by considering independent Gaussian variables:

PROPOSITION 1.27.– If $X_i \sim N(m_i, \sigma_i)$ and the variables $X_1, ..., X_k$ are independent, then $X = (X_1, ..., X_k)$ is a Gaussian vector and its covariance matrix $\mathbf{C}$ verifies:

$$C_{ij} = 0, \text{ se } i \neq j; \quad C_{ii} = \sigma_i^2 = V(X_i). \blacksquare$$

This result is a consequence of the following Lemma:

LEMMA 1.5.– Let $\mathbf{X} = (X_1, ..., X_k)$ be a vector formed of independent Gaussian variables $(X_i \sim N(m_i, \sigma_i))$. Then, for any $(\alpha_0, \alpha_1, ..., \alpha_n) \in \mathbb{R}^{n+1}$:

$$\alpha_0 + \sum_{i=1}^{n} \alpha_i X_i \text{ is Gaussian and has the distribution } N(m, \sigma);$$

$$m = \alpha_0 + \sum_{i=1}^{n} \alpha_i m_i \text{ and } \sigma^2 = \sum_{i=1}^{n} \alpha_i^2 \sigma_i^2. \blacksquare$$

PROOF.– Let $Y = \alpha_0 + \sum_{i=1}^{n} \alpha_i X_i$. The characteristic function of Y is:

$$\varphi(t) = E(\exp(itY)) = \exp(i\alpha_0 t) E\left(\exp\left(it \sum_{i=1}^{n} \alpha_i X_i\right)\right).$$

Given that the variables are independent:

$$\varphi(t) = \exp(i\alpha_0 t) \prod_{i=1}^{n} E(\exp(i\alpha_i t X_i)) = \exp(i\alpha_0 t) \prod_{i=1}^{n} \varphi_i(\alpha_i t),$$

where φ_i is the characteristic function of X_i. Thus,

$$\varphi(t) = \exp\left(imt - \frac{\sigma^2}{2}t^2\right); \; m = \alpha_0 + \sum_{i=1}^{n} \alpha_i m_i \text{ and } \sigma^2 = \sum_{i=1}^{n} \alpha_i^2 \sigma_i^2$$

and Y is Gaussian. ∎

PROOF.– OF THE PROPOSITION: It is an immediate consequence of the preceding lemma. ∎

We also have:

PROPOSITION 1.28.– Let $\mathbf{X} = (X_1, X_2)$ be a random vector such that $X_i \sim N(m_i, \sigma_i)$. Then, the variables X_1 and X_2 are independent if and only if the characteristic function of $\mathbf{X}$ is $\varphi(t) = \exp\left(i(t_1 m_1 + t_2 m_2) - \frac{1}{2}(\sigma_1^2 t_1^2 + \sigma_2^2 t_2^2)\right)$. ∎

PROOF.– X_1 and X_2 are independent if and only if $\varphi(t) = \varphi_{X_1}(t_1)\varphi_{X_2}(t_2)$, this establishes the result. ∎

COROLLARY 1.11.– Let $\mathbf{X} = (X_1, ..., X_k)$ be a random vector such that $X_i \sim N(m_i, \sigma_i)$. Then the components of $\mathbf{X}$ are independent if and only if the characteristic function of X is $\varphi(t) = \exp\left(i\sum_{i=1}^{k} t_i m_i - \frac{1}{2}\sum_{i=1}^{k} \sigma_i^2 t_i^2\right)$. ∎

PROOF.– It is an immediate consequence of the preceding result. ∎

COROLLARY 1.12.– Let $\mathbf{X} = (X_1, ..., X_k)$ be a Gaussian random vector. Then its components are independent if and only if $Cov(X_i, X_j) = 0$ for $i \neq j$. ∎

PROOF.– $\Longrightarrow$: this implication is immediate, since the independence implies that $Cov(X_i, X_j) = 0$ (Cf. preceding results). $\Longleftarrow$: Let φ be the characteristic function of $\mathbf{X}$ and $X_i \sim N(m_i, \sigma_i)$. If $Cov(X_i, X_j) = 0$ for

$i \neq j$, we have $\varphi(t) = \exp\left(i\sum_{i=1}^{k} t_i m_i - \frac{1}{2}\sum_{i=1}^{k} \sigma_i^2 t_i^2\right)$, so that X is a Gaussian random vector. ∎

COROLLARY 1.13.– Let $\mathbf{X} = (X_1, ..., X_k)$ be a random Gaussian vector and Y a random Gaussian variable such that $Cov(Y, X_i) = 0, 1 \leq i \leq k$. Then Y is independent of X and $Z = (Y, X_1, ..., X_k)$ is a Gaussian random vector. ∎

PROOF.– Let us denote by $\mathbf{m_A}$ the mean of $\mathbf{A}$ and by $\mathbf{C_A}$ the covariance matrix of $\mathbf{A}$. The preceding result shows that Y is independent of $\mathbf{X}$. Consequently, for $\tau = (t_0, \mathbf{t}) \in \mathbb{R} \times \mathbb{R}^k = \mathbb{R}^{k+1}$:

$$E\left(\exp\left(i\,(\tau, Z)_{k+1}\right)\right) = E\left(\exp(it_0 Y)\right) E\left(\exp\left(i\sum_{i=1}^{k} t_i X_i\right)\right),$$

so that:

$$\varphi_Z(\tau) = \varphi_Y(t_0)\,\varphi_X(\mathbf{t})$$

and

$$\varphi_Z(\tau) = \exp\left(it_0 \mathbf{m_Y} - \frac{1}{2}\sigma_Y^2 t^2 + i(\mathbf{t}, \mathbf{m_X})_k - \frac{1}{2}(\mathbf{t}, \mathbf{C_X t})_k\right),$$

i.e.,

$$\varphi_Z(\tau) = \exp\left(i\,(\tau, m_Z)_{k+1} - \frac{1}{2}(\tau, C_Z \tau)_{k+1}\right).$$

Thus, Z is a Gaussian vector. ∎

When the matrix $\mathbf{C}$ is invertible, the probability density of $\mathbf{X}$ is:

$$f(\mathbf{x}) = \frac{1}{(2\pi)^{k/2}\,|\det(\mathbf{C})|^{1/2}} \exp\left(-\frac{1}{2}\left(\mathbf{x}-\mathbf{m}, \mathbf{C}^{-1}(\mathbf{x}-\mathbf{m})\right)\right)$$

1.14. Samples of random variables

Let us consider a random variable $X \in L^2(\Omega, P)$ having the distribution F. This variable has a finite mean m, a finite standard deviation σ and a finite variance σ^2:

$$E(X) = m, \ V(X) = \sigma^2, \ \sigma(X) = \sigma.$$

In practice, the distribution of X and the values m and σ are often unknown and have to be determined by using a *sample* of X.

Naively, a *n-sample* (or *sample of size n*) of a random variable X is a set of n independent values of X: $(X_1, \ldots, X_n)$, where $X_i = X(\omega_i)$, for some $\omega_i \in \Omega$. Thus, a sample of X is a set of realizations of X (or variates from X). These values are used in order to estimate the unknown quantities, by using *estimators* such as, for instance, extremal likelihood, extremal entropy or minimum square ones. The theory of the estimators is outside the scope of this text and will not be studied here – the reader can find numerous references about this point in the literature – we limit ourselves to recall that the distribution F of X is estimated by using the empirical distribution of the sample and the mean is approximated by the empirical mean of the sample. It is interesting to notice that an n-sample may be seen as a vector defined on a *random finite population*, what connects the general approach to the finite population one: the unknown quantities are usually approximated by applying a finite population approach to the values observed in the sample.

There are five key results relating the properties of the samples of a variable X of finite mean and finite variance: the *weak law of large numbers*, the *strong law of large numbers*, the *central limit theorem*, the *law of large deviations* and the *Glivenko–Cantelli theorem*. The first two state that the empirical mean of the sample converges to the mean of X when the size of the sample increases. The third one says that the averages of independent identically distributed random variables is asymptotically normal distributed, the fourth states that the probability of having a large difference between the empirical mean of the sample and the mean of X rapidly decreases with the size of the difference and the size of the sample (it gives an exponentially decreasing upper bound for that probability). The last one establishes that the empirical distribution converges to F uniformly.

Formally, let us consider a sequence $\{X_n\}_{n \in \mathbb{N}} \subset L^2(\Omega, P)$ formed by *independent random variables* having *the same distribution F* as X (i.e. the distribution of X_n is F, $\forall\, n \in \mathbb{N}$. Since their distribution is same, all the elements of the sequence have the same mean, variance and standard deviation as X:

$$\forall n \in \mathbb{N}\ :\ E(X_n) = m,\ V(X_n) = \sigma^2,\ \sigma(X_n) = \sigma.$$

Let us introduce:

$$\psi(t) = \log(E(exp(tX))),$$
$$h(\eta) = \sup\{t(\eta + m) - \psi(\eta) : t \text{ such that } t\eta > 0\}$$

The empirical mean $\overline{X}_n$ is defined by:

$$\overline{X}_n = \frac{1}{n}\sum_{i=1}^{n} X_i.$$

The empirical distribution F_n is defined as:

$$F_n = \frac{1}{nd} sum_{i=1}^{n} 1_{(-\infty, x)}(X_i),\ 1_{(-\infty, x)}(s) = \begin{cases} 1, & \text{if } s < x \\ 0, & \text{otherwise} \end{cases}.$$

The key results are resumed in the following theorem:

THEOREM 1.5.–

1) $E(\overline{X}_n) = m$, $V(\overline{X}_n) = \frac{\sigma^2}{n}$, $\sigma(\overline{X}_n) = \frac{\sigma}{\sqrt{n}}$.

2) (weak law of large numbers) $\overline{X}_n \longrightarrow U$ p.

3) (large deviations law) let us consider $\varepsilon > 0$ p. Then:

$$P(\overline{X}_n - m \geq \varepsilon) \leq \exp(-nh(\varepsilon)),\ P(\overline{X}_n - m \leq -\varepsilon)$$
$$\leq \exp(-nh(-\varepsilon)),$$
$$P(|\overline{X}_n - m| \geq \varepsilon) \leq \exp(-nH(\varepsilon)),\ H(\varepsilon) = \min\{h(\varepsilon), h(-\varepsilon)\}.$$

4) (strong law of large numbers) If $H(\varepsilon) > 0$, for any $\varepsilon > 0$, then $\overline{X}_n \longrightarrow m$ a.s.

5) (central limit) Let:

$$Z_n = \frac{\overline{X}_n - m}{\frac{\sigma}{\sqrt{n}}}$$

Then $Z_n \longrightarrow N(0,1)$ D.

6) (Glivenko-Cantelli) $F_n \longrightarrow F$ p. ∎

1.15. Gaussian samples

Gaussian samples have particular properties which will be exploited in the following. We present here the essential elements which will be used in this text.

DEFINITION 1.25 (Gaussian sample).– Let $\mathbf{X} = (X_1, ..., X_k)$ be a random vector. We say that $\mathbf{X}$ is a *sample* from $N(m, \sigma)$ if and only if all its components are independent and have the same distribution $N(m, \sigma)$. ∎

A sample from $N(m, \sigma)$ is a Gaussian vector such that $\mathbf{m} = (m, ..., m) = m\mathbf{1}$ and $\mathbf{C} = \sigma^2 \mathbf{Id}$.

Let us recall that we can associate to a *linear transformation* $\mathbf{T} : \mathbb{R}^k \longrightarrow \mathbb{R}^k$ an *adjoint transformation* $\mathbf{T}^t$ defined by

$$(\mathbf{T}(\mathbf{x}), \mathbf{y})_k = (\mathbf{x}, \mathbf{T}^t(\mathbf{y}))_k, \ \forall \mathbf{x}, \mathbf{y} \in \mathbb{R}^k.$$

$\mathbf{T}^t$ is also linear: indeed, let us consider $\alpha \in \mathbb{R}$, $\mathbf{x}, \mathbf{y}, \mathbf{z} \in \mathbb{R}^k$. Then:

$$\left(\mathbf{x}, \mathbf{T}^t(\alpha \mathbf{y} + \mathbf{z})\right)_k = (\mathbf{T}(\mathbf{x}), \alpha \mathbf{y} + \mathbf{z})_k = \alpha(\mathbf{T}(\mathbf{x}), \mathbf{y})_k + (\mathbf{T}(\mathbf{x}), \mathbf{z})_k$$

and we have:

$$(\mathbf{x}, \mathbf{T}^t(\alpha \mathbf{y} + \mathbf{z}))_k = \alpha(\mathbf{x}, \mathbf{T}^t(\mathbf{y}))_k + (\mathbf{x}, \mathbf{T}^t(\mathbf{z}))_k = (\mathbf{x}, \alpha \mathbf{T}^t(\mathbf{y}) + \mathbf{T}^t(\mathbf{z}))_k.$$

Given that $\mathbf{x}$ is arbitrary, we obtain $\mathbf{T}^t(\alpha \mathbf{y} + \mathbf{z}) = \alpha \mathbf{T}^t(\mathbf{y}) + \mathbf{T}^t(\mathbf{z})$ and $\mathbf{T}^t$ is linear.

We have $\left(\mathbf{T}^t\right)^t = \mathbf{T}$, since:

$$\left(\mathbf{T}^t(\mathbf{x}), \mathbf{y}\right)_k = \left(\mathbf{y}, \mathbf{T}^t(\mathbf{x})\right)_k = (\mathbf{T}(\mathbf{y}), \mathbf{x})_k = (\mathbf{x}, \mathbf{T}(\mathbf{y}))_k.$$

The linear transformation $\mathbf{T}$ is *orthogonal* if and only if $\mathbf{T}^{-1} = \mathbf{T}^t$. In this case,

$$\|\mathbf{T}(\mathbf{x})\|_k^2 = (\mathbf{T}(\mathbf{x}), \mathbf{T}(\mathbf{x}))_k = (\mathbf{x}, \mathbf{T}^t(\mathbf{T}(\mathbf{x})))_k = (\mathbf{x}, \mathbf{x})_k = \|\mathbf{x}\|_k^2, \ \forall \mathbf{x} \in \mathbb{R}^k.$$

If **T** is orthogonal, then $\mathbf{T}^t$ is orthogonal, since $\left(\mathbf{T}^t\right)^t = \mathbf{T} = \left(\mathbf{T}^{-1}\right)^{-1} = \left(\mathbf{T}^t\right)^{-1}$.

If $\mathcal{B} = \{\mathbf{b}_1, ..., \mathbf{b}_k\}$ and $\mathcal{E} = \{\mathbf{e}_1, ..., \mathbf{e}_k\}$ are orthogonal basis of $\mathbb{R}^k$, then the transformation corresponding to a change of basis:

$$\mathbf{T}(x) = \boldsymbol{\alpha}; \quad \alpha_i = \sum_{j=1}^{k} x_j \left(\mathbf{e}_i, \mathbf{b}_j\right)_k \mathbf{e}_i$$

is orthogonal and

$$\mathbf{T}^t(y) = \boldsymbol{\beta}; \quad \beta_i = \sum_{j=1}^{k} y_j \left(\mathbf{b}_i, \mathbf{e}_j\right)_k \mathbf{b}_i :$$

It is enough to verify that:

$$(\boldsymbol{\alpha}, \mathbf{y})_k = \sum_{i,j=1}^{k} x_j \left(\mathbf{e}_i, \mathbf{b}_j\right)_k y_i = (x, \boldsymbol{\beta})_k \,,$$

so that:

$$(\mathbf{T}(\mathbf{x}), \mathbf{y})_k = (\boldsymbol{\alpha}, \mathbf{y})_k = (\mathbf{x}, \boldsymbol{\beta})_k = \left(\mathbf{x}, \mathbf{T}^t(y)\right)_k.$$

We have:

PROPOSITION 1.29.– Let $\mathbf{X} = (X_1, ..., X_k)$ be a sample from $N(0, \sigma)$. Then:

i) If $\mathbf{T} : \mathbb{R}^k \longrightarrow \mathbb{R}^k$ is an orthogonal transformation, then $\mathbf{T}(\mathbf{X})$ is a sample from $N(0, \sigma)$;

ii) The components of $\mathbf{X}$ in any orthonormal basis of $\mathbb{R}^k$ form a sample from $N(0, \sigma)$;

iii) Let $E_1 \oplus E_2 \oplus ... \oplus E_p$ be a decomposition of $\mathbb{R}^k$ in p orthogonal linear subspaces and $\mathbf{P}_i(\mathbf{X})$ be the orthogonal projection of $\mathbf{X}$ onto E_i. Then, each $\mathbf{P}_i(\mathbf{X})$ is a sample from $N(0, \sigma)$ and, in addition, $\mathbf{P}_i(\mathbf{X})$ and $\mathbf{P}_j(\mathbf{X})$ are independent for $i \neq j$. ∎

PROOF.– Let φ be the characteristic function of $N(0, \sigma)$. Given that the components of $\mathbf{X}$ are independent, we have:

$$\varphi_{\mathbf{X}}(t) = \prod_{i=1}^{k} \varphi(t_i) = \exp\left(-\frac{\sigma^2}{2}\|t\|_k^2\right).$$

Let $\mathbf{T}: \mathbb{R}^k \longrightarrow \mathbb{R}^k$ be an orthogonal transformation and $\mathbf{Y} = \mathbf{T}(\mathbf{X})$. We have:

$$(\mathbf{t}, \mathbf{Y})_k = (\mathbf{t}, \mathbf{T}(\mathbf{X}))_k = \left(\mathbf{T}^t(\mathbf{t}), \mathbf{X}\right)_k$$

so that:

$$\varphi_{\mathbf{Y}}(\mathbf{t}) = E\left(\exp\left(i(\mathbf{t}, \mathbf{Y})_k\right)\right) = E\left(\exp\left(i\left(\mathbf{T}^t(\mathbf{t}), X\right)_k\right)\right)$$

and

$$\varphi_{\mathbf{Y}}(\mathbf{t}) = \exp\left(-\frac{\sigma^2}{2}\left\|\mathbf{T}^t(\mathbf{t})\right\|_k^2\right) = \exp\left(-\frac{\sigma^2}{2}\|\mathbf{t}\|_k^2\right) = \varphi_{\mathbf{X}}(\mathbf{t}),$$

so that we have (i). (ii) follows from (i): The basis change transformation is orthogonal. Finally, (iii) follows from (ii): let $\mathcal{B}_i$ be an orthonormal basis of E_i, $i = 1, ..., p$. Then $\mathcal{B} = \mathcal{B}_1 \cup \mathcal{B}_2 \cup ... \cup \mathcal{B}_p$ is an orthonormal basis of $\mathbb{R}^k$ and the components of $\mathbf{X}$ in the basis $\mathcal{B}$ form a sample from $N(0, \sigma)$, so that we have the result. ■

THEOREM 1.6.– Let $(Y, X_1, ..., X_n)$ be a Gaussian vector. Let us denote $\mathbf{X} = (X_1, ..., X_n)$. Then, there exists $\boldsymbol{\alpha} = (\alpha_0, \alpha_1, ..., \alpha_n) \in \mathbb{R}^{n+1}$ such that:

$$E(Y \mid \mathbf{X}) = \alpha_0 + \sum_{i=1}^{n} \alpha_i X_i.$$

Moreover, $Y - E(Y \mid \mathbf{X})$ is Gaussian, has mean equal to zero and is independent of X and $E(Y \mid \mathbf{X})$. ■

PROOF.– let us consider the auxiliary linear subspace:

$$W = \left\{Z \in L^2(\Omega, P) : Z = \beta_0 + \sum_{i=1}^{n} \beta_i X_i; \ (\beta_0, \beta_1, ..., \beta_n) \in \mathbb{R}^{n+1}\right\}.$$

W is a finite-dimensional linear subspace ($\dim(W) = n+1$), so that the orthogonal projection $P_W(Y)$ of Y onto W exists and is uniquely determined. In addition, it verifies:

$$P_W(Y) \in S \text{ and } E((Y - P_W(Y))Z) = 0, \forall Z \in W.$$

Since $P_W(Y) \in S$, there exists $\boldsymbol{\alpha} = (\alpha_0, \alpha_1, ..., \alpha_n) \in \mathbb{R}^{n+1}$ such that:

$$P_W(Y) = \alpha_0 + \sum_{i=1}^{n} \alpha_i X_i.$$

It follows from the preceding proposition that both $P_W(Y)$ and $Y - P_W(Y)$ are Gaussian. Since $Z = 1 \in W$ (it correspond to $\beta_0 = 1$, $\beta_i = 0, i > 0$), we have:

$$E(Y - P_W(Y)) = E((Y - P_W(Y))1) = 0,$$

so that $Y - P_W(Y)$ has mean equal to zero. Since $X_i \in W$, we also have $E((Y - P_W(Y))X_i) = 0$, so that:

$$Cov(Y - P_W(Y), X_i) = \underbrace{E((Y - P_W(Y))X_i)}_{=0}$$
$$- \underbrace{E(Y - P_W(Y))}_{=0} E(X_i) = 0.$$

It yields from this equality that $Y - P_W(Y)$ is independent of X_i (given that both are Gaussian). Since i is arbitrary, $Y - P_W(Y)$ is independent from $\mathbf{X}$ and $(Y - P_W(Y), X_1, ..., X_n)$ is a Gaussian vector. We also have

$$Cov(Y - P_W(Y), P_W(Y)) =$$
$$\underbrace{E((Y - P_W(Y))P_W(Y))}_{=0} - \underbrace{E(Y - P_W(Y))}_{=0} E(P_W(Y)) = 0,$$

so that $Y - P_W(Y)$ is independent from $P_W(Y)$ (given that both are Gaussian). Let:

$$Z \in S = \{s \in L^2(\Omega, P) : s = \varphi(\mathbf{X}); \varphi : \mathbb{R}^n \longrightarrow \mathbb{R}\}.$$

We have:

$$E((Y - P_W(Y))(P_W(Y) - Z)) =$$
$$\underbrace{E((Y - P_W(Y))P_W(Y))}_{=0} - E((Y - P_W(Y))Z),$$

so that:

$$E((Y - P_W(Y))(P_W(Y) - Z)) = -E((Y - P_W(Y))Z)$$
$$\underset{\text{independent}}{=} -\underbrace{E(Y - P_W(Y))E(Z)}_{=0}$$

and, as a consequence, $E((Y - P_W(Y))(P_W(Y) - Z)) = 0$. Thus,

$$E\left((Y - Z)^2\right) = E\left((Y - P_W(Y) + P_W(Y) - Z)^2\right)$$

verifies:

$$E\left((Y - Z)^2\right) = E\left((Y - P_W(Y))^2\right) + E\left((P_W(Y) - Z)^2\right) \geq E\left((Y - P_W(Y))^2\right),$$

so that:

$$P_W(Y) = \arg\min\ \{\|s - Y\| : Y \in S\}$$

and $P_W(Y)$ is the orthogonal projection of Y onto S. Thus, $P_W(Y) = E(Y \mid \mathbf{X})$. ∎

We have also

PROPOSITION 1.30.– Let $\mathbf{U} = (U_1, U_2)$ be a pair of independent random variables having the same distribution $U(0,1)$. Let:

$$X_1 = \sqrt{-2\log(U_1)}\sin(2\pi U_2);\ X_2 = \sqrt{-2\log(U_1)}\cos(2\pi U_2).$$

Then $\mathbf{X} = (X_1, X_2)$ is a pair of independent random variables having the same distribution $N(0,1)$. ∎

PROOF.– The Jacobian of the transformation $\psi : \mathbb{R}^2 \longrightarrow \mathbb{R}^2$ such that $\psi(U_1, U_2) = (X_1, X_2)$ is:

$$J = \left| \det \begin{pmatrix} \dfrac{-1}{U_1\sqrt{-2\log(U_1)}} \sin(2\pi U_2) & 2\pi\sqrt{-2\log(U_1)} \cos(2\pi U_2) \\ \dfrac{-1}{U_1\sqrt{-2\log(U_1)}} \cos(2\pi U_2) & -2\pi\sqrt{-2\log(U_1)} \sin(2\pi U_2) \end{pmatrix} \right|,$$

that is,

$$J = \frac{2\pi}{U_1}.$$

In addition,

$$X_1^2 + X_2^2 = -2\log(U_1) \Longrightarrow U_1 = \exp\left(-\frac{1}{2}\|X\|_2^2\right)$$

and

$$\sin(2\pi U_2) = \frac{X_1}{\|X\|_2} \text{ and } \cos(2\pi U_2) = \frac{X_2}{\|X\|_2} \Longrightarrow U_2 = \frac{1}{2\pi}\arctan\left(\frac{X_1}{X_2}\right).$$

The probability density of $\mathbf{U}$ is:

$$f(\mathbf{u}) = \begin{cases} 1, & \text{if } \mathbf{u} \in (0,1) \times (0,1) \\ 0, & \text{otherwise.} \end{cases}$$

and the probability density of $\mathbf{X}$ is $g(\mathbf{x}) = f(\psi^{-1}(\mathbf{x}))/J$ (see [DE 92]), that is:

$$g(\mathbf{x}) = \frac{U_1}{2\pi} = \frac{1}{2\pi}\exp\left(-\frac{1}{2}\|\mathbf{x}\|_2^2\right).$$

It follows from this equality that $X = (X_1, X_2)$ is a pair of independent random variables having the distribution $N(0,1)$. ∎

1.16. Stochastic processes

Naively, stochastic process is a family of random variables depending on a parameter: $X = \{X(\lambda)\}_{\lambda \in \Lambda}$ and defined on the same universe Ω, i.e. $\forall n \in \mathbb{N} : X(\lambda) : \Omega \longrightarrow \mathbb{R}$. Λ is the family of *indexes* and we say that the stochastic process is *indexed by* Λ. Thus a stochastic process may simply be considered as a function $X : \Lambda \times \Omega \longrightarrow \mathbb{R}$.

We are mainly interested in stochastic processes indexed by time. Thus, we mainly consider applications $X : (a, b) \times \Omega \longrightarrow \mathbb{R}$, where $(a, b) \subset \mathbb{R}$, i.e.,

DEFINITION 1.26 (stochastic process).– Let $a, b \in \mathbb{R}$ be such that $a < b$. We say that X is a *stochastic process indexed by* (a, b) on the probability space (Ω, P) if and only

$$\forall t \in (a, b) : \omega \longrightarrow X(t, \omega) \text{ is a random variable on } \Omega. \blacksquare$$

Thus, for a given t, the random variable $X_t(\omega) = X(t, \omega)$ has a cumulative distribution given by:

$$P_t(x) = Pr(X_t < x) \qquad [1.25]$$

and a density of probability p_t. These quantities allow the evaluation of the mean of the process, given by

$$\mu_X(t) = E[X_t] = \int_{-\infty}^{\infty} x \, p_t(x) \, dx \qquad [1.26]$$

and its variance, given by

$$\sigma_X^2(t) = E[(X_t - \mu_X(t))^2] = E[X_t^2] - \mu_X(t)^2 \qquad [1.27]$$

The autocorrelation function of the process is

$$R_{XX}(s, t) = E[X_t \, X_s]. \qquad [1.28]$$

The covariance function of the process is

$$\begin{aligned} C(s,t) &= E\left[(X_s - \mu_X(s))(X_t - \mu_X(t))\right] \\ &= E[X_s X_t] - \mu_X(s)\mu_X(t) \\ &= R_{XX}(s,t) - \mu_X(s)\mu_X(t). \end{aligned} \qquad [1.29]$$

We have $C(t,t) = \sigma_X^2(t)$. All these definitions extend to the case of vectors.

1.17. Hilbertian structure

Stochastic processes possess Hilbertian properties analogous to those of random variables. For instance, we may consider the set $\mathcal{V} = \mathcal{E}_P(\ell, (a,b), L^2(\Omega, P))$, i.e. the set formed by the simple functions defined by partitions of (a,b) and taking their values in $L^2(\Omega, P)$. We have

$$\mathcal{V} = \left\{ X : (a,b) \times \Omega \longrightarrow \mathbb{R} : X(t,\omega) = \sum_{i=0}^{n-1} X_i(\omega) 1_{(t_i, t_{i+1})}(t) \; ; \right.$$
$$\left. t \in \mathfrak{Part}((a,b)), \; X_i \in L^2(\Omega, P) \right\}.$$

We may define a scalar product on $\mathcal{V}$:

$$(X, Y) = \int_{(a,b)} E(XY).$$

Indeed, $(\bullet, \bullet)$ is bilinear, symmetric and defined positive: when $X, Y \in \mathcal{V}$, we have

$$X(t,\omega) = \sum_{i=0}^{n_x-1} \overline{X_i} 1_{(x_i, x_{i+1})}(t) \quad \text{and} \quad Y(t,\omega) = \sum_{i=0}^{n_Y-1} \overline{Y_i} 1_{(y_i, y_{i+1})}(t)$$

where $x = (x_0, ..., x_{n_x}) \in \mathfrak{Part}((a,b))$ and $y = (y_0, ..., y_{n_y}) \in \mathfrak{Part}((a,b))$. For

$$n > 0 \text{ such that } h = \frac{b-a}{n} \leq \min\{\delta(x), \delta(y)\},$$

we have

$$XY = \sum_{i=0}^{n-1} X_i Y_i 1_{(a_i, a_{i+1})}(t) \;,\; a_i = a + (i-1)h \;,$$
$$X_i = X(a_i) \;,\; Y_i = Y(a_i)$$

so that

$$(X, Y) = h \sum_{i=0}^{n-1} E(X_i Y_i).$$

In particular,

$$\|X\|^2 = (X, X) = h \sum_{i=0}^{n-1} E\left([X_i]^2\right).$$

The completion of $\mathcal{V}$ for this scalar product is a Hilbert space denoted $V = L^2\left((a,b), L^2(\Omega, P)\right)$.

In the following, we shall use often the *extension principle*, which consists of extending a linear continuous map defined on $\mathcal{V}$ to the set formed by all the Cauchy sequences of its elements and to the sets V such that $\mathcal{V}$ is dense on V. The *extension principle* is based on the following theorems:

THEOREM 1.7.– Let $\mathcal{V}$ be a pre-Hilbertian space of scalar product $(\bullet, \bullet)$ and $I : \mathcal{V} \longrightarrow \mathbb{R}$ a continuous linear application on $\mathcal{V}$. Let

$$\mathbb{V} = \{\widetilde{v} = \{v_n\}_{n \in \mathbb{N}} \subset \mathcal{V} : \widetilde{v} \text{ is a Cauchy sequence for } (\bullet, \bullet)\},$$

and

$$\|\widetilde{v}\| = \lim_{n \longrightarrow +\infty} \|v_n\|.$$

Then

i) $\forall\, \widetilde{v} \in \mathbb{V} : \{I(v_n)\}_{n \in \mathbb{N}} \subset \mathbb{R}$ is a Cauchy sequence;

ii) There exists $\widetilde{I}(\widetilde{v}) \in \mathbb{R}$ such that $I(v_n) \longrightarrow \widetilde{I}(\widetilde{v})$ when $n \longrightarrow +\infty$.

iii) If $\widetilde{w} \in \mathbb{V}$ verifies $\|w_n - v_n\| \longrightarrow 0$ when $n \longrightarrow +\infty$, then $\widetilde{I}(\widetilde{w}) = \widetilde{I}(\widetilde{v})$.

iv) If $\alpha, \beta \in \mathbb{R}$; $\widetilde{v}, \widetilde{w} \in \mathbb{V}$, $\widetilde{u} = \alpha \widetilde{v} + \beta \widetilde{w}$, then $\widetilde{I}(\widetilde{u}) = \alpha \widetilde{I}(\widetilde{v}) + \beta \widetilde{I}(\widetilde{w})$.

v) Let $M \in \mathbb{R}$ verify $|I(v)| \leq M \|v\|$, $\forall\, v \in \mathcal{V}$. Then $\left|\widetilde{I}(\widetilde{v})\right| \leq M \|\widetilde{v}\|$, $\forall\, \widetilde{v} \in \mathbb{V}$. ∎

PROOF.– Let us observe that $\{\|v_n\|\}_{n \in \mathbb{N}} \subset \mathbb{R}$ is a Cauchy sequence, since

$$|\|v_m\| - \|v_n\|| \leq \|v_m - v_n\|$$

and

$$m, n \geq n(\varepsilon) \Longrightarrow \|v_m - v_n\| \leq \varepsilon \Longrightarrow |\|v_m\| - \|v_n\|| \leq \varepsilon.$$

Thus, there exists $m \in \mathbb{R}$ such that $\|v_n\| \longrightarrow m$ when $n \longrightarrow +\infty$.

i) $\{I(v_n)\}_{n \in \mathbb{N}} \subset \mathbb{R}$ is a Cauchy sequence, since

$$|I(v_m) - I(v_n)| = \|I(v_m - v_n)\| \leq M \|v_m - v_n\|$$

and $\{v_n\}_{n \in \mathbb{N}}$ is a Cauchy sequence.

ii) Given that $\mathbb{R}$ is complete, there exists $m \in \mathbb{R}$ such that

$$m = \lim_{n \longrightarrow +\infty} I(v_n).$$

The result is obtained by taking $\widetilde{I}(\widetilde{v}) = m$.

iii) If $\{w_n\}_{n \in \mathbb{N}} \subset \mathcal{V}$ verifies $\|w_n - v_n\| \longrightarrow 0$ when $n \longrightarrow +\infty$, we have

$$|I(w_n) - I(v_n)| = \|I(w_n - v_n)\| \leq M \|w_n - v_n\| \longrightarrow 0,$$

so that

$$\lim_{n \longrightarrow +\infty} I(w_n) = \lim_{n \longrightarrow +\infty} I(v_n)$$

and, as a consequence, $\widetilde{I}(\widetilde{w}) = \widetilde{I}(\widetilde{v})$.

iv) We have

$$I(u_n) = I(\alpha v_n + \beta w_n) = \alpha I(v_n) + \beta I(w_n) \longrightarrow \alpha \widetilde{I}(\widetilde{v}) + \beta \widetilde{I}(\widetilde{w}),$$

so that $\widetilde{I}(\widetilde{u}) = \alpha \widetilde{I}(\widetilde{v}) + \beta \widetilde{I}(\widetilde{w})$.

v) Since $\|v_n\| \longrightarrow \|\widetilde{v}\|$ when $n \longrightarrow +\infty$ and $|I(v_n)| \longrightarrow |\widetilde{I}(\widetilde{v})|$, we have

$$|I(v_n)| \leq M \|v_n\|, \forall n \in \mathbb{N} \Longrightarrow |\widetilde{I}(\widetilde{v})| \leq M \|v\|. \blacksquare$$

THEOREM 1.8.– Let $\mathcal{V}$ be a pre-Hilbertian space for the scalar product $(\bullet, \bullet)$ and $I : \mathcal{V} \longrightarrow \mathbb{R}$ a continuous linear application on $\mathcal{V}$. If V is a linear space such that $\mathcal{V}$ is dense on V, then I may be extended to V, i.e. there exists $I_V : V \longrightarrow \mathbb{R}$ linear continuous which coincides with I on $\mathcal{V}$. In addition, if $M \in \mathbb{R}$ verifies $|I(v)| \leq M \|v\|, \forall v \in \mathcal{V}$, then $|I_V(v)| \leq M \|v\|, \forall v \in V$. ∎

PROOF.– Let us recall that a linear map $I : \mathcal{V} \longrightarrow \mathbb{R}$ is continuous if and only if there exists $M \in \mathbb{R}$ tsuch that $|I(v)| \leq M \|v\|, \forall v \in \mathcal{V}$. Given that $\mathcal{V}$ is dense on V: for any $v \in V$, there exists $\widetilde{v} = \{v_n\}_{n \in \mathbb{N}} \subset \mathcal{V}$ such that $\|v_n - v\| \longrightarrow 0$ when $n \longrightarrow +\infty$. Let us consider

$$I_V(v) = \lim_{n \longrightarrow +\infty} I(v_n).$$

$\widetilde{v}$ is a sequence of Cauchy (since it converges). Thus, $I_V(v) = \widetilde{I}(\widetilde{v})$. It yields from the preceding theorem that the limit exists and is well-defined: if $\widetilde{w} = \{w_n\}_{n \in \mathbb{N}} \subset \mathcal{V}$ verifies $\|w_n - v\| \longrightarrow 0$ when $n \longrightarrow +\infty$, then $\{w_n\}_{n \in \mathbb{N}}$ is a sequence of Cauchy (given that it converges) and $\|w_n - v_n\| \longrightarrow 0$ when $n \longrightarrow +\infty$, so that $\widetilde{I}(\widetilde{v}) = \widetilde{I}(\widetilde{w})$ and, as a consequence,

$$\lim_{n \longrightarrow +\infty} I(v_n) = \lim_{n \longrightarrow +\infty} I(w_n).$$

$I_V : V \longrightarrow \mathbb{R}$ is linear: let $\alpha, \beta \in \mathbb{R}$; $v, w \in V$; $u = \alpha v + \beta w$; $\widetilde{v} = \{v_n\}_{n \in \mathbb{N}} \subset \mathcal{V}$, $\widetilde{w} = \{w_n\}_{n \in \mathbb{N}} \subset \mathcal{V}$, $\|v_n - v\| \longrightarrow 0$ and $\|w_n - w\| \longrightarrow 0$ when $n \longrightarrow +\infty$. Then $\widetilde{u} = \{u_n\}_{n \in \mathbb{N}} \subset \mathcal{V}$ given by $u_n = \alpha v_n + \beta w_n$ verifies $\widetilde{I}(\widetilde{u}) = \alpha \widetilde{I}(\widetilde{v}) + \beta \widetilde{I}(\widetilde{w})$. Thus, $I_V(w) = \alpha I_V(u) + \beta I_V(v)$. $I_V : V \longrightarrow \mathbb{R}$ is continuous : if $\widetilde{v} = \{v_n\}_{n \in \mathbb{N}} \subset \mathcal{V}$ verifies $\|v_n - v\| \longrightarrow 0$ when $n \longrightarrow +\infty$, then $\|v_n\| \longrightarrow \|v\|$, so that $\|\widetilde{v}\| = \|v\|$ and we have $\left|\widetilde{I}(\widetilde{v})\right| \leq M \|v\| \Longrightarrow |I_V(v)| \leq M \|v\|$. ∎

COROLLARY 1.14.– Let $\mathcal{V}$ be a pre-hilbertian space for the scalar product $(\bullet, \bullet)$ and $I : \mathcal{V} \longrightarrow \mathbb{R}$ a continuous linear map on $\mathcal{V}$. If V is the completion of $\mathcal{V}$ for $(\bullet, \bullet)$, then I extends to V. ∎

PROOF.– It is enough to notice that $\mathbb{V} = \{\{v_n\}_{n \in \mathbb{N}} \subset \mathcal{V} : \{v_n\}_{n \in \mathbb{N}}$ is a Cauchy sequence $(\bullet, \bullet) \}$ is dense on V. ∎

1.18. Wiener process

One of the main stochastice processes is the following one:

DEFINITION 1.27 (Unidimensional Wiener process).– Let $W = \{W(t)\}_{t \geq 0}$ be a stochastic process on the probability space (Ω, P). We say that W is a *unidimensional Wiener process* or *unidimensional brownian motion* if and only if:

i) $W(0) = 0$;

ii) $\forall (s, t) \in \mathbb{R}^2$ such that $0 \leq s \leq t$:

　a) $W(t) - W(s) \sim N(0, \sqrt{t-s})$;

　b) $W(t) - W(s)$ is independent of $W(b) - W(a)$, $\forall (a, b) \in \mathbb{R}^2$ such that $0 \leq a \leq b \leq s \leq t$.∎

Sometimes, the expression *univariate Wiener process* is used in order to refer to this stochastic process. We have

PROPOSITION 1.31.– Let $(s, t) \in \mathbb{R}^2$ be such that $0 \leq s \leq t$. Then

i) $E((W(t) - W(s))W(s)) = 0$;

ii) $E\left(W(s)^2\right) = s$;

iii) $E(W(t)W(s)) = s$;

iv) $E((W(t) - W(s))W(t)) = t - s$. ∎

PROOF.– Taking $a = 0$ and $b = s$, we have: $W(t) - W(s)$ is independent from $W(s) - W(0) = W(s) - 0 = W(s)$, so that

$$E((W(t) - W(s))W(s)) = \underbrace{E(W(t) - W(s))}_{=0} E(W(s)) = 0.$$

In an analogous way,

$$E\left(W(s)^2\right) = E\left((W(s) - W(0))^2\right) = V(W(s) - W(0)) = s - 0 = s.$$

Thus,

$$E(W(t)W(s)) = E((W(t) - W(s) + W(s))W(s)) =$$
$$\underbrace{E((W(t) - W(s))W(s))}_{=0} + \underbrace{E\left(W(s)^2\right)}_{=s},$$

so that $E(W(t)W(s)) = s$. Finally,

$$E((W(t) - W(s))W(t)) = \underbrace{E\left(W(t)^2\right)}_{=t} - \underbrace{E(W(t)W(s))}_{=s} = t - s. \blacksquare$$

THEOREM 1.9.– Let $(s,t) \in \mathbb{R}^2$ be such that $0 \leq s \leq t$; $k \in \mathbb{N}$ be such that $k > 0$; $t = (t_1, ..., t_k) \in \mathbb{R}^k$ be such that $0 \leq t_i \leq s$ ($1 \leq i \leq k$) and $t_i \leq t_{i+1}$ ($1 \leq i \leq k-1$); $\mathbf{X} = (W(t_1), ..., W(t_k))$, $Y = W(t) - W(s)$. Then $E(Y \mid \mathbf{X}) = 0. \blacksquare$

PROOF.– Let $a = 0$, $b = t_i : 0 \leq a \leq b \leq s$, so that $W(t) - W(s)$ is independent from $W(b) - W(a) = W(t_i)$. So, Y is independent from $\mathbf{X}$. As a consequence, $E(Y \mid \mathbf{X}) = E(Y) = 0. \blacksquare$

DEFINITION 1.28 (Multidimensional Wiener process).– Let $W = \{\mathbf{W}(t)\}_{t \geq 0}$ be a family of random vectors $\mathbb{R}^k$ on the probability space (Ω, P). We say that $\mathbf{W}$ is a *k-dimensional Wiener process* or *k-dimensional brownian motion* if and only if $\mathbf{W}(t) = (W_1(t), ..., W_k(t))$, where

(i) $\{W_i(t)\}_{t \geq 0}$ is an unidimensional Wiener process ($1 \leq i \leq k$);

(ii) The components of $\mathbf{W}$ are mutually independent, i.e. if $i \neq j$ then $W_i(t)$ is independent of $W_j(s)$, $\forall (s,t) \in \mathbb{R}^2$ tal que $s, t \geq 0. \blacksquare$

Sometimes, the expression *multivariate Wiener process* is used in order to refer to this stochastic process.

In order to simulate a Wiener process, discretization of time is requested. The reader may find in the literature many works on this topic. Here, we illustrate the simulation by using a simple Euler discretization combined to the random number generator `randn`):

Listing 1.21. *Simple simulation of a Wiener process*

```
function w = wiener(ndim, nstep, timestep)
%
% generates nstep steps of a Wiener process of  dimension ndim
%
% IN:
% ndim   : dimension  - type integer
% nstep: number of steps  - type integer
% timestep: the time step  - type double
%
% OUT:
% w : ndim x (nstep + 1) table  - type double
%     w(i,:) is the value of W((i-1)*timestep)
%
w = zeros(ndim, nstep+1);
aux = sqrt(timestep);
for i = 1: nstep
    w(i+1,:) = w(i,:) + aux*randn(1, ndim);
end;
return;
end
```

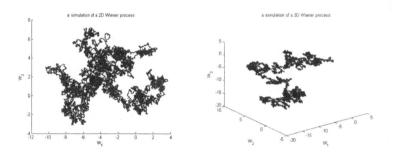

Figure 1.6. *Simulation of a Wiener process*

1.19. Ito integrals

We present in this section the definition of some stochastic integrals. We are mainly interested in

$$\int_0^T \varphi(W(t))\,dt \text{ and } \int_0^T \varphi(W(t))\,dW(t),$$

where $\varphi : \mathbb{R} \longrightarrow \mathbb{R}$ is a function and $W = \{W(t)\}_{t \geq 0}$ is a Brownian motion. The definitions below are usually referred as being those "in the sense of Ito". There are concurrent definitions, such as those "in the sense of Stratanovich", but they are outside the scope of this text.

1.19.1. *Integrals with respect to time*

In this section, we give a definition for

$$I(X) = \int_a^b X(t)\,dt.$$

Let $t = (t_0, ..., t_n) \in \mathfrak{Part}((a,b))$ be a n–partition of (a,b). Let us consider the finite sum

$$I(X, t) = \sum_{i=1}^n X(t_{i-1})(t_i - t_{i-1}).$$

We say that

$$\int_a^b X(t)\,dt = Y$$

when

$$I(X, t) \longrightarrow \int_a^b X(t)\,dt \text{ for } \delta(t) \longrightarrow 0,$$

i.e.,

DEFINITION 1.29.– Let $X \in L^2((a,b), L^2(\Omega, P))$. We say that $\int_a^b X(t)\,dt = Y \in L^2(\Omega, P)$ if and only if, for any $\varepsilon > 0$, there exists $\eta(\varepsilon) > 0$ such that

$$\forall t \in \mathfrak{Part}((a,b)) : \delta(t) \leq \eta(\varepsilon) \implies \|Y - I(X, t)\|_{L^2(\Omega, P)} \leq \varepsilon. \blacksquare$$

Let us recall that the scalar product and the norm of $L^2(\Omega, P)$ are given by

$$(X, Y)_{L^2(\Omega, P)} = E(XY) \; ; \; \|Z\|_{L^2(\Omega, P)} = E(Z^2).$$

Let $n > 0$ and consider

$$I_n(X) = h \sum_{i=1}^{n} X(a + (i-1)h) \; ; \; h = \frac{b-a}{n}.$$

When the integral exists, we also have

$$I_n(X) \longrightarrow \int_a^b X(t)\,dt \text{ for } n \longrightarrow +\infty.$$

The integral with respect to time is linear and continuous:

THEOREM 1.10.– let X, Y be two stochastic processes indexed by (a, b) such that $\int_a^b X(t)\,dt$ and $\int_a^b Y(t)\,dt$ both exist. Let $\alpha, \beta \in \mathbb{R}$. Then

$$\int_a^b (\alpha X(t) + \beta Y(t))\,dt = \alpha \int_a^b X(t)\,dt + \beta \int_a^b Y(t)\,dt.$$

In addition, there is a constant $C = C(a, b) \in \mathbb{R}$, independent from X such that

$$\forall X \in L^2((a, b), L^2(\Omega, P)) : \left\| \int_a^b X(t)\,dt \right\|_{L^2(\Omega, P)} \leq C \|X\|_{L^2((a,b), L^2(\Omega, P))}. \blacksquare$$

PROOF.– For any $t = (t_0, ..., t_n) \in \mathfrak{Part}((a, b))$,

$$I(\alpha X + \beta Y, t) = \alpha I(X, t) + \beta I(Y, t),$$

so that $I(\alpha X + \beta Y) = \alpha I(X) + \beta I(Y)$. By the extension principle, it is enough to consider $X \in \mathcal{E}_P(\ell, (a,b), L^2(\Omega, P))$. For such an X, we have

$$X(t, \omega) = \sum_{i=0}^{n_x - 1} \overline{X_i} 1_{(x_i, x_{i+1})}(t),$$

where $x = (x_0, ..., x_{n_x}) \in \mathfrak{Part}((a,b))$. Let $n > 0$, $h = \dfrac{b-a}{n}$, $a_i = a + (i-1)h$ and $X_i = X(a_i)$ ($1 \le i \le n+1$). Then, for $h \le \delta(x)$, we have

$$X(t, \omega) = \sum_{i=0}^{n-1} X_i 1_{(a_i, a_{i+1})}(t) = h^2 \sum_{i=1}^{n} X_i^2;$$

$$[I_n(X)]^2 = h^2 \left[\sum_{i=1}^{n} X_i\right]^2 = h^2 \sum_{i,j=1}^{n} X_i X_j \le h^2 \sum_{i,j=1}^{n} \frac{X_i^2 + X_j^2}{2}.$$

Thus,

$$[I_n(X)]^2 \le nh^2 \sum_{i=1}^{n} X_i^2 = (b-a) h \sum_{i=1}^{n} X_i^2$$

and

$$\|I_n(X)\|_{L^2(\Omega, P)}^2 = E\left([I_n(X)]^2\right) \le (b-a) h \sum_{i=1}^{n} E(X_i^2) =$$

$$(b-a) \int_{(a,b)} E(X^2) = (b-a) \|X\|_{L^2((a,b), L^2(\Omega, P))}^2,$$

that establishes the result. ∎

1.19.2. *Integrals with respect to a process*

In this section, we give a definition for

$$I(X, Y) = \int_a^b X(t)\, dY(t).$$

Let $t = (t_0, ..., t_n) \in \mathfrak{Part}((a,b))$ be a n–partition of (a,b). Let us consider the finite sum

$$I(X, Y, t) = \sum_{i=1}^{n} X(t_{i-1})(Y(t_i) - Y(t_{i-1})).$$

We say that

$$\int_a^b X(t)\, dY(t) = Z$$

when

$$I(X, Y, t) \longrightarrow Z \text{ for } \delta(t) \longrightarrow 0,$$

i.e.,

DEFINITION 1.30.– Let $X, Y \in L^2((a,b), L^2(\Omega, P))$. We say that $\int_a^b X(t)\, dY(t) = Z \in L^2(\Omega, P)$ if and only if, for any $\varepsilon > 0$, there exists $\eta(\varepsilon) > 0$ such that

$$\forall t \in \mathfrak{Part}((a,b)) : \delta(t) \leq \eta(\varepsilon) \implies \|Z - I(X, Y, t)\|_{L^2(\Omega, P)} \leq \varepsilon. \blacksquare$$

Let us consider $n > 0$ and, with $h = \frac{b-a}{n}$,

$$I_n(X, Y) = \sum_{i=1}^{n} X(a + (i-1)h)(Y(a+ih) - Y(a+(i-1)h)).$$

When the integral exists, we also have

$$I_n(X, Y) \longrightarrow \int_a^b X(t)\, dY(t) \text{ for } n \longrightarrow +\infty.$$

The integral with respect a process is bilinear:

PROPOSITION 1.32.– Let $X, Y, Z \in L^2\left((a,b), L^2(\Omega, P)\right)$ be such that $\int_a^b X(t)\, dY(t)$ and $\int_a^b X(t)\, dZ(t)$ there exist. Let $\alpha, \beta \in \mathbb{R}$. Then

$$\int_a^b X(t)\, d\left((\alpha Y(t) + \beta Z(t))\right) = \alpha \int_a^b X(t)\, dY(t) + \beta \int_a^b Y(t)\, dZ(t) \,. \blacksquare$$

PROOF.– For any $t = (t_0, ..., t_n) \in \mathfrak{Part}((a,b))$,

$$I(X, \alpha Y + \beta Z, t) = \alpha I(X, Y, t) + \beta I(X, Z, t) \,,$$

so that $I(X, \alpha Y + \beta Z) = \alpha I(X, Y) + \beta I(X, Z)$. $\blacksquare$

PROPOSITION 1.33.– Let $X, Y, Z \in L^2\left((a,b), L^2(\Omega, P)\right)$ be such that $\int_a^b X(t)\, dZ(t)$ and $\int_a^b Y(t)\, dZ(t)$ exist. let $\alpha, \beta \in \mathbb{R}$. Then

$$\int_a^b \left(\alpha X(t) + \beta Y(t)\right) dZ(t) = \alpha \int_a^b X(t)\, dZ(t) + \beta \int_a^b Y(t)\, dZ(t) \,. \blacksquare$$

PROOF.– For any $t = (t_0, ..., t_n) \in \mathfrak{Part}((a,b))$,

$$I(\alpha Y + \beta Z, Z, t) = \alpha I(X, Z, t) + \beta I(Y, Z, t) \,,$$

so that $I(\alpha X + \beta Y, Z) = \alpha I(X, Z) + \beta I(Y, Z)$. $\blacksquare$

1.19.3. *Integrals with respect to a Wiener process*

Integrals having the form

$$I_t(\varphi) = \int_a^a \varphi(W(t))\, dt \quad \text{and} \quad I_W(\varphi) = \int_a^b \varphi(W(t))\, dW(t)$$

are a particular case of the preceding ones, which corresponds to

$$I_t(\varphi) = I(\varphi(W)) \text{ and } I_W(\varphi) = I(\varphi(W), W)$$

We have

PROPOSITION 1.34.– Let $(W), \psi(W) \in L^2((a,b), L^2(\Omega, P))$. Then, for $\alpha, \beta \in \mathbb{R}$,

$$I_W(\alpha\varphi + \beta\psi) = \alpha I_W(\varphi) + \beta I_W(\psi).$$

In addition,

$$E(I_W(\varphi)) = 0,$$

$$\|I_W(\varphi)\|^2_{L^2(\Omega, P)} = \|\varphi(W)\|^2_{L^2((a,b), L^2(\Omega, P))},$$

$$(I_W(\varphi), I_W(\psi))_{L^2(\Omega, P)} = (\varphi(W), \psi(W))_{L^2((a,b), L^2(\Omega, P))}. \blacksquare$$

PROOF.– We have

$$I_W(\alpha\varphi + \beta\psi) = I(\alpha\varphi(W) + \beta\psi(W), W)$$
$$= \alpha I(\varphi(W), W) + \beta I(\psi(W), W) = \alpha I_W(\varphi) + \beta I_W(\psi).$$

For the rest of the equalities, the extension principle is applied: it is enough to show the result for φ, ψ such that $\varphi(W), \psi(W) \in \mathcal{E}_P(\ell, (a,b), L^2(\Omega, P))$. In this case,

$$\varphi(W)(t,\omega) = \sum_{i=0}^{n_x-1} \overline{\varphi_i} 1_{(x_i, x_{i+1})}(t) \text{ and } \psi(W)(t,\omega) = \sum_{i=0}^{n_Y-1} \overline{\psi_i} 1_{(y_i, y_{i+1})}(t)$$

where $x = (x_0, ..., x_{n_x}) \in \mathfrak{Part}((a,b))$ and $y = (y_0, ..., y_{n_y}) \in \mathfrak{Part}((a,b))$. For

$$n > 0 \text{ such that } h = \frac{b-a}{n} \leq \min\{\delta(x), \delta(y)\}, a_i = a + ih,$$
$$\varphi_i = \varphi(W_i), \psi_i = \psi(W_i), W_i = W(a_i),$$

we have

$$\varphi(W)(t,\omega) = \sum_{i=0}^{n-1} \varphi_i 1_{(a_i, a_{i+1})}(t) \text{ and } \psi(W)(t,\omega) = \sum_{i=0}^{n-1} \psi_i 1_{(a_i, a_{i+1})}(t) .$$

Thus,

$$I_W(\varphi) = \sum_{i=0}^{n-1} \varphi_i (W_{i+1} - W_i) \text{ and } I_W(\psi) = \sum_{i=0}^{n-1} \psi_i (W_{i+1} - W_i),$$

so that

$$E(I_W(\varphi)) = \sum_{i=0}^{n-1} E(\varphi_i (W_{i+1} - W_i)).$$

Since $\varphi_i = \varphi(W_i)$ is independent from $W_{i+1} - W_i$:

$$E(I_W(\varphi)) = \sum_{i=0}^{n-1} E(\varphi_i) \underbrace{E(W_{i+1} - W_i)}_{=0} = 0.$$

We also have

$$I_W(\varphi) I_W(\psi) = \sum_{i,j=0}^{n-1} \varphi_i \psi_j (W_{i+1} - W_i)(W_{j+1} - W_j).$$

For $i > j$, $(W_{i+1} - W_i)$ is independent from $(W_{j+1} - W_j)$, φ_i and ψ_j so that

$$E\left(\varphi_i \psi_j (W_{i+1} - W_i)(W_{j+1} - W_j)\right) = \underbrace{E(W_{i+1} - W_i)}_{=0} E\left(\varphi_i \psi_j (W_{j+1} - W_j)\right) = 0.$$

Thus,

$$E(I_W(\varphi) I_W(\psi)) = \sum_{i=0}^{n-1} E\left(\varphi_i \psi_i (W_{i+1} - W_i)^2\right) = h \sum_{i=0}^{n-1} E(\varphi_i \psi_i) = I_t(\varphi) = I(\varphi(W)).$$

Since $(W_{i+1} - W_i)$ is independent from φ_i and ψ_i:

$$E\left(I_W\left(\varphi\right) I_W\left(\psi\right)\right) = \sum_{i=0}^{n-1} E\left(\varphi_i \psi_i\right) \underbrace{E\left((W_{i+1} - W_i)^2\right)}_{= h} = h \sum_{i=0}^{n-1} E\left(\varphi_i \psi_i\right).$$

and we have

$$E\left(I_W\left(\varphi\right) I_W\left(\psi\right)\right) = \left(\varphi\left(W\right), \psi\left(W\right)\right)_{L^2((a, b), L^2(\Omega, P))},$$

i.e.,

$$\left(I_W\left(\varphi\right), I_W\left(\psi\right)\right)_{L^2(\Omega, P)} = \left(\varphi\left(W\right), \psi\left(W\right)\right)_{L^2((a, b), L^2(\Omega, P))}.$$

Taking $\psi = \varphi$ in this equality, we obtain

$$\|I_W\left(\varphi\right)\|_{L^2(\Omega, P)}^2 = \|\varphi\left(W\right)\|_{L^2((a, b), L^2(\Omega, P))}^2. \blacksquare$$

1.20. Ito Calculus

1.20.1. *Ito's formula*

We observe that

$$E\left(\frac{1}{2}\int_a^b W(t)\, dW(t)\right) = 0 \quad \text{and} \quad E\left([W(b)]^2 - [W(a)]^2\right) = b - a,$$

so that Ito integrals do not follow the standard calculus rules. We have

PROPOSITION 1.35 (Ito's formula).– Let $u \in \mathcal{D}\left(\mathbb{R}\right)$ (i.e., $u \in C^\infty\left(\mathbb{R}\right)$ and its support is compact) be such that $\int_a^b u'\left(W(t)\right) dW(t)$ and $\int_a^b u''\left(W(t)\right) dt$ both exist. Then

$$\int_a^b u'\left(W(t)\right) dW(t) = u\left(W(b)\right) - u\left(W(a)\right) - \frac{1}{2}\int_a^b u''\left(W(t)\right) dt. \blacksquare$$

PROOF.– Let us consider $n > 0$, $h = (b-a)/n$ and $a_i = a + (i-1)h$, $1 \leq i \leq n+1$. We have

$$u(W_{i+1}) - u(W_i) - u'(W_i)(W_{i+1} - W_i) - \tfrac{1}{2}u''(W_i)(W_{i+1} - W_i)^2 = \tfrac{1}{6}u'''(\xi_i)(W_{i+1} - W_i)^3,$$

where ξ_i is a random variable. Thus,

$$u(W(b)) - u(W(a)) - T_n - V_n - U_n = 0,$$

where

$$T_n = \sum_{i=1}^{n} u'(W_i)(W_{i+1} - W_i), \quad V_n = \sum_{i=1}^{n} u''(W_i)(W_{i+1} - W_i)^2,$$

$$U_n = \tfrac{1}{6}\sum_{i=1}^{n} u'''(\xi_i)(W_{i+1} - W_i)^3.$$

1) We have (by definition),

$$T_n \longrightarrow \int_a^b u'(W(t))\,dW(t) \text{ when } n \longrightarrow +\infty.$$

2) Let

$$S_n = \sum_{i=1}^{n}(W_{i+1} - W_i)^2 \; ; \; W_i = W(a_i).$$

Let us consider $Z_i = (W_{i+1} - W_i)^2 - h$. We have:

$$Z_i Z_j = (W_{i+1} - W_i)^2 (W_{j+1} - W_j)^2 - h(W_{i+1} - W_i)^2 - h(W_{j+1} - W_j)^2 + h^2.$$

For $i \neq j$, $W_{i+1} - W_i$ and $W_{j+1} - W_j$ are independent, so that

$$E\left((W_{i+1} - W_i)^2(W_{j+1} - W_j)^2\right) = \underbrace{E\left((W_{i+1} - W_i)^2\right)}_{=h}\underbrace{E\left((W_{j+1} - W_j)^2\right)}_{=h} = h^2$$

and

$$E(Z_i Z_j) = h^2 - h^2 - h^2 + h^2 = 0.$$

For $i = j$, we have

$$Z_i^2 = (W_{i+1} - W_i)^4 - 2h(W_{i+1} - W_i)^2 + h^2$$

$$E(Z_i^2) = \underbrace{M_4(W_{i+1} - W_i)}_{=3h^2} - 2h\underbrace{M_2(W_{i+1} - W_i)}_{=h} + h^2 = h^2.$$

Thus,

$$E\left((S_n - (b-a))^2\right) = E\left(\left(\sum_{i=1}^{n} Z_i\right)^2\right) =$$

$$\sum_{i=1}^{n} E(Z_i^2) = nh^2 = \frac{(b-a)^2}{n},$$

so that

$$E\left((S_n - (b-a))^2\right) \longrightarrow 0 \text{ when } n \longrightarrow +\infty.$$

3) Let

$$M_3 = \max\left\{|u'''(s)| : s \in \mathbb{R}\right\} < \infty.$$

Let $B_i = u'''(\xi_i)(W_{i+1} - W_i)^3$. We have

$$B_i^2 \le \frac{M_3^2}{36}(W_{i+1} - W_i)^6.$$

Since $W_{i+1} - W_i \sim N\left(0, \sqrt{h}\right)$:

$$E\left((W_{i+1} - W_i)^6\right) = h^3 E\left(\left(\frac{W_{i+1} - W_i}{\sqrt{h}}\right)^6\right) = \frac{(6)!}{3!2^3}h^3,$$

so that there exists a constant $C \in \mathbb{R}$ such that

$$E\left(B_i^2\right) \leq Ch^3.$$

Or,

$$U_n = \sum_{i=1}^{n} B_i,$$

so that

$$E\left(U_n^2\right) = \sum_{i,j=1}^{n} E\left(B_i B_j\right) \leq \sum_{i,j=1}^{n} \sqrt{E\left(B_i^2\right)}\sqrt{E\left(B_j^2\right)} \leq n^2 Ch^3 = C\frac{(b-a)^3}{n}$$

and

$$U_n \longrightarrow 0 \text{ when } n \longrightarrow +\infty.$$

4) Let

$$M_2 = \max\left\{\left|u''(s)\right| : s \in \mathbb{R}\right\} < \infty$$

and

$$A_i = u''(W_i)(W_{i+1} - W_i)^2 - u''(W_i)(a_{i+1} - a_i) = u''(W_i) Z_i.$$

For $i > j$, $W_{i+1} - W_i$ is independent of W_{i+1}, $W_{j+1} - W_j$ and W_j, so that Z_i is independent of $u''(W_i)$, $u''(W_j)$ and Z_j. Thus, for $i > j$,

$$E\left(A_i A_j\right) = \underbrace{E\left(Z_i\right)}_{=0} E\left(u''(W_i) u''(W_j) Z_j\right) = 0.$$

For $i = j$, we have

$$E\left(A_i^2\right) \leq M_2^2 E\left(Z_i^2\right) = M_2^2 h^2.$$

Thus,

$$E\left(\left(V_n - h\sum_{i=1}^{n} u''(W_i)\right)^2\right) = E\left(\left(\sum_{i=1}^{n} A_i\right)^2\right) = \sum_{i=1}^{n} E(A_i^2) \le M_2^2 nh^2 = M_2^2 \frac{(b-a)^2}{n},$$

and

$$E\left(\left(V_n - h\sum_{i=1}^{n} u''(W_i)\right)^2\right) \longrightarrow 0 \text{ when } n \longrightarrow +\infty,$$

so that

$$V_n \longrightarrow \int_a^b u''(W(t))\,dt \text{ when } n \longrightarrow +\infty.$$

5) Thus,

$$T_n + \frac{1}{2}V_n - U_n \longrightarrow \int_a^b u'(W(t))\,dW(t) + \frac{1}{2}\int_a^b u''(W(t))\,dt,$$

so that

$$u(W(b)) - u(W(a)) - \int_a^b u'(W(t))\,dW(t) - \frac{1}{2}\int_a^b u''(W(t))\,dt = 0$$

for any $u \in \mathcal{D}(\mathbb{R})$. ∎

Invoking the extension principle, this result applies to any u belonging to the adherence and completion of $D(\mathbb{R})$.

Ito's formula extends to the multidimensional situation where $u : \mathbb{R}^p \longrightarrow \mathbb{R}$:

$$\int_a^b \nabla u\left(W(t)\right) d\mathbf{W}(t) = u\left(\mathbf{W}(b)\right) - u\left(\mathbf{W}(a)\right) - \frac{1}{2} \int_a^b \Delta u\left(\mathbf{W}(t)\right) dt.$$

1.20.2. *Ito stochastic diffusions*

We write

$$dX_t = a\left(t, W_t\right) dW_t + b\left(t, W_t\right) dt$$

for the equality

$$X(t) - X(0) = \int_0^t a\left(s, W(s)\right) dW(s) + \int_0^t b\left(s, W(s)\right) ds.$$

and we say that the stochastic process X is a *stochastic diffusion*.

Ito's formula reads as

$$du\left(W_t\right) = u'\left(W_t\right) dW_t + \frac{1}{2} u''\left(W_t\right) dt.$$

We have:

PROPOSITION 1.36.– Let $X, Y \in L^2\left((a,b), L^2(\Omega, P)\right)$ be such that $dX_t = a_X(t, W_t) dW_t + b_X(t, W_t) dt$ and $dY_t = a_Y(t, W_t) dW_t + b_Y(t, W_t) dt$. Let $\alpha, \beta \in \mathbb{R}$. Then

$d(\alpha X_t + \beta Y_t)$
$= [\alpha a_X(t, W_t) + \beta a_Y(t, W_t)] dW_t + [\alpha b_X(t, W_t) + \beta b_Y(t, W_t)] dt.$ ∎

PROOF.– The result is a consequence of the linearity of Ito integrals. ∎

PROPOSITION 1.37.– Let $x, y : \mathbb{R} \longrightarrow \mathbb{R}$, $X = x(W)$, $Y = y(W)$ and $Z = XY$ be regular enough. Then

$$dZ_t = X_t dY_t + Y_t dX_t + x'(X_t) y'(W_t) dt. \blacksquare$$

PROOF.– Let $z(s) = x(s) y(s) = xy(s)$. From Ito's formula:

$$dZ_t = z'(W_t) dW_t + \frac{1}{2} z''(W_t) dt.$$

or,

$$z'(s) = x'(s) y(s) + x(s) y'(s)$$

and

$$z''(s) = x''(s) y(s) + 2x'(s) y'(s) + x(s) y''(s).$$

Thus,

$$dZ_t = [x'(W_t) Y_t + X_t y'(W_t)] dW_t + \frac{1}{2} [x''(W_t) Y_t + 2x'(W_t) y'(W_t) + X_t y''(W_t)] dt$$

i.e.,

$$dZ_t = X_t \underbrace{\left[y'(W_t) dW_t + \frac{1}{2} y''(W_t) dt \right]}_{dY_t} + Y_t \underbrace{\left[x'(W_t) dW_t + \frac{1}{2} x''(W_t) dt \right]}_{dX_t} + x'(W_t) y'(W_t) dt,$$

what establishes the result. ∎

COROLLARY 1.15.– If $dX_t = a_X(t, W_t) dW_t + b_X(t, W_t) dt$, $dY_t = a_Y(t, W_t) dW_t + b_Y(t, W_t) dt$ and $Z = XY$, then

$$dZ_t = X_t dY_t + Y_t dX_t + [a_X(t, W_t) a_Y(t, W_t)] dt. \blacksquare$$

PROOF.– The result is an immediate consequence of the preceding proposition.

PROPOSITION 1.38.– Let $x, y : \mathbb{R} \longrightarrow \mathbb{R}$, $X = x(W)$, $Z = y(x(W))$ be regular enough. Then

$$dZ_t = y'(X_t) dX_t + \frac{1}{2} y''(X_t) \left(x'(W_t) \right)^2 dt. \blacksquare$$

PROOF.– Let $z(s) = y(x(s)) = y \circ x(s)$. From Ito's formula:

$$dZ_t = z'(W_t) dW_t + \frac{1}{2} z''(W_t) dt.$$

Or,

$$z'(s) = [y(x(s))]' = y'(x(s)) x'(s)$$

and

$$z''(s) = [y'(x(s)) x'(s)]' = y''(x(s)) [x'(s)]^2 + y'(x(s)) x''(s).$$

Thus,

$$dZ_t = y'(X_t) x'(W_t) dW_t + \frac{1}{2} \left[y''(X_t) [x'(W_t)]^2 + y'(X_t) x''(W_t) \right] dt,$$

i.e.,

$$dZ_t = y'(X_t) \underbrace{\left[x'(W_t) dW_t + \frac{1}{2} x''(W_t) dt \right]}_{dX_t} + \frac{1}{2} y''(X_t) [x'(W_t)]^2 dt$$

and we have the result claimed. ∎

COROLLARY 1.16.– Let $Y(t) = F(X(t))$, where $dX_t = a_X(t, W_t) dW_t + b_X(t, W_t) dt$. Then

$$dY_t = F'(X_t) dX_t + \frac{1}{2} F''(X_t) (a_X(t, W_t))^2 dt . \blacksquare$$

PROOF.– It is an immediate consequence of the preceding proposition. ∎

When $F = F(t, x)$, $Y(t) = F(t, X(t))$ and $dX_t = A_t dW_t + B_t dt$, we have

$$dY_t = \frac{\partial F}{\partial t}(t, X_t) dt + \frac{\partial F}{\partial x}(t, X_t) dX_t + \frac{1}{2} \frac{\partial^2 F}{\partial x^2}(t, X_t) (A_t)^2 dt.$$

When $F = F(t, x, z)$, $Y(t) = F\left(t, X(t), \int_0^t g(s, X(s))\, ds\right)$ and $dX_t = A_t dW_t + B_t dt$, we have, by taking $P(t) = \left(t, X(t), \int_0^t g(s, X(s))\, ds\right)$,

$$dY_t = \frac{\partial F}{\partial t}(P_t)\, dt + \frac{\partial F}{\partial x}(P_t)\, dX_t + \frac{1}{2}\frac{\partial^2 F}{\partial x^2}(P_t)(A_t)^2\, dt + g(P_t)\frac{\partial F}{\partial z}(P_t).$$

In the multidimensional situation, where $\mathbf{F_x} = \mathbf{F}(t, \mathbf{x})$, $\mathbf{Y}(t) = \mathbf{F}(t, \mathbf{X}(t))$, $\mathbf{F} : \mathbb{R} \times \mathbb{R}^n \longrightarrow \mathbb{R}^m$, we have, for $1 \leq i \leq m$,

$$d(Y_i)_t = \frac{\partial F_i}{\partial t}(t, \mathbf{X}_t)\, dt + (\mathbf{A}d\mathbf{X}_t)_i + \frac{1}{2}(d\mathbf{X}_t)^t \mathbf{B}_i d\mathbf{X}_t\,,$$

where

$$A_{ij}(t, x) = \frac{\partial F_i}{\partial x_j}(t, x)\ ;\ B_{ijk}(t, x) = \frac{\partial F_i}{\partial x_j \partial x_k}(t, x).$$

It is usual to write these equalities by using the Einstein convention of sum about repeated indexes:

$$d(Y_i)_t = \frac{\partial F_i}{\partial t}(t, \mathbf{X}_t)\, dt + A_{ij} d(X_j)_t + \frac{1}{2} B_{ijk} d(X_j)_t d(X_k)_t.$$

The simulation of stochastic diffusions or stochastic differential equations request a discretization in time. Analogously to the Wiener processes, the reader may find in the literature many works on this topic. Here, we illustrate the simulation of Ito's diffusion $dX_t = a(t, W_t)dW_t + b(t, W_t)dt$, X_0 given, by using a simple Euler discretization:

Listing 1.22. *Simple simulation of* $dX_t = a(t, W_t)dW_t + b(t, W_t)dt$

```
function X = ito_diffusion(ndimw,nstep,timestep,a,b, X0)
%
% generates nstep steps of a the process
%    dX_t = a(t,W_t)dW_t + b(t,W_t)dt , X_0 = X0
%
% IN:
% ndimw: dimension of the Wiener process
```

```
% nstep: number of steps - type integer
% timestep: the time step - type double
% a: the first coefficient - type anonymous function
% b: the second coefficient - type anonymous function
% X0: the initial value - type double
%
% OUT:
% X : table - type double
%     X(i,:) is the value of X((i-1)*timestep)
%
ndim = length(X0);
X = zeros(ndim, nstep+1);
sqrdt = sqrt(timestep);
w = zeros(ndimw,1);
X(:,1) = X0;
Z = X;
t = 0;
for i = 1: nstep
    dw = sqrdt*randn(ndim,1);
    at = a(w,t);
    bt = b(w,t);
    X(:,i+1) = X(:,i) + at*dw + bt*timestep;
    t = t + timestep;
    w = w + dw;
end;
return;
end
```

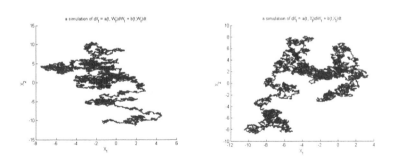

Figure 1.7. *Simulation results*

Analogously, the simulation of the stochastic differential equation $dX_t = a(t, X_t)dW_t + b(t, X_t)dt$, X_0 given, is implemented by a simple Euler discretization:

Listing 1.23. *Simple simulation of* $dX_t = a(t, X_t)dW_t + b(t, X_t)dt$

```
function X = ito_sde(ndimw,nstep,timestep,a,b, X0)
%
% generates nstep steps of a the process
%   dX_t = a(t,X_t)dX_t + b(t,X_t)dt  , X_0 = X0
%
% IN:
% ndimw: dimension of the Wiener process
% nstep: number of steps − type integer
% timestep: the time step − type double
% a: the first coefficient − type anonymous function
% b: the second coefficient − type anonymous function
% X0: the initial value − type double
%
% OUT:
% X : table − type double
%    X(i,:) is the value of X((i−1)*timestep)
%
ndim = length(X0);
X = zeros(ndim, nstep+1);
sqrdt = sqrt(timestep);
w = zeros(ndimw,1);
X(:,1) = X0;
Z = X;
t = 0;
for i = 1: nstep
    dw = sqrdt*randn(ndimw,1);
    at = a(X(:,i),t);
    bt = b(X(:,i),t);
    X(:,i+1) = X(:,i) + at*dw + bt*timestep;
    t = t + timestep;
    w = w + dw;
end;
return;
end
```

2

Maximum Entropy and Information

2.1. Construction of a stochastic model

The construction of stochastic simulations of a system is generally organized as follows:

1) First, we construct a *deterministic model* for the system.

2) In the second step, randomness is introduced by transforming the deterministic model into a *parametric stochastic model*: some parameters of the system are selected in order to be considered as random and their probabilistic models are generated, i.e. their respective probability density functions are determined.

3) A numerical method of stochastic simulation – most frequently the Monte Carlo method – is applied in order to generate information and facilitate statistical inferences about the system response.

The steps of the construction of deterministic and stochastic models of the system are essential in order to obtain realistic results in stochastic simulations. These two models are used in the stochastic simulation (for instance, by the Monte Carlo method) and directly influence the responses. Examples of construction of stochastic models can be found in [RIT 09, RIT 08, RIT 10b, RIT 10a, RIT 12].

The Monte Carlo method typically generates samples and uses them in order to generate information: on the one hand, samples of the parameters selected to be random are generated by using the pre-assigned probability distributions defined by the stochastic model and, on the other hand, each sample generates a variate from the system response by using the deterministic model. These deterministic results are collected and used for the generation of the targeted quantities. For instance, they may be aggregated in order to produce statistics and approximations of the probability distribution of the system response.

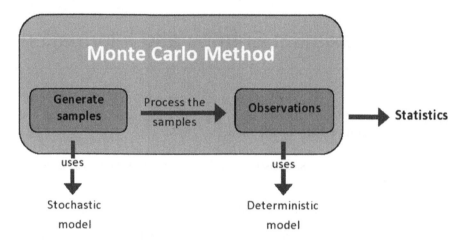

Figure 2.1. *Monte Carlo method*

Thus, the choice of the probability densities of the parameters selected to be random is a critical point, which requires some precaution.

These densities may be determined by various ways. For instance, experimental data may be used to construct histograms that represent the distribution of the parameters under analysis. An alternative way consists of determining a representation of these random parameters as functions of known random variables, by using the methods previously presented.

However, in many practical situations, performing experiments may be costly or impossible. In these situations, other arguments must be used in order to determine the probability density functions under analysis. In general, theoretical, physical or mathematical arguments may be used. Nevertheless, there is a systematic and popular procedure leading to the construction of such probability densities: the principle of maximum entropy (PME). In the following, we examine how this procedure may be applied and how it works in some simple situations.

2.2. The principle of maximum entropy

The PME is an efficient method for the generation of probabilistic models of scalar or vector parameters, and may be used for continuous or discrete random variables.

PME is a tool for the determination of probability density functions constrained by the respect of the statistical information available on the variables, such as, for instance, their support, their mean, their covariance matrix etc – eventually, several informations may be taken into account simultaneously: the result directly depends on the available information and changes by adding or subtracting information.

As its name indicates, the PME looks for a solution that maximizes the entropy subject to constraints defined by the available information. It reads as:

Among all probability distributions which satisfy the restrictions given by the available information, select the one that maximizes the entropy.

Naively, such an approach uses entropy as a measure of uncertainty and it may be interpreted as looking for the distribution that corresponds to the maximal uncertainty and the minimal quantity of information, i.e. assuming that the available information is minimal.

The idea of information directly leads to information theory and measures of *quantity of information*: when using PME, the popular approach consists of the use of *Shannon's entropy* (see [SHA 48]). The idea of maximizing the entropy was proposed by Jaynes (see [JAY 57a, JAY 57b]) in the context of statistical mechanics – field where the notion of entropy is close to Shannon's entropy.

In the following, we will illustrate the use of the PME by determining the distributions of discrete and continuous variables. We first consider discrete random variables taking a finite set of values, and then discrete variables taking an infinite set of values. Subsequently, we will present examples with continuous random variables and, finally, continuous random vectors.

2.2.1. *Discrete random variables*

Let us consider a discrete random variable X, defined on a finite set $\Omega = \{\omega_1, \cdots \omega_k\}$ and, thus, taking a finite number values $\mathbf{x} = \{x_1, \cdots, x_n\}$ ($n \leq k$).

Let us apply the PME in order to determine its distribution: in our first approach, we look for the values $p_i = P(X = x_i)$, $i = 1, \cdots, n$. These unknowns are grouped in a vector $\mathbf{p} = (p_1, \cdots, p_n)$. In this case, the Shannon's entropy is given by:

$$S(\mathbf{p}) = -\sum_{i=1}^{n} p_i \ln(p_i) \text{ (by convention } 0 \ln 0 = 0). \qquad [2.1]$$

Let us consider first the situation where no additional information is available: in this case, the only restrictions to be taken into account consist of imposing that the values of $\mathbf{p}$ define a probability, (i.e. these are non-negative values having their sum equal to one. Let C be the admissible set (i.e. the set of the values of which satisfy the restrictions): in this situation, we have

$$C = \{\mathbf{p} : p_1 + \cdots + p_n = 1 \qquad p_i \geq 0 \,\forall i \in \{1, \cdots, n\}\}. \qquad [2.2]$$

The PME states that

$$(\mathbf{p} = \arg\max \{S(\mathbf{q}) : \mathbf{q} \in C\}. \qquad [2.3]$$

Equation [2.3] defines a constrained optimization problem, which may be solved by introducing the Lagrange multipliers associated with C: let us consider $\boldsymbol{\lambda} = (\lambda_0, \cdots, \lambda_n)$ and the *Lagrangian*

$$\begin{aligned}L(\mathbf{p}, \boldsymbol{\lambda}) &= S(\mathbf{p}) - \lambda_0 \left(\sum_{i=1}^{n} p_i - 1\right) + \sum_{i=1}^{n} \lambda_i p_i \\ &= S(\mathbf{p}) - \sum_{i=1}^{n} (\lambda_0 + \lambda_i) p_i\end{aligned} \qquad [2.4]$$

With these notations, λ_0 is the Lagrange multiplier associated with the condition that the sum of the terms is equal to one and λ_i, with $i > 0$, is the Lagrange multiplier associated with the condition $p_i \geq 0$. From the general properties of Lagrange multipliers, we have

$$\lambda_i \geq 0 \, ; \, \lambda_i = 0, \text{ if } p_i > 0 \ (i > 0). \tag{2.5}$$

Let us denote

$$B = \{i \in \{1, \cdots, n\} : p_i > 0\} \, ; \, N = \{i \in \{1, \cdots, n\} : p_i > 0\}. \tag{2.6}$$

where B is the set of indexes such that $p_i > 0$ and let us denote by N the set of the indexes for which $p_i = 0$. Let n_B be the number of elements of B. We have

$$\sum_{i \in B} p_i = \sum_{i \in B} p_i + \sum_{i \in N} p_i = \sum_{i=1}^{n} p_i = 1. \tag{2.7}$$

so that $B \neq \emptyset$ – otherwise, the sum of the elements of $\mathbf{p}$ cannot be equal to one – and $n_B > 0$. Since

$$i \in B \Longrightarrow \lambda_i = 0,$$

we have:

$$\frac{\partial L}{\partial p_i}(\mathbf{p}, \boldsymbol{\lambda}) = -\ln(p_i) - p_i \frac{1}{p_i} - \lambda_0 = 0$$

$$= -\ln(p_i) - \lambda_0 = 1 \qquad i \in B \tag{2.8}$$

so that:

$$\ln(p_i) = 1 + \lambda_0 \quad \Longrightarrow \quad p_i = e^{-\lambda_0 - 1} \qquad i \in B. \tag{2.9}$$

Thus,

$$p_i = p, \ \forall i \in B \, ; \, p = e^{-\lambda_0 - 1}.$$

Since their sum is equal to one, we have:

$$1 = \sum_{i \in B} p_i = \sum_{i \in B} p = n_B \, p.$$

Thus,

$$p_i = \frac{1}{n_B} \qquad i \in B. \qquad [2.10]$$

So, if no information is available about the random variable X, the PME provides a set of solutions formed by all the uniform distributions defined on non-empty parts of **x**. If we add the supplementary information that all the values have a strictly positive probability, then the PME provides a single solution, which is the *uniform distribution* on the whole **x**.

In our second approach, we look for the probabilities of the elements of Ω and our unknowns are $p_i = P(\omega_i)$. Analogously to the preceding situation, the PME provides a set of solutions formed by all the uniform distributions defined on non-empty parts of Ω. If we add the supplementary information that all the values have a strictly positive probability, then the PME provides a single solution, which is the uniform distribution on the whole Ω. In this case, the distribution of X may not correspond to an uniform distribution. For instance, in the last situation, where all the elements have a strictly positive probability, $P(X = x_i) = card(X^{-1}(\{x_i\}))/k$, where $card(S)$ denotes the number of elements of S: if these subsets do not contain the same number of elements, X is not uniformly distributed.

This simple example shows that the results provided by PME are closely connected to the information used, *namely the support of the distribution.*

It is interesting to notice that, if **x** (or Ω) is a infinite countable set, then the result is the set of all the uniform distributions on finite non-empty parts of **x** (or Ω). In this case, assuming that all the elements have a strictly positive probability leads to a contradiction, since the condition $1 = \sum_{i \in B} p$ cannot be satisfied if p contains infinitely many elements. However, additional information about X may keep a solution possible (see example 2.2).

In the following, we limit ourselves to the situation where all the elements have a strictly positive probability. Thus, we do not consider Lagrange multipliers associated with the non-negativity condition of the probabilities: they are assumed to be null in the following.

Let us now consider a situation where more information is prescribed. Let us suppose that we have m informations of the form

$$\sum_{i=1}^{n} p_i g_{r\,i}(\mathbf{x}) = a_r \qquad r = 1, \cdots, m. \qquad [2.11]$$

In this case, the admissible set is (recall that we assume that all the p_i are strictly positive).

$$C = \left\{ \mathbf{p} : \sum_{i=1}^{n} p_i = 1 \,;\, \sum_{i=1}^{n} p_i g_{r\,i}(\mathbf{x}) = a_r \,,\, r = 1, \cdots, m \right\}. \qquad [2.12]$$

As in the preceding situation, let us introduce the Lagrange multipliers $\boldsymbol{\lambda} = (\lambda_0, \cdots, \lambda_m)$, where λ_0 is the Lagrange multiplier associated with the same condition as previously (the sum of the components of $\mathbf{p}$ is equal to one) and λ_i, with $i > 0$, is the Lagrange multiplier associated with the condition $\sum_{i=1}^{n} p_i g_{r\,i}(\mathbf{x}) = a_r$. Then, the Lagrangian reads as

$$L(\mathbf{p}, \boldsymbol{\lambda}) = S(\mathbf{p}) - \lambda_0 \left(\sum_{i=1}^{n} p_i - 1 \right) - \sum_{r=1}^{m} \lambda_r \left(\sum_{i=1}^{n} p_i g_{r\,i} - a_r \right) \qquad [2.13]$$

and the PME yields that :

$$\frac{\partial L}{\partial p_i} = 0 \quad \Longrightarrow \quad -\ln(p_i) - \lambda_0 - \sum_{r=1}^{m} \lambda_r \, g_{r\,i} = 1 \,, \qquad [2.14]$$

so that:

$$p_i = \exp\left(-\lambda_0 - 1 - \sum_{r=1}^{m} \lambda_r \, g_{r\,i}\right) \quad i = 1, \cdots, n. \qquad [2.15]$$

The multipliers are determined by replacing p_i [2.15] in [2.2] and [2.11]: we have

$$\sum_{i=1}^{n} \exp\left(-\lambda_0 - 1 - \sum_{j=1}^{m} \lambda_j g_{j\,i}\right) = 1 \qquad [2.16]$$

and

$$\sum_{i=1}^{n} g_{r\,i} \exp\left(-\lambda_0 - 1 - \sum_{j=1}^{m} \lambda_j g_{ji}\right) = a_r \quad r = 1, \cdots, m. \qquad [2.17]$$

It is interesting to note that these equations extend straightly to the situation where X takes infinitely many countable values by replacing n by $+\infty$ in equations [2.15]–[2.17]. Example 2.2 illustrates such a situation.

In the examples below, we illustrate the use of these equations in some particular situations.

EXAMPLE 2.1.– Let us consider a discrete random variable X taking a finite number n of values $\mathbf{x} = \{x_1, \cdots, x_n\}$. Let us assume that *the mean of X is known* and has the value μ. In this case, the admissible set is (recall that we assume that all the probabilities are strictly positive)

$$C = \left\{\mathbf{p} : \sum_{i=1}^{n} p_i = 1 \,;\, \sum_{i=1}^{n} x_i\, p_i = \mu\right\}. \qquad [2.18]$$

Applying the PME, we look for a solution $\mathbf{p} \in C$ such that $S(\mathbf{p}) \leq S(\mathbf{q})$, $\forall \mathbf{q} \in C$. This situation corresponds to

$$m = 1\,;\, g_{1\,i}(\mathbf{x}) = x_i\,;\, a_1 = \mu.$$

Thus, we have two Lagrange multipliers associated with C: $\boldsymbol{\lambda} = (\lambda_0, \lambda_1)$. The Lagrangian is

$$L(\mathbf{p}, \boldsymbol{\lambda}) = -\sum_{i=1}^{n} p_i \ln(p_i) - \lambda_0 \left(\sum_{i=1}^{n} p_i - 1\right) - \lambda_1 \left(\sum_{i=1}^{n} x_i\, p_i - \mu\right) \qquad [2.19]$$

and we have

$$\frac{\partial L}{\partial p_i} = 0 \quad \Longrightarrow \quad -\ln(p_i) - \lambda_0 - \lambda_1 x_i = 1. \quad [2.20]$$

Equation [2.15] shows that

$$p_i = \exp(-\lambda_0 - \lambda_1 x_i - 1) \quad i = 1, \cdots, n \quad [2.21]$$

The values of the Lagrange multipliers are determined by using equations [2.16]–[2.17]:

$$\sum_{i=1}^{n} \exp(-\lambda_0 - \lambda_1 x_i - 1) = 1 \quad \text{e} \quad \sum_{i=1}^{n} x_i \exp(-\lambda_0 - \lambda_1 x_i - 1) = \mu.$$

$$[2.22]$$

These equations may be solved by standard methods for nonlinear algebraic systems, such as Newton–Raphson. ∎

EXAMPLE 2.2.– Let X be a discrete random variable taking infinitely many values given by the natural numbers: $\mathbf{x} = \{0, 1, 2, \cdots\}$. Let us assume that *the mean of X is known* and has the value $\mu > 0$ – since we assume that all the p_i are strictly positive, the mean of such a variable must be strictly positive: this condition about the sign of μ appears as a condition for the existence of a solution in the developments below. Here, the admissible set is (recall that we still assume that all the probabilities are strictly positive)

$$C = \left\{ \mathbf{p} : \sum_{i=1}^{+\infty} p_i = 1 \,;\, \sum_{i=1}^{+\infty} x_i \, p_i = \mu \right\}. \quad [2.23]$$

In this situation, we have, analogous to example 2.1,

$$m = 1 \,;\, g_{1\,i}(\mathbf{x}) = x_i \,;\, a_1 = \mu$$

and we also have two Lagrange multipliers associated with C: $\boldsymbol{\lambda} = (\lambda_0, \lambda_1)$. The Lagrangian is analogous to the one of example 2.1:

$$L(\mathbf{p}, \boldsymbol{\lambda}) = -\sum_{i=0}^{+\infty} p_i \ln(p_i) - \lambda_0 \left(\sum_{i=0}^{+\infty} p_i - 1 \right) - \lambda_1 \left(\sum_{i=0}^{+\infty} i \, p_i - \mu \right). \quad [2.24]$$

We have

$$\frac{\partial L}{\partial p_i} = 0 \quad \Longrightarrow \quad -\ln(p_i) - \lambda_0 - \lambda_1 \, i = 1. \qquad [2.25]$$

which leads to equations analogous to equations [2.15]–[2.17], with n replaced by $+\infty$.

Since $x_i = i$, we have, from the analog of equation [2.15]:

$$p_i = \exp(-\lambda_0 - 1 - \lambda_1 \, i) \qquad i = 0, 1, 2, \cdots. \qquad [2.26]$$

Let us introduce

$$\alpha = \exp(-\lambda_0 - 1) \,;\, \beta = \exp(-\lambda_1).$$

We have $\alpha > 0$ and $\beta > 0$. In addition, p_i reads as

$$p_i = \alpha \, \beta^i.$$

By using the analogous of equation [2.16], we have

$$1 = \sum_{i=0}^{+\infty} p_i = \alpha \sum_{i=0}^{+\infty} \beta^i.$$

This equality implies that $\beta < 1$: if $\beta = 1$, then $\sum_{i=0}^{+\infty} \beta^i = +\infty$ and, since $\alpha > 0$, the equality leads to $1 = +\infty$, which is a contradiction. Thus, $0 < \beta < 1$ and we have

$$1 = \sum_{i=0}^{+\infty} p_i = \alpha \sum_{i=0}^{+\infty} \beta^i = \frac{\alpha}{1-\beta}. \qquad [2.27]$$

By using the analogous of equation [2.17], we have:

$$\mu = \sum_{i=0}^{\infty} i \, p_i = \alpha \sum_{i=0}^{+\infty} i \, \beta^i.$$

We observe that this equality cannot be satisfied if $\mu \leq 0$. Since $0 < \beta < 1$:

$$\sum_{i=0}^{+\infty} i\,\beta^i = \beta \sum_{i=0}^{+\infty} i\,\beta^{i-1} = \beta \frac{d}{d\beta} \sum_{i=0}^{+\infty} \beta^i = \frac{\beta}{(1-\beta)^2},$$

so that (using equation [2.27])

$$\mu = \frac{\alpha\beta}{(1-\beta)^2} = \frac{\beta}{1-\beta}. \qquad [2.28]$$

Thus,

$$\beta = \frac{\mu}{1+\mu} \quad\Longrightarrow\quad \lambda_1 = -\ln\left(\frac{\mu}{1+\mu}\right) \qquad [2.29]$$

and

$$\alpha = 1 - \frac{\mu}{1+\mu} = \frac{1}{1+\mu} \quad\Longrightarrow\quad \lambda_0 = -1 - \ln\left(\frac{1}{1+\mu}\right). \qquad [2.30]$$

So,

$$p_i = \left(\frac{1}{1+\mu}\right)\left(\frac{\mu}{1+\mu}\right)^i \qquad i = 0, 1, 2, \cdots. \blacksquare \qquad [2.31]$$

2.2.2. *Continuous random variables*

The PME is more frequently used for the determination of the distribution of continuous random variables. In the case of a continuous variable X having a density of probability $p : \mathbb{R} \longrightarrow \mathbb{R}$, Shannon's entropy is given by

$$S(p) = -\int_{\mathbb{R}} p(x) \ln p(x)\, dx \quad (\text{ by convention } 0 \ln 0 = 0). \qquad [2.32]$$

As previously observed, the support of the variable is an essential information: for a continuous variable, we may consider three situations: a finite interval (a, b), a semi-infinite interval – for instance, $[a, \infty)$, or the set of all the real numbers $(-\infty, \infty)$. Since we adopt the convention $0 \ln 0 = 0$, the integral in equation [2.32] reduces to an integral on the support. For instance, if the support is (a, b), we have:

$$S(p) = -\int_a^b p(x) \ln p(x)\, dx. \blacksquare \qquad [2.33]$$

Analogously to the discrete situation, let us examine the case of a variable X taking its values on a finite interval (a, b), for which no other information is available. In this case, the admissible set is :

$$C = \{p : (a, b) \longrightarrow \mathbb{R} : \int_a^b p(x)\, dx = 1\,;\, p(x) \geq 0 \text{ on } (a, b)\}. \qquad [2.34]$$

In this situation, we have two Lagrange multipliers $\boldsymbol{\lambda} = (\lambda_0,\, \lambda_1(x))$ such that:

$$L(p, \boldsymbol{\lambda}) = -\int_a^b p(x) \ln p(x)\, dx - \lambda_0 \left[\int_a^b p(x)\, dx - 1\right]$$
$$- \int_a^b \lambda_1(x) p(x)\, dx.$$

Thus:

$$L(p, \boldsymbol{\lambda}) = -\int_a^b h(p, \boldsymbol{\lambda})\, dx, \qquad [2.35]$$

where:

$$h(p, \boldsymbol{\lambda}) = p(x)[\ln p(x) + \lambda_0 + \lambda_1(x))]. \qquad [2.36]$$

As in the discrete situation, we have

$$\forall x \in (a, b) : \lambda_1(x) \geq 0\,;\, \lambda_1(x) = 0,\, \text{if } p(x) > 0.$$

Let us assume that $p(x) > 0$ on $(\alpha, \beta) \subset (a, b)$. From variational calculus, we have

$$\frac{\partial}{\partial p(x)} h(p, \boldsymbol{\lambda}) = 0 \text{ on } (\alpha, \beta). \qquad [2.37]$$

Thus:

$$p(x) = p_0 \,;\, p_0 = \exp(-\lambda_0 - 1) \text{ on } (a,b). \quad [2.38]$$

So, $p(x)$ is constant on its support. Let us denote by $A = supp(p) = \{x \in (a,b) : p(x) > 0\}$. We have

$$1 = \int_a^b p(x)\, dx = \int_A p_0 \, dx = p_0 \text{meas}(A).$$

so that

$$p(x) = \frac{1}{\text{meas}(A)} \text{ if } x \in A \,;\, p(x) = 0 \notin A.$$

So, analogously to the discrete situation, the PME furnishes as result the family of all the uniformly distributed variables having their support on $A \subset (a,b)$. By adding the supplementary information that $A = (a,b)$, we obtain a unique solution

$$p(x) = \mathbf{1}_{[a,b]} p_0 \,;\, \frac{1}{b-a}. \quad [2.39]$$

We observe that:

$$\frac{\partial^2}{\partial p(x)^2} h(p, \boldsymbol{\lambda}) > 0, \quad [2.40]$$

so that this extremal point corresponds to a maximum of S. Analogous to the discrete situation, we may also consider $\Omega = (\omega_a, \omega_b)$ and look for a probability on Ω: in this case, the solution will be the family of uniform distributions on subsets of Ω and analogously to the discrete situation, the distribution of X may result different from a uniform distribution. Finally, we observe that, as in the discrete case, considering unbounded sets, such as $(a, +\infty)$ or $\mathbb{R}$, results in the set of all the uniform distributions on bounded non-empty parts of the original set. Here yet, additional information about X may keep a solution possible (see examples 2.4, 2.5 and 2.6).

We observe again that the results furnished by PME are closely connected to the information used, *namely the support of the distribution*.

In the following, we limit ourselves to the situation where $p(x) > 0$ on the set under consideration. Thus, we do not consider Lagrange multipliers associated with the non-negativity condition of the probabilities: they are assumed to be null in the following.

It is common to consider situations involving information on statistics of the distribution to be determined, such as its mean or some moments: assume that the random variable X satisfies m restrictions written in the form:

$$\int_a^b p(x)g_r(x)\,dx = a_r \qquad r = 1, \cdots, m.$$

In this situation, the admissible set is

$$C = \{p : (a, b) \longrightarrow \mathbb{R} : \int_a^b p(x)\,dx = 1\,;$$

$$\int_a^b p(x)g_r(x)\,dx = a_r, r = 1, \cdots, m\}.$$

and we consider $m + 1$ Lagrange multipliers $\boldsymbol{\lambda} = (\lambda_0, \lambda_1, \cdots, \lambda_m)$ and the Lagrangian becomes:

$$L(p, \boldsymbol{\lambda}) = S(p) - \lambda_0 \left(\int_a^b p(x)\,dx - 1\right) \\ - \sum_{r=1}^m \lambda_r \left(\int_a^b p(x)g_r(x)\,dx - a_r\right) \qquad [2.41]$$

Equation [2.41] assumes the form:

$$L(p, \boldsymbol{\lambda}) = \int_a^b h(p, \boldsymbol{\lambda})\,dx, \qquad [2.42]$$

with:

$$h(p, \boldsymbol{\lambda}) = p(x)[\ln p(x) + \lambda_0 + \sum_{r=1}^m \lambda_r g_r(x)]. \qquad [2.43]$$

From variational calculus:

$$\frac{\partial}{\partial p(x)} h(p, \boldsymbol{\lambda}) = 0. \qquad [2.44]$$

and we have:

$$p(x) = 1_{[a,b]}(x) \exp\left(-\lambda_0 - 1 - \sum_{r=1}^{m} \lambda_r g_r(x)\right). \qquad [2.45]$$

The Lagrange multipliers are determined by using equation [2.45] in [2.41]: we have

$$\int_a^b \exp\left(-\lambda_0 - 1 - \sum_{r=1}^{m} \lambda_r g_r(x)\right) dx = 1 \qquad [2.46]$$

and

$$\int_a^b g_r(x) \exp\left(-\lambda_0 - 1 - \sum_{r=1}^{m} \lambda_r g_r(x)\right) dx = a_r, \, r = 1, \cdots, m. \qquad [2.47]$$

EXAMPLE 2.3.– Let us consider a continuous random variable X having as support (a, b), such that $E[X] = \mu$ and $E[X^2] = \eta^2$. In this case, we have $m = 2$, $g_1(x) = x$, $g_2(x) = x^2$, $a_1 = \mu$, $a_2 = \eta^2$. Thus, we consider $\boldsymbol{\lambda} = (\lambda_0, \lambda_1, \lambda_2)$. From equation [2.45]:

$$p(x) = 1_{[a,b]}(x) \exp\left(-\lambda_0 - 1 - \lambda_1 x - \lambda_2 x^2\right). \qquad [2.48]$$

It is convenient to rewrite this equality as

$$p(x) = \alpha \exp\left(-\beta(x - \gamma)^2\right), \qquad [2.49]$$

where

$$\alpha = \exp\left(-\lambda_0 - 1 + \frac{\lambda_1^2}{4\lambda_2}\right), \, \beta = \lambda_2, \, \gamma = \frac{\lambda_1}{2\lambda_2}.$$

The values of the multipliers are determined by solving equations [2.46]–[2.47] for the unknowns α, β, γ. These equations read as

$$\alpha \int_a^b \exp\left(-\beta(x-\gamma)^2\right) dx = 1$$

$$\alpha \int_a^b x \exp\left(-\beta(x-\gamma)^2\right) dx = \mu. \quad [2.50]$$

$$\alpha \int_a^b x^2 \exp\left(-\beta(x-\gamma)^2\right) dx = \eta^2$$

Equation [2.50] may be solved by methods adapted to the solution of nonlinear algebraic equations, such as Newton–Raphson. ∎

EXAMPLE 2.4.– Let us consider a continuous random variable X supported by $(0, \infty)$ and having a known mean $E[X] = \mu > 0$. In this case, we have $m = 1$, $g_1(x) = x$, $a_1 = \mu$.

From equation [2.45]:

$$p(x) = \exp\left(-\lambda_0 - 1 - \lambda_1 x\right). \quad [2.51]$$

Let us denote $\alpha = \exp\left(-\lambda_0 - 1\right)$. Then $p(x) = \alpha \exp\left(-\lambda_1 x\right)$ and we have, from equation [2.46]:

$$1 = \alpha \int_0^\infty \exp\left(-\lambda_1 x\right) dx = \frac{\alpha}{\lambda_1},$$

so that $\alpha = \lambda_1$ (note that $\lambda_1 > 0$: otherwise, the equality is impossible). Thus,

$$\mu = \lambda_1 \int_0^\infty x \exp\left(-\lambda_1 x\right) dx = \frac{1}{\lambda_1}, \implies \lambda_1 = \frac{1}{\mu}$$

so that

$$p(x) = 1_{[0,\infty)}(x) \frac{1}{\mu} e^{\frac{-x}{\mu}}. \quad \blacksquare \quad [2.52]$$

EXAMPLE 2.5.– Let us consider a continuous random variable X supported by $(0, \infty)$ for which the following additional information is available: $E[X] = \mu > 0$ and $E[\ln(X)] = \sigma$. This situation corresponds to $m = 2$, $g_1(x) = x$, $g_2(x) = \ln(X)$, $a_1 = \mu$, $a_2 = \eta$.

From equation [2.45]:

$$p(x) = \exp(-\lambda_0 - 1 - \lambda_1 x - \lambda_2 \ln(x)). \qquad [2.53]$$

Let us denote $\alpha = \exp(-\lambda_0 - 1)$. Then $p(x) = \alpha x^{-\lambda_2} \exp(-\lambda_1 x)$. This expression may be compared with the classical Γ distribution:

$$p(x) = \frac{x^{a-1} exp\left(-\frac{x}{\theta}\right)}{\theta^a \Gamma(a)}.$$

Thus, the solution of equations [2.46]–[2.47] leads to a *Gamma distribution*:

$$p(x) = 1_{]0,+\infty]}(x) \frac{1}{\mu} \left(\frac{1}{\delta^2}\right)^{\frac{1}{\delta^2}} \frac{1}{\Gamma(1/\delta^2)} \left(\frac{x}{\mu}\right)^{\frac{1}{\delta^2}-1} \exp\left(\frac{x}{\delta^2 \mu}\right), \qquad [2.54]$$

with ($\sigma = ...$):

$$\delta = \frac{\sigma}{\mu} \; ; \; \Gamma(a) = \int_0^\infty t^{a-1} \exp(-t)dt.$$

$\Gamma(\bullet)$ is referred to as Gamma function. ■

EXAMPLE 2.6.– Let us consider a continuous random variable X having as support $(-\infty, \infty)$ for which two moments are given: $E[X] = \mu$ and $E[X^2] = \eta^2$. The situation is similar to the situation considered in example 2.3: $m = 2$, $g_1(x) = x$, $g_2(x) = x^2$, $a_1 = \mu$, $a_2 = \eta^2$ (only the support is different). As in example 2.5, we have:

$$p(x) = \alpha \exp(-\beta(x-\gamma)^2), \qquad [2.55]$$

where α, β and γ are the solutions of equations [2.46]–[2.47], which read as

$$\alpha \int_{-\infty}^{\infty} \exp\left(-\beta(x-\gamma)^2\right) dx = 1$$

$$\alpha \int_{-\infty}^{\infty} x \exp\left(-\beta(x-\gamma)^2\right) dx = \mu. \qquad [2.56]$$

$$\alpha \int_{-\infty}^{\infty} x^2 \exp\left(-\beta(x-\gamma)^2\right) dx = \eta^2$$

Since (see [KAP 92]):

$$\int_{-\infty}^{\infty} \exp\left(-\beta(x-\gamma)^2\right) dx = \sqrt{\frac{\pi}{\beta}}, \quad \text{for } \beta > 0, \qquad [2.57]$$

we have:

$$\alpha = \frac{1}{\sqrt{2\pi}\sigma}, \qquad \beta = \frac{1}{2\sigma^2}, \qquad c = \mu, \qquad [2.58]$$

with $\sigma^2 = \eta^2 - \mu^2$. Thus,

$$p(x) = \frac{1}{\sqrt{2\pi}\sigma} \exp\left[-\frac{1}{2}\frac{(x-\mu)^2}{\sigma^2}\right]. \qquad [2.59]$$

So, the PME furnishes as solution a Gaussian density. ∎

In the following, we summarize some classical results derived from the PME.

Continuous variables with bounded support (a, b):

We recall below some classical results furnished by the PME under the assumption that X is a continuous random variable having as support (a, b):

1) If no additional information is given, the PME furnishes as solution the *uniform distribution* on (a, b):

$$p(x) = \mathbf{1}_{[a,b]}(x) \frac{1}{b-a}. \qquad [2.60]$$

2) If the mean of X is given: $E[X] = \mu$, then the solution is a *truncated exponential distribution*:

$$p(x) = 1_{[a,b]}(x) \, \exp\left(-\lambda_0 - \lambda_1 x\right), \qquad [2.61]$$

where:

$$e^{-\lambda_0} \int_a^b e^{-\lambda_1 x} \, dx = 1 \quad \text{e} \quad e^{-\lambda_0} \int_a^b x \, e^{-\lambda_1 x} \, dx = \mu. \qquad [2.62]$$

3) If $a = 0$, $b = 1$, and we have $E[\ln(X)] = k_1$ and $E[\ln(1-X)] = k_2$, then the PME furnishes a *beta distribution*:

$$p(x) = 1_{[0,1]}(x) \, \frac{1}{B(m,n)} x^{(m-1)} (1-x)^{n-1}, \qquad [2.63]$$

where B is the *beta function*:

$$B(m,n) = \frac{\Gamma m \Gamma n}{\Gamma m + n} \qquad [2.64]$$

and $\Gamma(m) = \int_0^\infty t^{m-1} \exp(-t) dt$ is the *Gamma function*. The values of m and n are determined from the equations $E[\ln(X)] = k_1$ and $E[\ln(1-X)] = k_2$:

$$\frac{1}{B(m,n)} \int_0^1 x^{(m-1)} (1-x)^{n-1} \ln(x) \, dx = E[\ln(X)] = k_1, \qquad [2.65]$$

$$\frac{1}{\beta(m,n)} \int_0^1 x^{(m-1)} (1-x)^{n-1} \ln(1-x) \, dx = E[\ln(1-X)] = k_2. \qquad [2.66]$$

Continuous variables with support $(0, \infty)$:

Here, we collect some classical results furnished by the PME under the assumption that X is a continuous random variable having as support $(0, \infty)$:

1) If no additional information about X is given, the PME does not admit a solution.

2) If the mean of X is known, $E[X] = \mu$, the solution is the *exponential distribution*:

$$p(x) = 1_{[0,\infty)}(x) \, \frac{1}{\mu} e^{\frac{-x}{\mu}}. \qquad [2.67]$$

3) If the available information consists of the mean of X and the mean of $\ln(X)$; $E[X] = \mu$ and $E[\ln(X)] = q$, the solution is the *Gamma distribution*:

$$p(x) = 1_{[0,+\infty)}(x) \frac{1}{\mu} \left(\frac{1}{\delta^2}\right)^{\frac{1}{\delta^2}} \frac{1}{\Gamma(1/\delta^2)} \left(\frac{x}{\mu}\right)^{\frac{1}{\delta^2}-1} \exp\left(\frac{x}{\delta^2 \mu}\right), \qquad [2.68]$$

where Γ is the *Gamma function* ($\Gamma(a) = \int_0^\infty t^{a-1} \exp(-t) dt$) and $\delta = \frac{\sigma}{\mu}$.

4) If $E[\ln(X)] = k_1$ e $E[\ln(1-X)] = k_2$, the PME provides as result:

$$p(x) = 1_{]0,+\infty)}(x) \frac{1}{B(m,n)} x^{(m-1)} (1+x)^{-(n+m)}, \qquad [2.69]$$

where B is the *beta function* [2.64] and the values of m and n are determined from the equations $E[\ln(X)] = k_1$ and $E[\ln(1-X)] = k_2$. Thus:

$$\frac{1}{B(m,n)} \int_0^\infty x^{(m-1)} (1+x)^{-(n+m)} \ln(x) \, dx = E[\ln(X)] = k_1, \qquad [2.70]$$

$$\frac{1}{\beta(m,n)} \int_0^\infty x^{(m-1)} (1+x)^{-(n+m)} \ln(1-x) \, dx = E[\ln(1-X)] = k_2.$$

[2.71]

Continuous variables having by support $(-\infty, \infty)$:

Here, we recall some classical results furnished by the PME under the assumption that X is a continuous random variable having as support $(-\infty, \infty)$:

1) If no additional information about X is given, the PME does not admit a solution.

2) If the only information is the mean of X, the PME does not admit a solution.

3) If $E[X] = \mu$ and $E[X^2] = \alpha^2$, the PME furnishes a Gaussian density:

$$p(x) = \frac{1}{\sqrt{2\pi}\sigma} \exp\left[-\frac{1}{2}\frac{(x-\mu)^2}{\sigma^2}\right]. \blacksquare \qquad [2.72]$$

2.2.3. *Random vectors*

In this section, we consider a random vector $\mathbf{X} = (X_1, \cdots, X_n)$ of dimension n formed by n random variables X_i, $i = 1, \ldots, n$. The cumulative density function of such a vector is:

$$\begin{aligned} P : \mathbb{R}^n &\longmapsto [0,1] \\ \mathbf{x} &\longmapsto P(X_1 < x_1,\ X_2 < x_2, \cdots,\ X_n < x_n) \end{aligned} \qquad [2.73]$$

where $\mathbf{x} \in \mathbb{R}^n$. In the following, we denote $d\mathbf{X} = dx_1 \cdots dx_n$. Its density of probability p is given by:

$$p(\mathbf{x}) = \frac{\partial^n}{\partial x_1\, \partial x_2 \cdots \partial x_n} P(X_1 < x_1,\ X_2 < x_2, \cdots,\ X_n < x_n), \quad [2.74]$$

sometimes referred to as the joint distribution of the vector $\mathbf{X} = (X_1, \cdots, X_n)$.

The Shannon's entropy of a random vector $\mathbf{X}$ is:

$$S(p) = -\int_{\mathbb{R}^n} p(\mathbf{x}) \ln p(\mathbf{x})\, d\mathbf{x}. \qquad [2.75]$$

Analogously to the preceding situations, p verifies:

$$p(\mathbf{x}) \geq 0, \qquad \int_{\mathbb{R}^n} p(\mathbf{x})\, d\mathbf{x} = 1. \qquad [2.76]$$

Analogously to the preceding situations, we denote by λ_0 the Lagrange multiplier associated with the condition $\int_{\mathbb{R}^n} p(\mathbf{x})\, d\mathbf{x} = 1$ and we limit the following to the situation where the support of $\mathbf{X}$ is given and the Lagrange multiplier associated with the condition $p(\mathbf{x}) \geq 0$ is identically null. In addition, we consider that this support is the whole space $\mathbb{R}^n$.

Let us consider the situation where m supplementary informations on $\mathbf{X}$ are given under the form:

$$\int_{\mathbb{R}^n} p(\mathbf{x})\mathbf{g}_r(\mathbf{x})\,d\mathbf{x} = \mathbf{a}_r \in \mathbb{R}^{\nu_r} \qquad r = 1,\cdots,m. \qquad [2.77]$$

with $\nu_1,\cdots,\nu_m$ being all strictly positive integers.

Let C be the space of the maps $\mathbf{X} \longmapsto p(\mathbf{X})$ from $\mathbb{R}^n$ onto $\mathbb{R} \geq 0$, having all the same support $k_n \subset \mathbb{R}^n$ (eventually, $k_n = \mathbb{R}^n$) and satisfying equations [2.76]–[2.77]:

$$C = \{p : \mathbb{R}^n \longrightarrow \mathbb{R}^{\geq 0} : supp(p) = k_n \,;\, p \text{ satisfies } [2.76]\text{–}[2.77]\}.$$

The PME reads as

$$p = \arg\max\{S(q) : q \in C\}. \qquad [2.78]$$

As previously, we introduce the Lagrange multipliers $\boldsymbol{\lambda} = (\lambda_0,\cdots,\lambda_m)$, where λ_0 is the Lagrange multiplier associated with equation [2.76] and λ_i, with $i > 0$, is the Lagrange multiplier associated with equation [2.77] with $r = i$. Then, the Lagrangian reads as

$$L(p,\boldsymbol{\lambda}) = S(p) - \lambda_0 \left(\int_{\mathbb{R}^n} p(\mathbf{x})\,d\mathbf{x} - 1\right)$$
$$- \sum_{r=1}^{m} <\boldsymbol{\lambda}_r\,,\,\int_{\mathbb{R}^n} p(\mathbf{X})\mathbf{g}_r(\mathbf{x})\,d\mathbf{x} - \mathbf{a}_r>_{\mathbb{R}^{\nu_r}} \qquad [2.79]$$

where $<\mathbf{u},\mathbf{v}>_{\mathbb{R}^{\nu_r}} = u_1 v_1 + \cdots + u_{\nu_r} v_{\nu_r}$ denotes the euclidean scalar product in $\mathbb{R}^{\nu_r}$.

Equation [2.79] may be written as:

$$L(p,\boldsymbol{\lambda}) = -\int_{\mathbb{R}^n} h(p(\mathbf{x}),\boldsymbol{\lambda})\,d\mathbf{x}, \qquad [2.80]$$

with:

$$h(p, \boldsymbol{\lambda}) = p(\mathbf{x})\left[\ln p(\mathbf{x}) + \lambda_0 + \sum_{r=1}^{m} <\boldsymbol{\lambda}_r , \mathbf{g}_r(\mathbf{x})>_{\mathbb{R}^{\nu_r}}\right].$$ [2.81]

From variational calculus:

$$\frac{\partial}{\partial p(\mathbf{x})} h(p, \boldsymbol{\lambda}) = 0,$$ [2.82]

which gives:

$$p(\mathbf{x}) = 1_{k_n}(\mathbf{x}) \exp\left(-\lambda_0 - \sum_{r=1}^{m} <\boldsymbol{\lambda}_r , \mathbf{g}_r(\mathbf{x})>_{\mathbb{R}^{\nu_r}}\right).$$ [2.83]

So, analogously to the preceding situations, we have:

$$\int_{k_n} \exp\left(-\lambda_0 - \sum_{r=1}^{m} <\boldsymbol{\lambda}_r , \mathbf{g}_r(\mathbf{x})>_{\mathbb{R}^{\nu_r}}\right) d\mathbf{x} = 1.$$ [2.84]

$$\int_{k_n} \mathbf{g}_r(\mathbf{x}) \exp\left(-\lambda_0 - \sum_{r=1}^{m} <\boldsymbol{\lambda}_r , \mathbf{g}_r(\mathbf{x})>_{\mathbb{R}^{\nu_r}}\right) d\mathbf{x} = \mathbf{a}_r.$$ [2.85]

EXAMPLE 2.7.– Let us consider a random vector $\mathbf{X}$ having dimension n and supported by $k_n = [a_1, b_1] \times \cdots \times [a_n, b_n]$. If no other information is available, equation [2.83] shows that

$$p(\mathbf{x}) = p_0 \, ; \, p_0 = 1_{k_n} \exp\left(-\lambda_0\right),$$ [2.86]

so that $p(\mathbf{x})$ is constant. Using [2.84]:

$$\int_{\mathbb{R}^n} 1_{k_n} p_0 \, d\mathbf{X} = 1$$ [2.87]

and we have

$$p_0 = \frac{1}{\operatorname{meas}(k_n)}.$$

Thus, the solution is the uniform distribution:

$$p(\mathbf{x}) = 1_{k_n}(\mathbf{x}) \frac{1}{\text{meas}(k_n)} . \blacksquare \qquad [2.88]$$

EXAMPLE 2.8.– Let us consider $\mathbf{X}$ having dimension n, support $k_n = (0, +\infty) \times \cdots \times (0, +\infty) \subset \mathbb{R}^n$ and a known mean $E[\mathbf{X}] = \boldsymbol{\mu}_\mathbf{X} \in k_n$. This situation corresponds to $m = 1$, $\nu_1 = n$, $\mathbf{a}_1 = \boldsymbol{\mu}_\mathbf{X}$ and $\mathbf{g}_r(\mathbf{x}) = \mathbf{x}$. The solution furnished by the PME is

$$p(\mathbf{x}) = p_{X_1}(x_1) \times \cdots \times p_{X_n}(x_n), \qquad [2.89]$$

where p_{X_j} is an *exponential density* [2.52]:

$$p_{X_j}(x_j) = 1_{(0,\infty)}(x_j) \frac{1}{\mu_{X_j}} e^{\frac{-x_j}{\mu_{X_j}}}, \qquad j = 1, \cdots, n. \qquad [2.90]$$

In this case, the PME furnishes as solution independent random variables. ∎

EXAMPLE 2.9.– Let us consider $\mathbf{X}$ of dimension n, with support $k_n = \mathbb{R}^n$, known mean $E[\mathbf{X}] = \boldsymbol{\mu}_\mathbf{X} \in k_n$, and a given covariance matrix $\mathbf{C}$ (a real symmetric positive definite matrix such that $Cov(X_i, X_j) = C_{ij}$). In this situation, the PME furnishes as solution the Gaussian distribution:

$$p(\mathbf{x}) = \frac{1}{\sqrt{(2\pi)^n \det(\mathbf{C})}} \cdot \exp\left\{-\tfrac{1}{2}\langle \mathbf{C}^{-1}(\mathbf{X} - \boldsymbol{\mu}_\mathbf{X}), (\mathbf{x} - \boldsymbol{\mu}_\mathbf{X})\rangle_{\mathbb{R}^n}\right\}. \blacksquare \qquad [2.91]$$

2.2.4. Random matrices

In the previous sections, we have used the PME for determining the distributions of random variables and vectors in situations where some information about these variables is given. We observe that the method extends straightly to the determination of the distribution of random matrices (see [SOI 00]).

Let us denote by:

- $\mathbb{M}_n = \mathcal{M}(n,n)$: the set of all the square real matrices $n \times n$;
- $\mathbb{M}_n^S = \mathcal{M}^S(n,n)$: the set of all the real symmetric matrices $n \times n$;
- $\mathbb{M}_n^{+0} = \mathcal{M}^{+0}(n,n)$: the set of all the real symmetric matrices $n \times n$ which are, in addition, positive semi-definite;
- $\mathbb{M}_n^+ = \mathcal{M}^+(n,n)$: the set of all the real symmetric matrices $n \times n$ which are, in addition, positive definite;
- $\mathbb{M}_n^D = \mathcal{M}^D(n,n)$: the set of all the real diagonal matrices $n \times n$.

We have

$$\mathbb{M}_n^D (\subset \mathbb{M}_n^+ \subset \mathbb{M}_n^{+0} \subset \mathbb{M}_n^S \subset \mathbb{M}_n \quad [2.92]$$

The norm of a matrix is given by:

$$\|A\| = \sup_{\|x\| \leq 1} \|[A]x\| \quad , \quad x \in \mathbb{R}^n. \quad [2.93]$$

and its Frobenius norm (or Hilbert–Schmidt norm) is:

$$\|A\|_F^2 = \operatorname{tr}\left\{[A]^T[A]\right\} = \sum_{j=1}^n \sum_{k=1}^n [A]_{jk}^2 \quad [2.94]$$

We have $\|A\| \leq \|A\|_F \leq \sqrt{n}\,\|A\|$.

Let us introduce the element of volume $\tilde{d}A_n$ given by:

$$\tilde{d}\mathbf{A_n} = 2^{n(n-1)/4} \prod_{1 \leq i \leq j \leq n} dA_{nij}. \quad [2.95]$$

Let $\mathbf{A_n}$ be a random matrix taking its values on $\mathbb{M}_n^+$, which has as density of probability ($\bullet$), such that

$$\mathbf{A_n} \longmapsto p(\mathbf{A_n}), \quad \text{from } \mathbb{M}_n^+(\mathbb{R}) \text{ to } \mathbb{R}^+ = [0, +\infty) \quad [2.96]$$

Analogously to the preceding situations, the PME may be used in order to generate the distribution p. For instance, let us suppose that the following information is known about $\mathbf{A_n}$:

1) $\mathbf{A_n} \in \mathbb{M}_n^+$ and $\int_{\mathbb{M}_n^+} p_{\mathbf{A_n}}(\mathbf{A_n}) \, \tilde{d}\mathbf{A_n} = 1$;

2) the mean of $\mathbf{A_n}$ is given:

$$E(\mathbf{A_n}) = \int_{\mathbb{M}_n^+} \mathbf{A_n} p(\mathbf{A_n}) \, \tilde{d}\mathbf{A_n} = \underline{\mathbf{A_n}}. \qquad [2.97]$$

3) $\int_{\mathbb{M}_n^+} \ln(\det \mathbf{A_n}) p(\mathbf{A_n}) \, \tilde{d}\mathbf{A_n} = v$, with $|v| < +\infty$.

Analogously to the preceding situations, Shannon's entropy reads as:

$$S(p) = -\int_{\mathbb{M}_n^+} p(\mathbf{A_n}) \ln p(\mathbf{A_n}) \, \tilde{d}\mathbf{A_n} \qquad [2.98]$$

and the admissible set C is formed by all the densities of probability p from $\mathbb{M}_n^+$ on $\mathbb{R}^+$ verifying all these conditions.

In order to apply the PME, some algebraic manipulations are requested. Let us decompose $\underline{\mathbf{A}}_n$ from the Cholesky decomposition:

$$\underline{\mathbf{A}}_n = \underline{\mathbf{L}}_{\mathbf{A_n}}{}^T \underline{\mathbf{L}}_{\mathbf{A_n}}, \qquad [2.99]$$

where $\underline{\mathbf{L}}_{\mathbf{A_n}}$ is an upper triangular matrix. Then, the PME furnishes:

$$\mathbf{A_n} = \underline{\mathbf{L}}_{\mathbf{A_n}}{}^T \mathbf{G_n} \underline{\mathbf{L}}_{\mathbf{A_n}}, \qquad [2.100]$$

where $\mathbf{G_n}$ is a random matrix taking its values on $\mathbb{M}_n^+$ having mean equal to the identity matrix $\mathbf{Id_n}$:

$$\underline{\mathbf{G}}_n = E(\mathbf{G_n}) = \mathbf{Id_n}. \qquad [2.101]$$

The probability density of $\mathbf{G_n}$ is denoted by $q(\bullet)$. We have:

$$q(\mathbf{G_n}) = 1_{\mathbb{M}_n^+}(\mathbf{G_n}) \times C_{G_n} \times (\det \mathbf{G_n})^{(n+1)\frac{(1-\delta_A^2)}{2\delta_A^2}}$$
$$\times \exp\left\{-\frac{(n+1)}{2\delta_A^2} \mathrm{tr}(\mathbf{G_n})\right\}, \qquad [2.102]$$

where C_{G_n} is a normalizing constant:

$$C_{G_n} = \frac{(2\pi)^{-n(n-1)/4}\left(\frac{n+1}{2\delta_A^2}\right)^{n(n+1)(2\delta_A^2)^{-1}}}{\left\{\prod_{j=1}^{n}\Gamma\left(\frac{n+1}{2\delta_A^2}+\frac{1-j}{2}\right)\right\}}. \qquad [2.103]$$

δ_A is a dispersion parameter given by:

$$\delta_A = \left\{\frac{E(\|\mathbf{G_n}-\underline{\mathbf{G_n}}\|_F^2)}{\|[\underline{\mathbf{G_n}}]\|_F^2}\right\}^{\frac{1}{2}}, \qquad [2.104]$$

with $0 < \delta < \sqrt{\frac{n+1}{n+5}}$.

In addition, the matrix $\mathbf{G_n}$ is positive definite and may be decomposed by Cholesky:

$$\mathbf{G_n} = \mathbf{L_n}^T \mathbf{L_n}, \qquad [2.105]$$

where $\mathbf{L_n}$ is an upper triangular matrix taking its values on $\mathbb{M}_n$. With this decomposition, the elements of $\mathbf{L_n}$ are independent and characterized by:

– if $j < j'$: $\mathbf{L_n}_{jj'} = \sigma_n U_{jj'}$, where $\sigma_n = \delta_A(n+1)^{-1/2}$ and $U_{jj'}$ is a real Gaussian random variable having mean equal to 0 and variance equal to 1;

– if $j = j'$: $\mathbf{L_n}_{jj} = \sigma_n\sqrt{2V_j}$, where V_j is a real-positive random variable Gamma distributed, having as density of probability:

$$p_{V_j}(v) = \mathbf{1}_{\mathbb{R}^+}(v)\frac{1}{\Gamma\left(\frac{n+1}{2\sigma_A^2}+\frac{1-j}{2}\right)}v^{\left(\frac{n+1}{2\delta_A^2}-\frac{1+j}{2}\right)}e^{-v}. \qquad [2.106]$$

2.3. Generating samples of random variables, random vectors and stochastic processes

As mentioned in the beginning of the chapter, once deterministic and stochastic models (e.g. using PME) have been generated, the Monte Carlo method can be applied in order to furnish information and statistics about the response of the system (Figure 2.1).

However, the implementation of the Monte Carlo method involves the generation of samples, i.e. variates from a given probability distribution. For instance, assume that the PME has furnished a density of probability for the parameters chosen to be considered as random: the practical application of the Monte Carlo method needs the generation of variates from these distributions.

So, we need to provide methods for the generation of random variables, vectors or matrices, as for stochastic processes and fields.

This topic has been extensively treated in the literature, and a large number of works consider the construction of *random number generator* (RNG). In general, some qualities are expected from an RNG:

1) rapidity, for the generation of large numbers of variates;

2) controlled reproducibility, in order to make possible comparisons and debugging;

3) knowledge of its probabilistic and statistical properties;

4) an extremely large period;

5) pseudo-randomization, namely for the targeted applications.

Early in the development of generators, many hybrid techniques mixing analog and digital systems have been used, for instance, electronic circuits generating white noise. However, these methods had shown some disadvantages: not so rapid as necessary and not easily reproducible, they also tended to produce bias and requested highly specialized equipment. Therefore, these techniques were replaced over the years. Today, virtually all generators are based on algorithms.

Since computers are expected to be deterministic devices, it may seem impossible to use them as random generators. In fact, computer-generated random samples are only pseudo-random (see [SHO 09]), since they are obtained through deterministic algorithms. However, they appear as random and reproduce random distributions – the main generators available on most of the computers have been subjected to rigorous testing in order to confirm their quality.

In [KNU 98], the author describes his personal attempt to build a random generator of random variables. He proposed an algorithm that worked with 10-digit decimal numbers, generating a sequence of values $\{x_i\}_i \in \mathbb{N}$ from an initial value.

Despite the apparent complication, Knuth found that his algorithm could quickly converge to a fixed point. In addition, the tests showed that, even when using different initial values, the sequence of numbers quickly began to repeat itself (so, the period was relatively small).

The algorithm of Knuth is an example that complexity in the construction of the algorithm does not guarantee that the resulting random generator will have all the expected qualities. Frequently, the complexity hides the period, i.e the rapidity with which the numbers generated repeat. Trivially, we can say that: *random samples should not be generated by a method chosen at random* (see [KNU 98]).

Many methods for generating samples of random variables and vectors have been developed over the recent years. Examples include the popular generators based on linear congruence, the inverse transform method, or the Markov chains (Markov Chain Monte Carlo (MCMC)), among others.

This chapter will discuss the construction of RNG. The readers interested in this topic may consult [ROS 06, DEV 86] and [RUB 08].

We focus on the generation of samples from stochastic processes and fields, for which the generation process is a bit more complicated, since they are infinite-dimensional objects. In the following, we present the Karhunen–Loève approach for the generation of realizations of stochastic processes.

2.4. Karhunen–Loève expansions and numerical generation of variates from stochastic processes

The Monte Carlo method requests the generation of samples formed by variates from a random variable, vector or process: Monte Carlo approaches use random sampling as a tool to produce observations which may be used in order to perform statistical analysis or inference, with the aim of extracting information about quantities of interest (see [SHO 09]).

The generation of large samples as requested by the Monte Carlo approach may result expensive. The Karhunen–Loève expansion provides a parameterization of a stochastic process that enables its approximation through a finite-dimensional process, i.e. by a random vector. Such an approximation can be used, for example, to reduce the computational cost of generating variates from the process under consideration.

2.4.1. *Karhunen–Loève expansions*

Let us consider a stochastic process X, indexed by time and defined on a time interval $T = (a, b)$. We denote by μ_X its mean and by $C(s, t)$ its covariance function (see section 1.16). We consider the following eigenvalue problem:

$$\text{find } (\lambda, \psi) \text{ such that } \psi \neq 0 \text{ and } \int_T C(t, s)\psi(s)\, ds = \lambda\, \psi(t), \forall t \in T \quad [2.107]$$

The eigenvectors ψ_i are normalized in order to have a mean square norm equal to the unity. Since C is positive, the eigenvalues are always positive $\lambda_i \geq 0, \forall i \in \mathbb{N}$. We assume that this problem has a countable set of solutions $\{(\lambda_i, \psi_i) : i \in \mathbb{N}\}$, which are positive *and may be ordered decreasingly*: $0 \leq \lambda_{i+1} \leq \lambda_i, \forall i \in \mathbb{N}$. In general, the inequality is not strict (see, for instance, example 2.11).

We define a sequence $\{Z_i\}_{i \in \mathbb{N} \geq 1}$ of random variables by

$$Z_i = \frac{1}{\lambda_i} \int_T [X_t - \mu_X(t)]\, \psi_i(t)\, dt \qquad \forall i \in \mathbb{N}. \blacksquare \quad [2.108]$$

We have

$$E[Z_i] = 0 \quad \text{and} \quad E[Z_i\, Z_j] = \delta_{ij} \qquad \forall i, j \in \mathbb{N}, \quad [2.109]$$

The Karhunen–Loève expansion of X is:

$$X_t = \mu_X(t) + \sum_{i=1}^{\infty} \sqrt{\lambda_i}\, \psi_i(t)\, Z_i, \quad [2.110]$$

Thus, the Karhunen–Loève expansion furnishes a particular way for the decomposition of a stochastic process into an infinite sum of time-dependent terms (the eigenfunctions ψ_i) and a sequence of random variables Z_i. In practice, Karhunen–Loève expansions are approximated by finite sums involving d terms:

$$X_t \approx \mu_X(t) + \sum_{i=1}^{d} \sqrt{\lambda_i}\, \psi_i(t)\, Z_i. \qquad [2.111]$$

Obviously, such a truncation requests some reflection about the choice of d – the number of terms to be used in order to obtain a given quality. This question may be answered by considering one of the main characteristics of Karhunen–Loève expansions: the value of λ_i also gives its relative contribution to the expansion: the smaller the eigenvalue, the smaller the contribution of the term (recall that both ψ_i and Z_i are normalized). Since the eigenvalues are arranged in decreasing order, the contribution decreases with i. Thus, we may evaluate the error by considering the mean square norm of the remaining eigenvalues $i > d$. Examples of applications can be found in [TRI 05, BEL 06, BEL 09, SAM 07, SAM 08, SAM 10, DOR 12, CAT 09, MAU 12].

In the following, we illustrate Karhunen–Loève expansions for some particular stochastic processes.

EXAMPLE 2.10.– Let us consider a stochastic process X_t, indexed by the parameter $t \in T = [-b, b]$ and having as covariance function C the exponential one:

$$C(t_1, t_2) = \exp\left(\frac{-|t_1 - t_2|}{a}\right) \quad , \quad a > 0. \qquad [2.112]$$

In this situation, the eigenfunctions ψ_i and the eigenvalues λ_i are obtained by solving equation [2.107], which reads as:

$$\int_T \exp{-c|t_1 - t_2|}\psi_i(t_2)\, dt_2 = \lambda_i\, \psi_i(t_1), \qquad [2.113]$$

where $c = 1/a$. Equation [2.113] may be written as:

$$\int_{-b}^{t_1} \exp -c(t_1 - t_2)\psi_i(t_2)\, dt_2 + \int_{t_1}^{b} \exp -c(t_2 - t_1)\psi_i(t_2)\, dt_2 = \lambda_i\, \psi_i(t_1).$$
[2.114]

By differentiating equation [2.114] twice with respect to t_1 (and using some algebraic manipulations), we obtain a differential equation for ψ_i, which leads to the following eigenvalues λ_i (see [XIU 10]):

$$\lambda_i = \begin{cases} \dfrac{2a}{1 + a^2\omega_i^2}, & \text{for } i = 2, 4, 6, \cdots \\ \dfrac{2a}{1 + a^2 v_i^2}, & \text{for } i = 1, 3, 5, \cdots \end{cases}$$
[2.115]

and eigenfunctions ψ_i (see [XIU 10]):

$$\psi_i(t) = \begin{cases} \sin(\omega_i t)\Big/\sqrt{b - \dfrac{\sin(2\omega_i b)}{2\omega_i}}, & \text{for } i = 2, 4, 6, \cdots \\ \sin(v_i t)\Big/\sqrt{b + \dfrac{\sin(2v_i b)}{2v_i}}, & \text{for } i = 1, 3, 5, \cdots \end{cases}$$
[2.116]

where ω_i and v_i satisfy the equations below:

$$\begin{cases} a\omega + \tan(\omega b) = 0, & \text{for } i = 2, 4, 6, \cdots \\ 1 - a v \tan(vb) = 0, & \text{for } i = 1, 3, 5, \cdots \end{cases}$$
[2.117]

In Figure 2.2, we exhibit the first 20 eigenvalues λ_i for different values of a. ∎

EXAMPLE 2.11.– Let us consider a stochastic process X_t, indexed by a parameter t defined by $t \in T = [-b, b]$ and having as covariance function C the orthonormal one:

$$C(t_1, t_2) = \delta(t_1 - t_2).$$
[2.118]

In this case, equation [2.107] admits as a solution any family $\{(\lambda_i, \psi_i) : i \in \mathbb{N}\}$, such that $\{\psi_i : i \in \mathbb{N}\}$ is a family of orthonormal functions and $\lambda_i = 1, \forall i \in \mathbb{N}$. In this case, the eigenvalues are not strictly decreasing. ∎

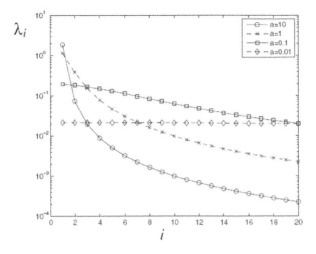

Figure 2.2. *First 20 eigenvalues λ_i determined for the exponential covariance function with different values of a*

NOTE 2.1.– Karhunen–Loève expansion of Gaussian stochastic processes: a particular situation often found and, thus, having a particular interest is the one where both X_t and $\{Z_i : i \in \mathbb{N}\}$ are Gaussian. In this case, equation [2.110] corresponds to the sum of uncorrelated normalized Gaussian variables – thus, to independent Gaussian variables. So, the Karhunen–Loève expansion corresponds to the approximation of a Gaussian process by sums of independent Gaussian variables with time-dependent coefficients. ∎

2.4.2. *Numerical determination of Karhunen–Loève expansions*

In the previous examples, we have considered analytical expressions of the terms involved in the Karhunen–Loève expansion, derived from the analytical expression of the covariance function of the stochastic process X. In this section, we consider the situation where such an expression is not available and we have only observations connected to a time discretization.

For instance, let us assume that $T = (a, b)$ and consider that a *finite number* n of time instants is given: $\tau = (t_1, t_{2,;} \cdots, t_n)^t$ such that $a = t_1 < t_2 < \cdots < t_n = b$. In practice, data are often sampled at fixed time intervals and we have $t_{j+1} - t_j = \Delta t$, $\Delta t = (b-a)/(n-1)$.

For each t_j in this set, we consider $X_j = X_{t_j}$: associated with the time instants, we have the random variables $X_1, X_2, \cdots, X_n$. A realization of the stochastic process at these instants consists of a vector of variates: $\mathbf{X}_m = ((X_1)_m, \cdots, (X_n)_m)^t$, where $1 \leq m \leq M$ is an integer that indexes the realizations: M is the number of available realizations.

For M and n large enough, approximated realizations of X_t, as estimations of the mean, variance, autocorrelation and covariance of the stochastic process X, may be obtained (see [WIR 95]). For instance, we may use an interpolation of the discrete values $\mathbf{X}_m$ in order to generate an approximated generalization $X_m(t)$. An example is provided by the simple linear interpolation

$$X_m(t) \approx (X_j)_m + (t_{j+1} - t_j)\frac{(X_j)_m - (X_j)_m}{t_{j+1} - t_j} \qquad t_j \leq t_{j+1}. \quad [2.119]$$

The mean μ_X may be estimated by $\hat{\mu}_X$ defined as follows:

$$\hat{\mu}_X(t_j) = \frac{1}{M}\sum_{m=1}^{M}(X_j)_m \qquad j = 1, \cdots, n. \qquad [2.120]$$

These estimates generate a vector $\hat{\boldsymbol{\mu}} = (\hat{\mu}_X(t_1), \cdots, \hat{\mu}_X(t_n))^t$, which is an estimation of $\boldsymbol{\mu}_T = (\hat{\mu}_X(t_1), \cdots, \hat{\mu}_X(t_n))^t$. Analogously to the realizations of the stochastic process itself, an estimation $\hat{\mu}_X(t)$ is obtained from these values by using these discrete values in order to generate a function defined on T; for instance, we may use an interpolation or other approximation procedures.

An analogous procedure may be used in order to obtain an estimation $\hat{\sigma}_X^2$ of σ_X^2 (see 1.27):

$$\hat{\sigma}_X^2(t_j) = \frac{1}{M}\sum_{m=1}^{M}((X_j)_m - \hat{\mu}_X(t_j))^2 \qquad j = 1, \cdots, n. \qquad [2.121]$$

Analogously to the mean, an estimation $\hat{\sigma}_X^2(t)$ is obtained from these values by using these discrete values in order to generate a function defined on T.

The situation is entirely analogous for the autocorrelation or the covariance. We generate

$$\hat{R}(t_j, t_k) = \frac{1}{M} \sum_{m=1}^{M} \left((X_j)_m \cdot (X_k)_m\right) ; \qquad [2.122]$$

$$\hat{C}(t_j, t_k) = \frac{1}{M} \sum_{m=1}^{M} \left((X_j)_m - \hat{\mu}_X(t_j)\right) \cdot \left((X_k)_m - \hat{\mu}_X(t_k)\right). \qquad [2.123]$$

These quantities may be used for the construction of continuous functions $\hat{R}(s,t)$ and $\hat{C}(s,t)$, which estimate $R(s,t)$ and $C(s,t)$, respectively.

2.4.2.1. *Estimating the eigenvalues and eigenfunctions of the covariance matrix*

In practice, we may determine estimations of the eigenvalues and eigenvectors of the covariance matrix *without going through the construction* $\hat{C}(s,t)$ defined in the previous section: let us represent the M realizations $\mathbf{X}_m$, $1 \leq m \leq M$ by using two $n \times M$ matrices $\mathbf{X}$ and $\boldsymbol{\mu}$, such that the jth column of $\mathbf{X}$ is $\mathbf{X}_j$ and all the columns of $\boldsymbol{\mu}$ are equal to $\boldsymbol{\mu}_\tau$:

$$\mathbf{X} = \begin{pmatrix} \mathbf{X}_1 \, \mathbf{X}_2 \, \cdots \, \mathbf{X}_M \end{pmatrix}, \boldsymbol{\mu} = \begin{pmatrix} \boldsymbol{\mu}_\tau \, \boldsymbol{\mu}_\tau \, \cdots \, \boldsymbol{\mu}_\tau \end{pmatrix} ; \qquad [2.124]$$

i.e.:

$$\mathbf{X} = \begin{pmatrix} (X_1)_1 & (X_1)_2 & \cdots & (X_1)_M \\ \vdots & \vdots & \vdots & \vdots \\ (X_n)_1 & (X_n)_2 & \cdots & (X_n)_M \end{pmatrix}, \boldsymbol{\mu} = \begin{pmatrix} \hat{\mu}_X(t_1) & \cdots & \hat{\mu}_X(t_1) \\ \vdots & \vdots & \vdots \\ \hat{\mu}_X(t_n) & \cdots & \hat{\mu}_X(t_n) \end{pmatrix}. \qquad [2.125]$$

Then, we consider

$$\mathbf{X_0} = \mathbf{X} - \boldsymbol{\mu}. \qquad [2.126]$$

and may estimate the covariance function $C(s,t)$ by using the matrix

$$\hat{\mathbf{C}} = \frac{1}{M} \mathbf{X_0} \, \mathbf{X_0}^t, \qquad [2.127]$$

which gives the estimates of $C(t_i, t_j)$ for pairs (t_i, t_j) from τ:

$$\hat{\mathbf{C}} = \begin{bmatrix} \hat{C}(t_1, t_1) & \cdots & \hat{C}(t_1, t_n) \\ \vdots & \ddots & \vdots \\ \hat{C}(t_n, t_1) & \cdots & \hat{C}(t_n, t_n) \end{bmatrix}. \qquad [2.128]$$

This matrix is used as follows: equation [2.107] is approximated as

$$\hat{\mathbf{C}}\hat{\boldsymbol{\psi}} \, \Delta t = \hat{\lambda}\hat{\boldsymbol{\psi}}, \qquad [2.129]$$

The solution of equation [2.129] provides n eigenvalues $\hat{\lambda}_i$ and their associated eigenvectors $\hat{\boldsymbol{\psi}}_i$ (normalized to the unity norm). These quantities estimate the n first eigenvalues λ_i and the values $(\boldsymbol{\psi}_i)_\tau = (\psi_i(t_1), \cdots, \psi_i(t_n))^t$, which may be used for the construction of an estimate $\hat{\psi}_i$ of ψ_i on T, for $1 \leq i \leq n$ (for instance, by using an interpolation procedure).

EXAMPLE 2.12.– Let us consider the stochastic process:

$$X_t = A_1 \, t + A_2, \qquad [2.130]$$

where $\mathbf{A} = (A_1, A_2)^t$ is a Gaussian vector having the mean and covariance given by:

$$\{\mu_\mathbf{A}\} = \begin{bmatrix} 1 \\ 2 \end{bmatrix} \qquad [C_\mathbf{A}] = \begin{bmatrix} 1 & 1/2 \\ 1/2 & 1 \end{bmatrix}. \qquad [2.131]$$

In this situation, X_t is a linear combination of the components of a Gaussian vector – thus, X_t is a Gaussian variable and, so, the process is Gaussian. As a result, for any $t_j \in T$, the probability density $p_j = p_{t_j}$ is Gaussian. Such a process may be represented as follows:

$$X_t = \boldsymbol{\alpha}^t \mathbf{A} \; ; \; \boldsymbol{\alpha} = \begin{pmatrix} t \\ 1 \end{pmatrix}, \qquad [2.132]$$

We have:

$$\mu_X(t) = \boldsymbol{\alpha}^t \mu_\mathbf{A} = t + 2 \qquad [2.133]$$

$$\sigma_X^2(t) = \alpha^t \mathbf{C_A} \alpha = t^2 + t + 1. \qquad [2.134]$$

As a result, the probability density p_{t_j} is known :

$$p_{t_j}(x) = \frac{1}{\sqrt{2\pi}\sqrt{t_j^2 + t_j + 1}} e^{\frac{(x-(t_j+2))^2}{2(t_j^2+t_j+1)}}. \qquad [2.135]$$

Consequently, this process is *first-order defined*.

Using the probability density given in equation [2.135], we may generate realizations: for instance, we have generated 10^4 realizations on the interval $T = (0, 10)$, with $\Delta t = 0.1$. These data have been used for the estimation $\hat{C}$ using equation [2.127] and for the determination of the approximated eigenvalues which satisfy equation [2.129].

The results are shown in Figure 2.3. ∎

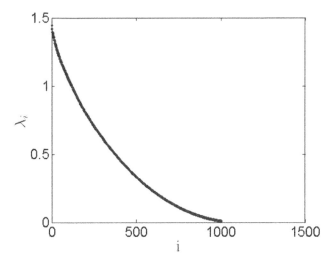

Figure 2.3. *Eigenvalues λ_i estimated from the covariance matrix $\hat{C}$ of the stochastic process [2.130]*

EXAMPLE 2.13.– Let us consider the stochastic process:

$$\mathcal{X}(t,\omega) = A_1 \cos(\omega t) + A_2 \sin(\omega t), \qquad [2.136]$$

where $\mathbf{A} = (A_1, A_2)^t$ is a Gaussian vector having the mean and covariance given by equation [2.131]. Analogously to the preceding example, X_t is Gaussian and the process is Gaussian. We have:

$$X_t = \alpha^t \mathbf{A} \; ; \; \alpha = \begin{pmatrix} \cos(\omega t) \\ \sin(\omega t) \end{pmatrix}. \quad [2.137]$$

In this case, the mean and the covariance are:

$$\mu_X(t) = \alpha^t \mu_\mathbf{A} = \cos(\omega t) + 2\sin(\omega t) \quad [2.138]$$

$$\sigma_X^2(t) = \alpha^t \mathbf{C_A} \alpha = \cos^2(\omega t) + \cos(\omega t)\sin(\omega t) + \sin^2(\omega t). \quad [2.139]$$

Here, the probability density p_{t_j} is:

$$p_{t_j}(x) = \frac{1}{\sqrt{2\pi}\sqrt{\sigma_X^2(t_j)}} e^{\frac{(x-\mu_X(t_j))^2}{2\sigma_X^2(t_j)}} \quad [2.140]$$

and this process is also *first-order defined*.

As in the preceding situation, the probability density given in equation [2.140] has been used in order to generate 10^4 realizations on the interval $T = (0, 10)$, with $\Delta t = 0.1$. The data collected have been used for the estimation $\hat{C}$ using equation [2.127] and for the determination of the approximated eigenvalues which satisfy equation [2.129].

The results are shown in Figure 2.4. ∎

2.4.2.2. Using the estimations for the generation of an approximated Karhunen–Loève expansion

Once the estimates of $\hat{\lambda}_i$ and $\hat{\psi}_i$n are obtained, we have to generate the elements Z_i to be used in the Karhunen–Loève expansion. Namely, we must determine their distributions of probability. From equation [2.108], we may generate an approximated realization of Z_i by using:

$$(Z_i)_m = \frac{1}{\lambda_i} < (\mathbf{X}_m - \hat{\mu}) \, , \, (\boldsymbol{\psi_i})_\tau >_{\mathbb{R}^n}. \quad [2.141]$$

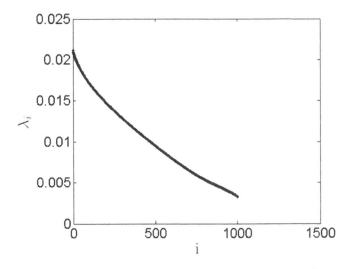

Figure 2.4. *Eigenvalues λ_i estimated from the covariance matrix $\hat{C}$ of the stochastic process [2.136]*

We may collect the realizations of Z_i in a vector $\mathbf{Z_i} = ((Z_i)_1, \cdots, (Z_i)_M)^t$. We have:

$$\mathbf{Z_i} = \frac{1}{\lambda_i}\mathbf{X_0}^t \cdot (\boldsymbol{\psi_i})_T, \qquad [2.142]$$

where the matrix $\mathbf{X_0}$ is defined in equation [2.126] and has dimension $(n \times M)$. This equality shows that the samples $\mathbf{X}_m$, $1 \leq m \leq M$, provide M realizations of Z_i: for M large enough, we may use these values in order to determine the distribution of Z_i. For instance, we may generate a histogram in order to get an approximate probability density.

Once this step is performed, we may consider an approximate Karhunen–Loève expansion involving d terms ($d < n$), given by:

$$X_t^{(KL)} = \hat{\mu}(t) + \sum_{i=1}^{d} \sqrt{\hat{\lambda}_i}\, \hat{\psi}_i(t)\, Z_i. \qquad [2.143]$$

Thus, for a given stochastic process X, the procedure consists of generating a sample $\mathbf{X}$; estimating the mean $\hat{\mu}$ and the covariance matrix $\hat{C}$

(equations [2.120] and [2.123] or [2.127]); determining the eigenvalues $\hat{\lambda}_i$ and their associated eigenvectors $(\psi_i)_\tau$ (equation [2.129]); generating the continuous approximations, the continuous $\hat{\psi}_i$ and $\hat{\mu}$; finding the variables Z_i and their probability densities (for instance, by using equation [2.142]).

EXAMPLE 2.14.– Assume that we are interested in the generation of realizations from the stochastic process $\mathcal{X}$, indexed by the parameter $t \in T = (0, 10)$, such that:

1) $X_0 = 0$;

2) $X_t, t \geq 0$, is a stochastic process that has independent increments;

3) the increments $Y = X_{t_2} - X_{t_1}$, with $t_2 > t_1 \geq 0$, are Gamma distributed, have a mean $\mu_Y = m(t_2 - t_1)$ and a coefficient of variation $\delta_Y = \frac{\delta}{\sqrt{t_2 - t_1}}$), with $m > 0$ and $\delta > 0$ given.

We have generated the approximated Karhunen–Loève expansion by using the procedure indicated and a sample of 10^4 variates from X on $T = (0, 10)$, obtained with different values of Δt.

The results obtained for the first 20 eigenvalues $\hat{\lambda}_i$ are shown in Figure 2.5.

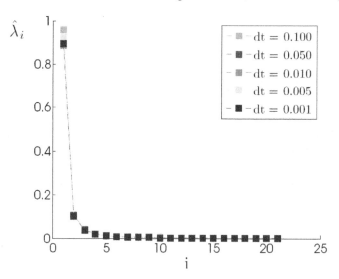

Figure 2.5. *Estimations of the eigenvalues $\hat{\lambda}_i$ of the matrix $\hat{C}$ for different values of Δt. For a color version of the figure, see www.iste.co.uk/souzadecursi/quantification.zip*

The estimated probability densities of the variables Z_i are shown in Figure 2.6 – they were obtained by using the histograms associated with equation [2.142].

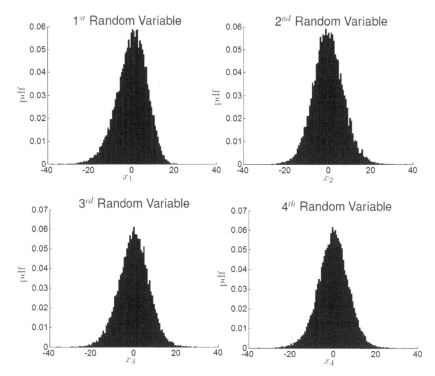

Figure 2.6. *Estimations for the densities of probability of the first four random variables Z_i in the Karhunen–Loève expansion. For a color version of the figure, see www.iste.co.uk/souzadecursi/quantification.zip*

Then, samples from Z_i were generated by using the MCMC method, which provided a sample from X.

We have studied the influence of d from two different approaches. In the first approach, the covariance matrix $\hat{\mathbf{C}}_\mathbf{d}$ has been estimated from the samples. Then, we have considered the error

$$Error = \frac{\left|\hat{\mathbf{C}}_\mathbf{d}\right|_{Fr} - \left|\hat{\mathbf{C}}\right|_{Fr}}{\left|\hat{\mathbf{C}}\right|_{Fr}}.$$

This expression has been evaluated for different values of d. The results are shown in Figure 2.7. As we can observe, the error [2.145] decreases when d increases.

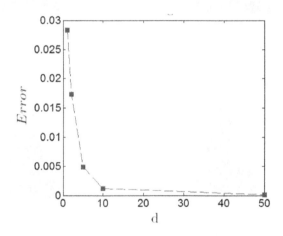

Figure 2.7. *Error [2.145] for different numbers of terms d*

In the second approach, we compare, for a given fixed t, the probability density of X_t and the histogram of $X_t^{(KL)}$.

Figures 2.8–2.9 exhibit results for $t = 2$ and $d = 1$, 2, 5, 10. We observe that, when d increases, the histogram of $X_t^{(KL)}$ becomes closer to the correct result p_{X_2}. ∎

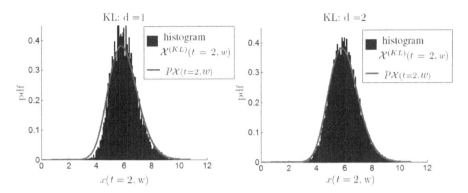

Figure 2.8. *Densities $p_{\mathcal{X}(t=2,\omega)}$ and histograms of $\mathcal{X}^{KL}(t=2,\omega)$ for $d=1$ and $d=2$ terms. For a color version of the figure, see www.iste.co.uk/souzadecursi/quantification.zip*

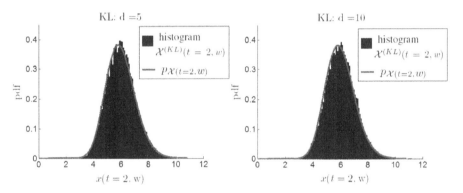

Figure 2.9. *Densities $p_{\mathcal{X}(t=2,\omega)}$ and histograms of $X^{KL}(t=2,\omega)$ for $d=5$ and $d=10$ terms. For a color version of the figure, see www.iste.co.uk/souzadecursi/quantification.zip*

3

Representation of Random Variables

When considering random variables, the terminology *representation* is generally used in the sense of *approximation*. Naively, the problem of determining a representation for a random variable **X** using another random variable **U** may stand as follows:

MODEL PROBLEM 3.1 (informal).– Let **X** and **U** be two *random vectors* (eventually having different dimensions) on the probability space (Ω, P). Let $S \neq \varnothing$ be a set of functions of **U**. Determine an element **PX** from S such that **PX** is the best approximation of **X** on S, i.e. **PX** is the element of S that is the closest 1 to **X**. ∎

The main objective of a representation is to use **PX** instead of **X** in a procedure of design, in the evaluation of the probability of crucial events. Ideally, we desire to get a representation (i.e. an approximation) such that **X** − **PX** is equivalent to **0** (i.e. it is negligible and we have **X** = **PX**, almost everywhere (a.e.)). In general, this aim cannot be achieved and we look for an approximation such that **PX** may replace **X** in practice, i.e. such that the difference **X** − **PX** is small enough in a sense corresponding to our purposes.

It is interesting to note that this formulation leads to an optimization problem, since we must determine the element of S that is the closest to **X** according to a proximity criterion: this is generally achieved by introducing a function that measures the distance between the random variables and determining its minimum on S. In order to obtain a precise formulation, it is necessary to introduce formally precise definitions of subspace S and of the distance – this is also necessary in order to provide algorithms for the numerical approximation.

3.1. Approximations based on Hilbertian properties

When considering square summable variables ($\mathbf{X} \in \left[L^2\left(\Omega, P\right)\right]^k$ and $S \subset \left[L^2\left(\Omega, P\right)\right]^k$), we may exploit the *Hilbertian structure* of this set (see sections 1.6, 1.11) in order to characterize the representation $\mathbf{PX}$ as an *orthogonal projection* (see definition 1.5). In this case, the scalar product and the norm are those of $\left[L^2\left(\Omega, P\right)\right]^k$:

$$(\mathbf{Z}, \mathbf{Y}) = E\left(\mathbf{Z}.\mathbf{Y}\right), \quad \|\mathbf{Z}\| = \sqrt{(\mathbf{Z}, \mathbf{Z})}.$$

The *orthogonal projection* is the solution of the following problem.

PROBLEM 3.1.– Let $\mathbf{X} = (X_1, \ldots, X_k)$ and $\mathbf{U} = (U_1, \ldots, U_m)$ be two *random vectors* taking their values on $\mathbb{R}^k$ and $\mathbb{R}^m$, respectively. Let $S \neq \varnothing$, $S \subset \left[L^2\left(\Omega, P\right)\right]^k$ be a set of functions of $\mathbf{U}$. Find

$$\mathbf{Z} \in S \text{ and } \|\mathbf{Z} - \mathbf{X}\| = \min\left\{\|\mathbf{s} - \mathbf{X}\| : \mathbf{s} \in S\right\}. \blacksquare \qquad [3.1]$$

The error in the approximation is measured by the distance between $\mathbf{X}$ and S: $\text{dist}\left(\mathbf{X}, S\right) = \|\mathbf{X} - \mathbf{PX}\|$.

As previously observed, we have to solve an optimization problem: find the argument $\mathbf{Z}$ that minimizes $\|\mathbf{Z} - \mathbf{X}\|$ on S. With this formulation, the theoretical framework about orthogonal projections may be applied, namely, the main fundamental result given by the *orthogonal projection theorem* and its consequences (see proposition 1.8). For instance, if S is a closed linear subspace S, then $\mathbf{PX}$ is uniquely determined and the numerical determination of $\mathbf{PX}$ may be performed by using the proposition 1.9. We have:

$$\mathbf{PX} \in S \text{ and } \mathbf{X} - \mathbf{PX} \perp S,$$

so that:

$$\mathbf{PX} \in S \text{ and } (\mathbf{X} - \mathbf{PX}, \mathbf{s}) = 0, \forall \, \mathbf{s} \in S.$$

Indeed, if $dim(S) = n$, we may consider a basis $\mathcal{B} = \{\mathbf{e}_i : 1 \leq i \leq n\}$ of $L^2(\Omega, P)$ of the linear space S and take unknowns as the coefficients of the expression of $\mathbf{PX}$ in this basis. In the case of a closed subset S, the representation is exact (i.e. $\mathbf{X} - \mathbf{PX} = 0$) if and only if $\mathbf{X} \in S$ (since $\mathbf{X} = \mathbf{PX} \in S$). The last equation may be used in order to generate equations leading to the complete determination of the coefficients of $\mathbf{PX}$, by taking successively $\mathbf{s} = \mathbf{e}_i$ for $i = 1, \ldots, n$.

In practice, S is often generated by considering a basis $\mathcal{B} = \{\varphi_i : 1 \leq i \leq N_X\}$ ($N_X > 0$) of $L^2(\Omega, P)$ (instead $\left[L^2(\Omega, P)\right]^k$) and taking S as being the subspace of dimension $n = k \times N_X$ given by

$$S = \left[\{\varphi_1(\mathbf{U}), \ldots, \varphi_{N_X}(\mathbf{U})\}\right]^k = \left\{\mathbf{Z} = \sum_{i=1}^{N_X} \mathbf{z}_i \varphi_i(\mathbf{U}) : \mathbf{z}_i \in \mathbb{R}^k, 1 \leq i \leq N_X\right\}, \quad [3.2]$$

Then, an element $\mathbf{Z}(\mathbf{U}) = \sum_{i=1}^{N_X} \mathbf{z}_i \varphi_i(\mathbf{U})$ is defined by the coefficients $\mathbf{z} = (\mathbf{z}_1, \ldots, \mathbf{z}_n)$, which may be ranged in a matrix $\mathbf{z} \in \mathcal{M}(N_X, k)$ having the coefficients of $\mathbf{z}_j$ as $j - th$ line, i.e.:

$$z = (z_{ij}), \mathbf{z_i} = (z_{i1}, \ldots, z_{ik}) \in \mathbb{R}_t^k \quad [3.3]$$

Let us introduce:

$$\boldsymbol{\varphi}(\mathbf{U}) = (\varphi_1(\mathbf{U}), \ldots, \varphi_n(\mathbf{U})). \quad [3.4]$$

Then, we have:

$$\mathbf{Z}(\mathbf{U}) = \boldsymbol{\varphi}(\mathbf{U}).\mathbf{z}, \quad [3.5]$$

Since $\mathbf{PX} = \sum_{i=1}^{N_X} \mathbf{x}_i \varphi_i \in S$, we may also write:

$$\boldsymbol{PX}(\mathbf{U}) = \boldsymbol{\varphi}(\mathbf{U}).\mathbf{x}, \quad [3.6]$$

where $\mathbf{x} \in \mathcal{M}(N_X, k)$ has the coefficients of $\mathbf{x_i}$ as $j - th$ line, i.e.:

$$x = (x_{ij}), \mathbf{x_i} = (x_{i1}, \ldots, x_{ik}) \in \mathbb{R}^k. \quad [3.7]$$

The unknowns to be determined are the coefficients forming **x**. Since **X** − **PX**⊥S, we have:

$$(\mathbf{X} - \mathbf{PX}, \mathbf{Z}) = 0, \forall \mathbf{Z} \in S.$$

This equation reads as:

$$E\left((\mathbf{X} - \mathbf{PX})^t \mathbf{Z}\right) = 0, \forall \mathbf{Z} \in S,$$

so that:

$$E\left((\mathbf{PX})^t \mathbf{Z}\right) = E\left(\mathbf{X}^t \mathbf{Z}\right), \forall \mathbf{Z} \in S.$$

By using equations [3.5]–[3.6], we obtain:

$$E\left(\mathbf{x}^t \varphi^t \varphi \mathbf{z}\right) = E\left(\mathbf{X}^t \varphi \mathbf{z}\right), \forall \mathbf{z} \in \mathbf{z} \in \mathcal{M}(k, N_X),$$

which shows that:

$$\mathbf{x}^t E\left(\varphi^t \varphi\right) \mathbf{z} = E\left(\mathbf{X}^t \varphi\right) \mathbf{z}, \forall \mathbf{z} \in \mathbf{z} \in \mathcal{M}(k, N_X),$$

Since **z** is arbitrary, we have:

$$\mathbf{x}^t E\left(\varphi^t \varphi\right) = E\left(\mathbf{X}^t \varphi\right). \tag{3.8}$$

Equation [3.8] forms a linear system that may be used in order to determine the coefficients **x**. Examples and the Matlab implementation of this approach are given in the following.

If S is not a finite-dimensional subspace, but it admits a *Hilbertian basis* $\mathcal{F} = \{\mathbf{e_i} : i \in \mathbb{N}^*\}$, we may still consider $\mathbf{PX} = \sum_{i=1}^{+\infty} u_i \mathbf{e_i}$ and to determine the coefficients $\mathbf{u} = (u_1, u_2, \ldots) \in \mathbb{R}^\infty$. From the mathematical standpoint, all the coefficients may be determined, but the approximation is, in practice, limited to a maximum number of terms n, which corresponds to the situation where S is a finite-dimensional subspace. This situation extends to the case where $\mathcal{F}$ is a *countable total family*.

3.1.1. *Using the conditional expectation in order to generate a representation*

As previously observed in sections 1.6.5 and 1.11.1.4, the most general approximation of **X** as a function of **X** is provided by the *conditional mean* or *conditional expectation*. In this case, S i.e. taken as the set of *all the functions of* **U** *that are square summable*:

$$S = \{\mathbf{Z} : \mathbf{Z} = \varphi(\mathbf{U}), \varphi : \mathbb{R}^p \longrightarrow \mathbb{R}^q, \mathbf{Z} \in L^2(\Omega, P)\}$$

The solution is the *conditional mean* of **X** with respect to **U** (sections 1.6.5, 1.11.1.4), $\mathbf{PX} = E(\mathbf{X}|\mathbf{U}) = \mathbf{g}(\mathbf{U})$, given by:

$$\mathbf{g}(\mathbf{u}) = \int \mathbf{x} f(\mathbf{x} \mid \mathbf{U} = \mathbf{u})\, d\mathbf{u},$$

where $f(\mathbf{x} \mid \mathbf{U} = \mathbf{u})$ is the *conditional distribution* of **X** with respect to **u**:

$$f(\mathbf{x} \mid \mathbf{U} = \mathbf{u}) = f(\mathbf{x}, \mathbf{u}) \Big/ \int_{\mathbb{R}} f(\mathbf{x}, \mathbf{u})\, d\mathbf{u} = f(\mathbf{x}, \mathbf{u}) \Big/ f_{\mathbf{X}}(\mathbf{x}).$$

Here, f is the density of the pair $(\mathbf{X}, \mathbf{U})$ (i.e. $P(\mathbf{x} \in d\mathbf{x}, \mathbf{U} \in d\mathbf{u}) = f(\mathbf{x}, \mathbf{u})\, d\mathbf{x}d\mathbf{u}$). The conditional mean provides a lower bound for the error $\|\mathbf{X} - \mathbf{PX}\|$: $E(\mathbf{X}|\mathbf{UX})$ is the best approximation of **X** by a function of **U** in the sense of the norm $\|\bullet\|$, so that $\|\mathbf{X} - \varphi(\mathbf{U})\| \geq \|\mathbf{X} - E(\mathbf{X}|\mathbf{U})\|$ for any approximation $\mathbf{Z} = \varphi(\mathbf{U}) \in S$.

EXAMPLE 3.1.– Let us consider $\Omega = (0,1)$ and $P((a,b)) = b - a$ (i.e. $P(du) = \ell(du)$ where ℓ is the Lebesgue measure). Let $U(\omega) = \sqrt{\omega}$ and $X(\omega) = \omega^2$. The cumulative function of the pair (U, X) is:

$$F(u, x) = P(U < u, X < x) = P\left(\sqrt{\omega} < u, \omega^2 < x\right)$$
$$= P\left(\omega < u^2, \omega < \sqrt{x}\right),$$

so that:

$$F(u, x) = \min\left\{u^2, \sqrt{x}\right\}.$$

The joint probability density is:

$$f(u, x) = \frac{\partial^2}{\partial u\, \partial x} F(u, x).$$

This derivative must be evaluated in its variational form: at a given point (u, x), we consider the ball B_ε of radius $\varepsilon > 0$ and the set $D(B_\varepsilon)$ formed by the compact support functions B_ε. Then, we have:

$$\left\langle \frac{\partial^2}{\partial u\, \partial x} F(u, x), \phi \right\rangle = \int_{B_\varepsilon} F \frac{\partial^2}{\partial u\, \partial x} \phi\, du dx, \forall \phi \in D(B_\varepsilon).$$

Let:

$$B_\varepsilon^< = \{(u, x) \in B_\varepsilon : u^2 < \sqrt{x}\};\ B_\varepsilon^> = \{(u, x) \in B_\varepsilon : u^2 > \sqrt{x}\};$$

$$\Sigma = \{(u, x) : u^2 = \sqrt{x}, 0 < u,\ x < 1\},\ \Sigma_\varepsilon = \{(u, x) \in B_\varepsilon : u^2 = \sqrt{x}\}.$$

We have:

$$(u, x) \in B_\varepsilon^< \Longrightarrow f(u, x) = 0$$

and, in an analogous manner:

$$(u, x) \in B_\varepsilon^> \Longrightarrow f(u, x) = 0$$

so that f is concentrated on the curve Σ: we have $f = A(u, x)\, \delta_\Sigma$.

So, by using that $u^2 = \sqrt{x} \iff u^4 = x$, we have:

$$f(x\,|\,U = u) = A(u, x)\, \delta_\Sigma(u, x) \Big/ \int_\mathbb{R} A(u, x)\, \delta_\Sigma dx = \begin{cases} 0, \text{ if } x \neq u^4, \\ 1, \text{ if } x = u^4. \end{cases}$$

Thus:

$$E(X\,|\,U = u) = u^4 \text{ and } E(X\,|\,U) = U^4.$$

$A(u, x)$ can be determined: indeed, let us introduce the unitary normal oriented outwards the region $B_\varepsilon^<$:

$$n = \begin{pmatrix} n_u \\ n_x \end{pmatrix},\ n_u = \frac{N_u}{\sqrt{N_u^2 + N_x^2}},\ n_x = \frac{N_x}{\sqrt{N_u^2 + N_x^2}},$$
$$N_u = 2u,\ N_x = -\frac{1}{2\sqrt{x}}.$$

We have, at any point $(u, x) \in \Sigma$:

$$\int_{B_\varepsilon} F \frac{\partial^2}{\partial u \, \partial x} \phi \, dudx = \int_{B_\varepsilon^<} u^2 \frac{\partial^2}{\partial u \, \partial x} \phi + \int_{B_\varepsilon^>} \sqrt{x} \frac{\partial^2}{\partial u \, \partial x} \phi,$$

or:

$$\int_{B_\varepsilon(u)} u^2 \frac{\partial^2}{\partial u \, \partial x} \phi = \int_{B_\varepsilon(u)} \operatorname{div}\left(u^2 \begin{pmatrix} \partial \phi/\partial x \\ 0 \end{pmatrix} - N_u \begin{pmatrix} 0 \\ \phi \end{pmatrix}\right),$$

so that (Green's theorem):

$$\int_{B_\varepsilon(u)} u^2 \frac{\partial^2}{\partial u \, \partial x} \phi = \int_{\Sigma_\varepsilon(u)} \left[u^2 \begin{pmatrix} \partial \phi/\partial x \\ 0 \end{pmatrix}.n - N_u \begin{pmatrix} 0 \\ \phi \end{pmatrix}.n \right],$$

and

$$\int_{B_\varepsilon(u)} u^2 \frac{\partial^2}{\partial u \, \partial x} \phi = \int_{\Sigma_\varepsilon(u)} \left[n_u u^2 \partial \phi/\partial x - n_x N_u \phi \right].$$

In an analogous manner:

$$\int_{B_\varepsilon(u)} \sqrt{x} \frac{\partial^2}{\partial u \, \partial x} \phi = \int_{B_\varepsilon(u)} \operatorname{div}\left(\sqrt{x} \begin{pmatrix} \partial \phi/\partial x \\ 0 \end{pmatrix}\right)$$

$$= -\int_{\Sigma_\varepsilon(u)} \sqrt{x} \begin{pmatrix} \partial \phi/\partial x \\ 0 \end{pmatrix}.n,$$

and we have:

$$\int_{B_\varepsilon(u)} \sqrt{x} \frac{\partial^2}{\partial u \, \partial x} \phi = -\int_{\Sigma_\varepsilon(u)} \left[n_u \sqrt{x} \partial \phi/\partial x \right],$$

Thus, we also have:

$$\int_{B_\varepsilon} F \frac{\partial^2}{\partial u \, \partial x} \phi \, dudx = \int_{\Sigma_\varepsilon(u)} \left[\underbrace{n_u (u^2 - \sqrt{x}) \partial \phi/\partial x}_{=0} + n_u N_x \phi \right]$$

$$= \int_{\Sigma_\varepsilon(u)} n_u N_x \phi = n_u N_x \delta_\Sigma(\phi),$$

and $A = n_u N_x = n_x N_u$. ∎

Example 3.1 is implemented by a Matlab program that generates a sample of the pair (U, X) and uses this sample in order to determine, on the one hand, the empirical cumulative distribution function and, on the other hand, the associated density. These quantities are obtained by using the programs presented in section 1.5. The results obtained from two samples of 10,000 variates are shown in Figure 3.1 below. The first sample is random, obtained by using rand, while the second is composed of equally spaced values of ω. The value of parameter h is $h = 0.02$ that coincides with the step used in the equally spaced abscissa.

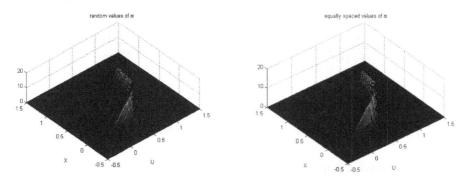

Figure 3.1. *Empirical density from a sample of 10,000 realizations of the pair with $h = 0.02$. For a color version of the figure, see www.iste.co.uk/souzadecursi/quantification.zip*

3.1.2. *Using the mean for the approximation by a constant*

The simplest approximation for the random variable $\mathbf{X}$ consists of determining a constant corresponding to the element of the linear subspace:

$$S = \left\{ \mathbf{Z} \in L^2(\Omega) : \mathbf{Z} \text{ is constant: } \mathbf{Z}(\omega) = \mathbf{s} \in \mathbb{R}^k, \forall \omega \in \Omega \right\}$$

which is the closest to $\mathbf{X}$. As shown in the preceding (section 1.11.1.2), the solution is the *mean* or *expectation* of $\mathbf{X}$: $\mathbf{PX} = \mathbf{m} = E(\mathbf{X})$. The error dist $(\mathbf{X}, S)$ in the approximation is the standard deviation of $\mathbf{X}$.

EXAMPLE 3.2.– Let us consider $\Omega = (0,1)$ and $P((a,b)) = b - a$ (i.e. $P(d\omega) = \ell(d\omega)$, where ℓ is the Lebesgue measure. Let $U(\omega) = \sqrt{\omega}$ and $X(\omega) = \omega^2$. We have:

$$E(U) = \int_\Omega U(\omega) P(d\omega) = \int_0^1 \sqrt{\omega}\, d\omega = \frac{2}{3}$$

and

$$E(X) = \int_\Omega X(\omega) P(d\omega) = \int_0^1 \omega^2\, d\omega = \frac{1}{3}.$$

The variances are:

$$V(U) = \int_\Omega (U(\omega) - E(U))^2 P(d\omega) = \int_0^1 \left(\sqrt{\omega} - \frac{2}{3}\right)^2 d\omega = \frac{1}{18}$$

and

$$V(X) = \int_\Omega (X(\omega) - E(X))^2 P(d\omega) = \int_0^1 \left(\omega^2 - \frac{1}{3}\right)^2 d\omega = \frac{4}{45}.$$

The result may also be obtained by using the probability density of the variable **U**. For instance, the cumulative function of **U** is:

$$F_U(u) = P(U < u) = \begin{cases} 0, & \text{if } u \leq 0 \\ u^2, & \text{if } 0 < u \leq 1, \\ 1, & \text{if } u > 1 \end{cases}$$

so that its probability density is:

$$f_U(u) = \begin{cases} 0, & \text{if } u < 0 \text{ or } u > 1 \\ 2u, & \text{if } 0 < u < 1 \end{cases}.$$

Thus:

$$E(U) = \int_0^1 u f_U(u)\, du = \int_0^1 2u^2\, du = \frac{2}{3}$$

and

$$V(U) = \int_0^1 \left(u - \frac{2}{3}\right)^2 f_U(u) du = \int_0^1 2u\left(u - \frac{2}{3}\right)^2 du = \frac{1}{18}$$

In an analogous manner:

$$F_X(x) = P(X < x) = \begin{cases} 0, \text{ if } x \leq 0 \\ \sqrt{x}, \text{ if } 0 < x \leq 1 \\ 1, \text{ if } x > 1 \end{cases},$$

$$f_X(x) = \begin{cases} 0, \text{ if } x < 0 \text{ or } x > 1 \\ \frac{1}{2\sqrt{x}}, \text{ if } 0 < x < 1 \end{cases}.$$

Thus:

$$E(X) = \int_0^1 x f_X(x) = \int_0^1 \frac{1}{2}\sqrt{x} dx = \frac{1}{3}$$

and

$$V(X) = \int_0^1 \left(x - \frac{2}{3}\right)^2 f_X(x) dx = \int_0^1 \frac{1}{2\sqrt{x}}\left(x - \frac{1}{3}\right)^2 dx = \frac{4}{45}. \blacksquare$$

Example 3.2 may be implemented in Matlab by using the programs introduced in section 1.10.1.

3.1.3. *Using the linear correlation in order to construct a representation*

As shown in section 1.11.1.3, the best approximation of **X** by an affine function of **U** is the element of the linear subspace:

$$S = \left\{ \mathbf{s} \in L^2\left[(\Omega)\right]^k : \mathbf{s} = \boldsymbol{\alpha}\mathbf{U} + \boldsymbol{\beta}; \boldsymbol{\alpha} \in \mathcal{M}(k, m), \boldsymbol{\beta} \in \mathbb{R}^k \right\},$$

which is the closest to **X**. The solution is **X** = **AU**+**B**, where **A** is determined by solving, for $i = 1, \ldots, k$:

C.**a** = **b**,

with $\mathbf{a} = (a_1, \ldots, a_m)^t \in \mathcal{M}(m, 1)$ such that $a_j = A_{ij}$, $\mathbf{C} \in \mathcal{M}(m, m)$ such that $C_{rj} = Cov(U_j, U_r)$ and $\mathbf{b} = (b_1, \ldots, b_m)^t \in \mathcal{M}(m, 1)$ such that $b_r = Cov(X_i, U_r)$. After the determination of $\mathbf{A}$, $\mathbf{B}$ is given by:

$$\mathbf{B} = \mathbf{X} - \mathbf{A}.E(\mathbf{U}).$$

In the situation where $k = m = 1$, we have $X = aU + b$, with:

$$a = \frac{Cov(X, U)}{V(U)} \; ; \; b = E(X) - aE(U).$$

and the error may be determined by using the *linear correlation coefficient*:

$$\rho(X, U) = \frac{Cov(X, U)}{\sqrt{V(X)V(U)}}.$$

We have:

$$\text{dist}(\mathbf{X}, S) = \sqrt{V(Y)\left(1 - [\rho(X, Y)]^2\right)}.$$

EXAMPLE 3.3.– Let us consider $\Omega = (0, 1)$ and $P((a, b)) = b - a$ (i.e. $P(d\omega) = \ell(d\omega)$ where ℓ is the Lebesgue measure). Let $U(\omega) = \sqrt{\omega}$ and $X(\omega) = \omega^2$. We have:

$$E(U) = \frac{2}{3}, V(U) = \frac{1}{18}; \; E(X) = \frac{1}{3}, V(X) = \frac{4}{45}.$$

In addition:

$$E(UX) = \int_\Omega U(\omega)X(\omega) P(d\omega) = \int_0^1 \omega^2 \sqrt{\omega} d\omega = \frac{2}{7},$$

so that:

$$Cov(U, X) = E(UX) - E(U)E(X) = \frac{2}{7} - \frac{1}{3} \times \frac{2}{3} = \frac{4}{63}.$$

Thus:

$$a = \frac{Cov(U, X)}{V(U)} = \frac{4/63}{1/18} = \frac{8}{7};$$

$$b = E(X) - aE(U) = \frac{1}{3} - \frac{8}{7} \times \frac{2}{3} = -\frac{3}{7};$$

$$\rho(U, X) = \frac{Cov(U, X)}{\sqrt{V(U)V(X)}} = \frac{2\sqrt{10}}{7} \approx 0.903508;$$

$$\|X - aU - b\| = \sqrt{V(X)\left(1 - [\rho(U, X)]^2\right)} = \frac{2}{7\sqrt{5}} \approx 0.127775.$$

Indeed, if:

$$J(a, b) = \int_\Omega (X(\omega) - aU(\omega) - b)^2 P(d\omega) = \int_\Omega (\omega^2 - a\sqrt{\omega} - b)^2 d\omega,$$

we have:

$$J(a, b) = \frac{1}{5} + \frac{a^2}{2} - \frac{2b}{3} + b^2 + \frac{4}{21}a(-3 + 7b),$$

so that:

$$\frac{\partial J}{\partial a} = a + \frac{4}{3}b - \frac{4}{7},$$

and

$$\frac{\partial J}{\partial b} = \frac{2}{3}(2a + 3b - 1).$$

Thus, the minimum of J is attained in (a, b), solution of the linear system is:

$$a + \frac{4}{3}b = \frac{4}{7} \; ; 2a + 3b = 1.$$

So, $a = \frac{8}{7}$ and $b = -\frac{3}{7}$. Let us note that:

$$E(U^2) = \int_\Omega U(\omega)^2 P(d\omega) = \int_0^1 \omega d\omega = \frac{1}{2}$$

while:

$$E(U) = \frac{2}{3}, E(X) = \frac{1}{3}, E(UX) = \frac{2}{7}.$$

Thus, the equations

$$aE(U^2) + bE(U) = E(UX) \; ; \; aE(U) + b = E(X)$$

may be written as:

$$\frac{1}{2}a + \frac{2}{3}b = \frac{2}{7} \iff a + \frac{4}{3}b = \frac{4}{7}$$

and

$$\frac{2}{3}a + b = \frac{1}{3} \iff 2a + 3b = 1. \blacksquare$$

Example 3.3 may be implemented in Matlab by using the programs introduced in section 1.10.2.

3.1.4. *Polynomial approximation*

A classical (and often useful) expansion consists of approximating $\mathbf{X}$ by a polynomial function of $\mathbf{U}$. In this case, we must consider a basis $\mathcal{B} = \{\varphi_i(\mathbf{U}): 1 \leq i \leq N_X\}$ containing N_X linearly independent polynomials in the variable $\mathbf{U}$ (such as, for instance the polynomials 1; U_r for $1 \leq i \leq m$; $U_r U_s$ for $1 \leq r, s \leq m$, etc.) and, as previously introduced, take as unknowns the coefficients $\mathbf{x} = (\mathbf{x_1}, \ldots, \mathbf{x_n})$, with $\mathbf{x_i} \in \mathbb{R}^k$ of the expression of $\mathbf{PX}$. For instance, when we are interested in a polynomial where the maximal degree in component U_i is $d_i - 1 \geq 0$, we set $\mathbf{d} = (d_1, \ldots, d_m)$, $\boldsymbol{\alpha} = (\alpha_1, \ldots, \alpha_m)$ and we consider:

$$A(\mathbf{d}) = \{\boldsymbol{\alpha} \in \mathbb{R}^m : 1 \leq \boldsymbol{\alpha} \leq \mathbf{d} \text{ (i.e. } 1 \leq \alpha_i \leq d_i \text{ for } 1 \leq i \leq m)\}$$

and

$$\mathcal{B} = \left\{U_1^{\alpha_1 - 1} U_2^{\alpha_2 - 1} \ldots U_m^{\alpha_m - 1} : \boldsymbol{\alpha} \in A(\mathbf{d})\right\} .$$

We observe that, in this case, term U_i may take any exponent going from 0 to $d_i - 1$ so that $N_X = \prod_{i=1}^{m} d_i$. In order to apply the procedure exposed, it is necessary to introduce a map that indexes the elements $\boldsymbol{\alpha} \in A(\mathbf{d})$ into a single index. For instance, we may consider *Morton ordering* or, more simply, an *index map* given *recursively* by:

$$index(\alpha_1, \ldots, \alpha_m) = index(\alpha_1, \ldots, \alpha_{m-1}) + (\alpha_m - 1)$$
$$*index(d_1, \ldots, d_{m-1}), index(\alpha_1) = \alpha_1.$$

Such a map is implemented in Matlab as follows:

Listing 3.1. *Index map*

```
function ind = index_map(alfa,d)
if length(d) == 1
    ind = alfa(1);
elseif length(d) == 2
    ind = alfa(1) + (alfa(2)-1)*d(1);
else
    ind = index_map(alfa(1:end-1),d(1:end-1)) + (alfa(end) - 1)
        *index_map(d(1:end-1),d(1:end-1));
end;
return;
end

function alfa = inverse_index_map(ind,d)
alfa = zeros(size(d));
iind = ind;
if length(d) > 1
    for j = 2: length(d)
        jj = length(d) - j + 2;
        m = index_map(d(1:jj-1),d(1:jj-1));
        alfa(jj) = floor(iind/m) + 1;
        ii = mod(iind,m);
        if ii > 0
            iind = ii;
        else
            alfa(jj) = alfa(jj) - 1;
            iind = iind - (alfa(jj) - 1)*m;
        end;
    end;
end;
if iind > 0
    alfa(1) = iind;
else
    alfa(1) = d(1);
```

```
        alfa (2) = alfa (2) - 1;
end;
return;
end
```

By using these maps, we associate to a multi-index $(\alpha_1, \ldots, \alpha_m)$ an unique index $0 \leq i \leq N_X$ and conversely. Let us introduce

$$\varphi_i(\mathbf{U}) = U_1^{\alpha_1-1} U_2^{\alpha_2-1} \ldots U_m^{\alpha_m-1}, i = index(\alpha_1, \ldots, \alpha_m).$$

Then,

$$S = \left\{ \mathbf{s} \in \left[L^2(\Omega)\right]^k : \mathbf{s} = \sum_{i=1}^{N_X} \mathbf{s}_i \varphi_i(\mathbf{U}); \mathbf{s}_i \in \mathbb{R}^k \right\},$$

so that% and

$$\left(\sum_{j=1}^{N_X} x_j \varphi_j(\mathbf{U}), \varphi_i(\mathbf{U}) \right) = (X, \varphi_i(\mathbf{U})), \ \forall\ 1 \leq i \leq N_X.$$

This last equation provides a linear system for the determination of the coefficients x_i.

A particular situation of interest is the one where $\mathbf{U}$ is a *Gaussian vector* (see section 1.13.4. A vector is said to be Gaussian if, for instance, all its components U_i are Gaussian and independent). In this case, the approach may be connected with Wiener's one (see [WIE 38, CAM 47, GHA 91]), which has generated the approximation by *polynomial chaos*.

This approach may be implemented in Matlab by using the programs of section 3.1.5.

EXAMPLE 3.4.– Let us consider $\Omega = (0,1)$ and $P((a,b)) = b - a$ (i.e. $P(d\omega) = \ell(d\omega)$ where ℓ is the Lebesgue's measure). Let $U(\omega) = \sqrt{\omega}$ and $X(\omega) = \omega^2$. In order to approximate X by a polynomial function of U, having degree d, we must determine:

$$PX = \sum_{j=0}^{d} x_j U^j,$$

where the $d+1$ coefficients $x_0, x_1, ..., x_n$ verify:

$$\left(\sum_{j=0}^{d} x_j U^j, U^i\right) = \left(X, U^i\right), \ 0 \leq i \leq d,$$

i.e.:

$$\sum_{j=0}^{n} x_j E\left(U^{i+j}\right) = E\left(XU^i\right), \ 0 \leq i \leq d.$$

By setting $\mathbf{X} = (x_0, ..., x_d)$, we have:

$$\mathbf{AX} = \mathbf{B},$$

where:

$$A_{ij} = E\left(U^{i+j}\right) = \int_{\Omega} [U(\omega)]^{i+j} P(d\omega) = \int_0^1 \omega^{(i+j)/2} d\omega$$
$$= \frac{2}{2+i+j}, \ 0 \leq i, j \leq d$$

and

$$B_i = E\left(XU^i\right) = \int_{\Omega} X(\omega) [U(\omega)]^i P(d\omega) = \int_0^1 \omega^{(i+4)/2} d\omega$$
$$= \frac{2}{6+i}, \ 0 \leq i \leq d.$$

The solution provided by Matlab for different values of d is shown below:

d	X
0	(0.3333)
1	(−0.4286, 1.1429)
2	(0.2143, − 1.4286, 2.1429)
3	(−0.0397, 0.4762, − 1.6667, 2.2222)
4	(0.0000, − 0.0000, 0.0000, − 0.0000, 1.0000)
5	(0.0000, − 0.0000, 0.0000, − 0.0000, 1.0000, − 0.0000)
6	(0.0000, − 0.0000, 0.0000, − 0.0000, 1.0000, − 0.0000, 0.0000)

We observe that:

– for $d = 0$, it coincides with the approximation of X by a constant: the exact result is $PX = \frac{1}{3} \approx 0.3333$,

– for $d = 1$, it is equivalent to approximate X by an affine function of U: the exact result is $PX = -\frac{3}{7} + \frac{8}{7}U$, which corresponds to $\mathbf{X} = \left(-\frac{3}{7}, \frac{8}{7}\right) \approx (-0.4286, 1.1429)$.

– for $d \geq 4$, the result is $PX = U^4 = E(X \mid U)$, which corresponds to the exact 1.

3.1.5. *General finite-dimensional approximations*

A usual situation is that where S is a finite-dimensional subspace, i.e. a subspace having a basis formed by a finite number n of elements:

$$F = \{\varphi_1(U), ..., \varphi_n(U)\}$$

such that, for $k = 1$ (extension for $k > 1$ is obtained by applying the results below to each component):

$$S = \left\{ s \in L^2(\Omega) : \mathbf{s} = \sum_{i=1}^{n} s_i \varphi_i(U) ; s_i \in \mathbb{R}, \ 1 \leq i \leq n \right\}.$$

In this case:

$$PX = \sum_{j=1}^{n} x_j \varphi_j(U) \in S$$

and

$$\left(\sum_{j=1}^{n} x_j \varphi_j(U), \varphi_i(U) \right) = (X, \varphi_i(U)), \ 1 \leq i \leq n.$$

Thus, $\mathbf{X} = (x_1, ..., x_n)^t$ is the solution of the following linear system:

$$\mathbf{AX} = \mathbf{B}, \ A_{ij} = (\varphi_j(U), \varphi_i(U)), \ B_i = (X, \varphi_i(U)).$$

We observe that this approximation is *linear*: is $\alpha \in \mathbb{R}$, we have $P(X_1 + \alpha X_2) = PX_1 + \alpha PX_2$. Indeed, if:

$$(B_1)_i = (X_1, \varphi_i(U)), \; (B_2)_i = (X_2, \varphi_i(U)), \; B_i = (B_1)_i + \alpha (B_2)_i$$

and

$$\mathbf{AX_1} = \mathbf{b_1}, \; \mathbf{AX_2} = \mathbf{B_2},$$

then:

$$\mathbf{X} = \mathbf{X_1} + \alpha \mathbf{X_2}$$

satisfies:

$$\mathbf{AX} = \mathbf{AX_1} + \alpha \mathbf{AX_2} = \mathbf{B_1} + \alpha \mathbf{B_2} = \mathbf{B},$$

and we have $P(X_1 + \alpha X_2) = PX_1 + \alpha PX_2$.

If the basis is *orthogonal*, i.e.:

$$(\varphi_i(U), \varphi_j(U)) = 0, \text{ if } i \neq j, \; (\varphi_i(U), \varphi_i(U)) > 0.$$

then:

$$A_{ij} = 0, \text{ if } i \neq j, \; A_{ii} > 0.$$

In this case, the solution is given by:

$$x_i = B_i / A_{ii} = (X, \varphi_i(U)) / (\varphi_i(U), \varphi_i(U)), \; 1 \leq i \leq n.$$

If the basis is *orthonormal*, i.e.:

$$(\varphi_i(U), \varphi_j(U)) = \begin{cases} 1, \text{ if } i = j, \\ 0, \text{ if } i \neq j. \end{cases}$$

then:

$$x_i = B_i = (X, \varphi_i(U)), \; 1 \leq i \leq n.$$

We observe that any basis $G = \{\psi_1(U), ..., \psi_n(U)\}$ may be orthonormalized using the Gram–Schmidt's orthogonalization procedure:

$$\phi_1 = \psi_1 \ ; \ \varphi_1 = \phi_1/\|\phi_1\| \ ;$$

$$k > 1 : \phi_k = \psi_k - \sum_{i=1}^{k-1} (\psi_k, \varphi_i) \varphi_i \ ; \ \varphi_k = \phi_k/\|\phi_k\|.$$

This approach may be implemented in Matlab as follows: assume that, on the one hand, $\varphi_k(U)$ is provided by a subprogram phi(k,U) while, on the other hand, the scalar product of two random variables Y and Z is provided by a subprogram scalprod(Y,Z). For instance, $Y = Y(\omega)$, $Z = Z(\omega)$ and the density of ω is $f_\omega(\omega)$, defined on (a, b), scalprod(Y,Z) must return the value of $\int_a^b Y(\omega) Z(\omega) f_\omega(\omega) d\omega$. In this case, the coefficients x are provided by the code

Listing 3.2. *Scalar products calculated by integration*

```
function  x = expcoef(phi, N_X, X, U,f_omega,a,b)
%
%   determines the coefficients of the expansion.
%    by integrating the variables
%
% IN:
% phi : basis function phi(k,U) - type anonymous function
% N_X : order of the expansion - type integer
% X : the random variable to approach - type anonymous function
% U : the random variable argument of the expansion - type
       anonymous function
% f_omega : density of omega - type anonymous function
% a,b : bounds of integration - type double
%
% OUT:
% x: table 1 x N_X of coefficients - type array of double
%
[A, B] = variational_matrices(phi, N_X, X, U,f_omega,a,b);
x = A \ B;
x = x'; % transpose to 1 x N_x vector
return;
end

function [A, B] = variational_matrices(phi, N_X, X, U, f_omega,
     a,b)
%
```

```
% creates the matrices of the linear system by using
%   by integrating the variables
%
% IN:
% phi : basis function phi(k,U) - type anonymous function
% N_X : order of the expansion - type integer
% X : the random variable to approach - type anonymous function
% U : the random variable argument of the expansion - type
%     anonymous function
% f_omega : density of omega - type anonymous function
% a,b : bounds of integration - type double
%
% OUT:
% A: table N_X x N_X of scalar products - type array of double
%    A(i,j) = (phi(i,U), phi(j,U))
% B : table N_x x 1 of scalar products - type array of double
%    B(i) = (X, phi(i,U))
%
A = zeros(N_X, N_X);
B = zeros(N_X, 1);
for i = 1: N_X
    Y = @(om) phi(i,U(om));
    A(i,i) = scalprod(Y,Y,f_omega,a,b);
    Z = @(om) X(om);
    B(i) = scalprod(Y, Z,f_omega,a,b);
    for j = i+1:N_X
        Z = @(om) phi(j,U(om));
        aux = scalprod(Y,Z,f_omega,a,b);
        A(i,j) = aux;
        A(j,i) = aux;
    end;
end;
return;
end

function v = scalprod(Y,Z,f_omega,a,b)
%
%   evaluates the scalar product (Y, Z) by integration
%
% IN:
% Y : a random variable   - type anonymous function
% U : a random variable   - type anonymous function
% f_omega : density of omega - type anonymous function
% a,b : bounds of integration - type double
%
% OUT
% v : value of the scalar product - type double
```

```
%
f1 = @(om) Y(om)*Z(om)*f_omega(om);
v = numint(f1,a,b);
return;
end
```

If a sample $\{(Y_i, Z_i) : 1 \leq i \leq ns\}$ of the pair (Y, Z) is available, scalprod(Y,Z) must return the empirical mean $(\sum_{i=1}^{ns} Y_i Z_i)$ of the product YZ. Then, we may use the code:

Listing 3.3. *Scalar products evaluated from a sample*
```
function x = expcoef(phi, N_X, Xs, Us)
%
%   determines the coefficients of the expansion
%   using a sample (U_i, X_i) from (X,U)
%
% IN:
% phi : basis function phi(k,U) - type anonymous function
% N_X : order of the expansion - type integer
% Xs : table of values of X - type array of double
% Us : table of values of U - type array of double
% (Us(i),Xs(i)) is a variate from (U,X)
%
% OUT:
% x: table 1 x N_X of coefficients - type array of double
%
[A, B] = variational_matrices(phi, N_X, Xs, Us);
x = A \ B;
x = x'; % transpose to 1 x N_x vector
return;
end

function [A, B] = variational_matrices(phi, N_X, Xs, Us)
%
% creates the matrices of the linear system by using
% scalar products evaluated from a sample
%
% IN:
% phi : basis function phi(k,U) - type anonymous function
% N_X : order of the expansion - type integer
% Xs : table of values of X - type array of double
% Us : table of values of U - type array of double
% (Us(i),Xs(i)) is a variate from (U,X)
%
% OUT:
% A: table N_X x N_X of scalar products - type array of double
```

```
%      A(i,j) = (phi(i,U), phi(j,U))
% B :  table N_x x 1 of scalar products - type array of double
%      B(i) = (X, phi(i,U))
%
A = zeros(N_X, N_X);
B = zeros(N_X, 1);
for i = 1: N_X
    f1 = @(U) phi(i,U);
    Y = map(f1, Us,1);
    A(i,i) = scalprod(Y,Y);
    Z = Xs;
    B(i) = scalprod(Y, Z);
    for j = i+1:N_X
        f1 = @(U) phi(j,U);
        Z = map(f1, Us, 1);
        aux = scalprod(Y,Z);
        A(i,j) = aux;
        A(j,i) = aux;
    end;
end;
return;
end

function v = scalprod(Y,Z)
%
%   evaluates the scalar product (Y, Z) by using a sample
%
% IN:
% Y : table of values of Y - type array of double
% Z : table of values of Z - type array of double
% (Y(i),Z(i)) is a variate from (Y,Z)
%
% OUT
% v : value of the scalar product - type double
%
v = mean(Y.*Z);
return;
end
```

Once the coefficients have been obtained, the values of the projection PX at the values of a sample Us of ns variates from U are provided by the code:

Listing 3.4. *Scalar products evaluated from a sample*

```
function PX = projection(c,Us,phi)
%
%   determines the projection PX at the points Us
%   assumes that Us contains ns points of dimension k
%   each line of Us is a point
%   assumes that c contains N_X vectors of dimension n
%   each line of c is a vector.
%
% IN:
% c : table n x N_X of the coefficients of the expansion -
     type array of double
% Us : table ns x k of the sample points - type array of double
% phi : basis function phi(k,U) - type anonymous function
%
% OUT:
% PX: table ns x n of the values of the projection - type array
     of double
%
ns = size(Us,1);
n = size(c,1);
N_X = size(c,2);
PX = zeros(n,ns);
for i = 1: N_X
    c_i = c(:,i);
    for j = 1: ns
        PX(:,j) = PX(:,j) + c_i*phi(i,Us(j,:));
    end;
end;
PX = PX'; % transpose PX to   ns x n table
return;
end
```

EXAMPLE 3.5.– Let us consider $\Omega = (0,1)$ and $P((a,b)) = b - a$ (i.e. $P(d\omega) = \ell(d\omega)$ where ℓ is the Lebesgue's measure). Let $U(\omega) = \sqrt{\omega}$ and $X(\omega) = \omega^2$. In the last example, we have considered the approximation of X by a polynomial of degree d in U, which corresponds to $\varphi_i(U) = U^{i-1}$, $0 \leq i \leq d+1$. Other families may be considered, such as, for instance:

$$\varphi_1(U) = 1; \varphi_{2k}(U) = \sin(kU),$$
$$\varphi_{2k+1}(U) = \cos(kU), 1 \leq k \leq d \ (n = 2d+1).$$

In this case, we have:

$$B_1 = \frac{1}{3}.$$

$$B_{2k} = -\tfrac{2}{k^6}\left(k\left(120 - 20k^2 + k^4\right)\cos(k) - 5\left(24 - 12k^2 + k^4\right)\sin(k)\right) \quad (k \geq 1),$$

$$B_{2k+1} = \tfrac{2}{k^6}\left(-120 + 5\left(24 - 12k^2 + k^4\right)\cos(k) + k\left(120 - 20k^2 + k^4\right)\sin(k)\right) \quad (k \geq 1),$$

$$A_{11} = 1,$$

$$A_{1,2k} = \frac{2}{k^2}\left(\sin(k) - k\cos(k)\right) \quad (k \geq 1),$$

$$A_{1,2k+1} = \frac{2}{k^2}\left(cos(k) + ksin(k) - 1\right) \quad (k \geq 1),$$

$$A_{2k,2k} = \frac{1 + 2k^2 - \cos(2k) - 2k\sin(2k)}{4k^2} \quad (k \geq 1),$$

$$A_{2k,2k+1} = \frac{\sin(2k) - 2k\cos(2k)}{4k^2} \quad (k \geq 1),$$

$$A_{2k+1,2k+1} = \frac{2k^2 + \cos(2k) + 2k\sin(2k) - 1}{4k^2} \quad (k \geq 1),$$

and, for $k \neq p$,

$$A_{2k,2p} = \tfrac{2}{(k^2-p^2)^2}\left(p\left(k^2 - p^2\right)\cos(p)\sin(k) + k\left(p^2 - k^2\right)\cos(k)\sin(p)\right.$$
$$\left. + \left(k^2 + p^2\right)\sin(k)\sin(p) + 2kp\left(\cos(k)\cos(p) - 1\right)\right),$$

$$A_{2k,2p+1} = \tfrac{2}{(k^2-p^2)^2}\left(k\left(p^2 - k^2\right)\cos(p)\cos(k) + p\left(p^2 - k^2\right)\sin(k)\sin(p)\right.$$
$$\left. - 2kp\cos(k)\sin(p) + \left(k^2 + p^2\right)\cos(p)\sin(k)\right),$$

$$A_{2k+1,2p+1} = \tfrac{2}{(k^2-p^2)^2}\left(k\left(k^2 - p^2\right)\cos(p)\sin(k) + p\left(p^2 - k^2\right)\cos(k)\sin(p)\right.$$
$$\left. + \left(k^2 + p^2\right)\left(\cos(p)\cos(k) - 1\right) + 2kp\sin(k)\sin(p)\right).$$

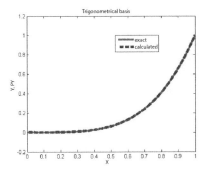 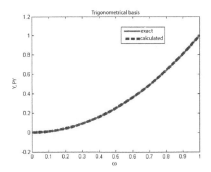

Figure 3.2. *Approximation using a trigonometrical basis. For a color version of the figure, see www.iste.co.uk/souzadecursi/quantification.zip*

The result obtained for $d = 6$ is shown in Figure 3.2.

It is also interesting to compare the cumulative probability function and the density of probability (see Figure 3.3). ∎

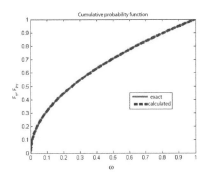 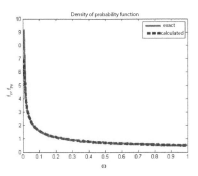

Figure 3.3. *Approximation using a trigonometrical basis. For a color version of the figure, see www.iste.co.uk/souzadecursi/quantification.zip*

EXAMPLE 3.6.– We stress that the existence of a dependence between the variables U and X is essential. Indeed, if the variables are independent, then:

$$B_i = (X, \varphi_i(U)) = E(X, \varphi_i(U)) = E(X) E(\varphi_i(U)),$$

so that the solution of the linear system $\mathbf{AX} = \mathbf{B}$ is:

$$\mathbf{X} = E(X) \overline{\mathbf{X}},$$

where:

$$\mathbf{A}\overline{\mathbf{X}} = \mathbf{B}, \ \overline{B}_i = E\left(\varphi_i\left(U\right)\right).$$

Thus:

$$PX = E(X) P(\mathbf{1}),$$

where **1** is a constant function taking the value 1 everywhere. ∎

3.1.6. *Approximation using a total family*

A first generalization of the preceding situation is obtained by considering a subspace S possessing a *total family* F, i.e. a countable family F such that its finite linear span (the set of the finite linear combinations of elements from F) is dense on S see [DE 08]. We consider $k = 1$ and extension for $k > 1$ is obtained by applying the procedure to each component):

$$F = \{\varphi_n\}_{n \in \mathbb{N}}; \ [F] = \left\{Z \in L^2\left(\Omega, P\right) : Z = \sum_{i=1}^{k} a_i \varphi_{n_i}\right\}; \ S = \overline{[F]}.$$

Then:

$$\forall \, \varepsilon > 0 : \exists \, U_\varepsilon \in [F] \text{ tal que } \|X - U_\varepsilon\| \leq \varepsilon.$$

3.1.6.1. *Case of an increasing family of subspaces*

The first popular situation corresponding to total families is the one where $[F]$ is parameterized by one dimension, i.e. $[F]$ is the reunion of increasing finite-dimensional subspaces:

$$[F] = \bigcup_{n=0}^{+\infty} [F_n]; \ \dim([F_k]) = d_k < \infty; \ d_{k+1} \geq d_k, \ \forall k \geq 0; \ [F_k] \subset [F_{k+1}].$$

In this case, we may define $S_k = \overline{[F_k]} = \overline{[F_k]}$ and we have:

$$S = \overline{\bigcup_{n=0}^{+\infty} S_k}; \ \dim(S_k) = d_k < \infty; \ d_{k+1} \geq d_k, \ \forall k \geq 0; \ S_k \subset S_{k+1}.$$

For an arbitrary Z, we denote $P_k Z$ the projection of Z on S_k and $c_k(Z) = \|Z - P_k Z\|$. Then, we have:

$$\|Z - PZ\| \leq c_{k+1}(Z) \leq c_k(Z), \ \forall k \in \mathbb{N},$$

so that the sequence $\{c_k(Z)\}_{k \in \mathbb{N}}$ is decreasing and lower bounded. Thus:

$$c_k(Z) \longrightarrow c(Z) \text{ para } k \longrightarrow +\infty.$$

For $Z \in S$, we have:

$$\forall \varepsilon > 0 : \exists Z_\varepsilon \in [F] \text{ such that } \|Z - Z_\varepsilon\| \leq \varepsilon.$$

Since $[F] = \bigcup_{n=0}^{+\infty} [F_n]$, we have $Z_\varepsilon \in S_k$ for some $k \in \mathbb{N}$. Thus:

$$c_k(Z) = \|Z - P_k(Z)\| \leq \|Z - Z_\varepsilon\| \leq \varepsilon.$$

As a result:

$$c_n(Z) \leq \varepsilon \text{ for any } n \geq k$$

and $c(Z) \leq \varepsilon$. Since ε is arbitrary, it follows that $c(Z) = 0$. So:

$$\forall Z \in S : P_k(Z) \longrightarrow Z \text{ when } k \longrightarrow +\infty.$$

Since $X - PX$ is orthogonal to S and $S_k \subset S$, we have $X - PX \perp S_k$ for any $k \in \mathbb{N}$, so that:

$$\forall k \in \mathbb{N} : P_k(X - PX) = 0.$$

Then, the linearity of the projection operator on finite-dimensional subspaces shows that $P_k X = P_k(PX)$. Since $PX \in S$, we have:

$$P_k X = P_k(PX) \longrightarrow PX \text{ for } k \longrightarrow +\infty.$$

So:

$$PX = \lim P_k X \text{ and } \|X - PX\| = \lim \|X - P_k X\|.$$

Thus, $P_k X$ provides an approximation of PX. In practice, we do not generate subspaces such that $S_k \subset S_{k+1}$, but:

$$\forall k : S_i \subset \bigcup_{n=k+1}^{+\infty} S_n \text{ for } i \leq k,$$

a condition that is satisfied when $\forall k : \exists n > 0$ such that $S_k \subset S_{k+n}$.

$P_k X$ may be determined in an analogous manner to the one used in the case of finite dimensional subspaces: if

$$F_k = \{\varphi_1(U), ..., \varphi_{d_k}(U)\}$$

is a basis of S_k, then:

$$P_k X = \sum_{j=1}^{d_k} x_{j,k} \varphi_j(U) \in S_k$$

and $\mathbf{X}_k = (x_{1,k}, ..., x_{d_k,k})^t$ is the solution of the linear system:

$$\mathbf{A}_k \mathbf{X}_k = \mathbf{B}_k, \quad (A_k)_{ij} = (\varphi_j(U), \varphi_i(U)), \quad (B_k)_i = (X, \varphi_i(U)).$$

3.1.6.2. *Case of an orthogonal family*

The second popular situation is the case where the family F is *orthogonal*, i.e.:

$$(\varphi_m(U), \varphi_n(U)) = 0, \text{ if } m \neq n.$$

In this case, we define:

$$F_k = \{\varphi_0(U), \varphi_1(U), ..., \varphi_n(U)\}$$

and $[F]$ is the reunion of an increasing family of finite-dimensional subspaces:

$$[F] = \bigcup_{n=0}^{+\infty} [F_n]; \quad \dim([F_k]) = k+1 < \infty; \quad d_{k+1} \geq d_k, \forall k \geq 0; \quad [F_k] \subset [F_{k+1}].$$

In addition:

$$(A_k)_{ij} = (\varphi_j(U), \varphi_i(U)) = 0, \text{ if } i \neq j,$$

so that:

$$P_k X = \sum_{j=0}^{k} x_j \varphi_j(U) \in S_k ,$$

where:

$$x_i = B_i/A_{ii} = (X, \varphi_i(U))/(\varphi_i(U), \varphi_i(U)), \ 0 \leq i \leq k.$$

In this situation, classical orthogonal families associated with square summable functions are often used. For instance, we may use a Fourier series associated with a *trigonometrical basis* or *orthogonal polynomials*. We underline that the orthogonality is connected to the distribution of the variable U: let f_X be the density of probability of U and its range (i.e. the set of the values of U) be denoted by D. Then:

$$(\varphi_i, \varphi_j) = E\left(\varphi_i(U) \varphi_j(U)\right) = \int_D \varphi_i(u) \varphi_j(u) f_U(u) du,$$

so that f_U appears as the *weight function associated with the family*. In general, $f_U = Aw(u)$, where A is a normalizing constant and w is the usual weight associated with the orthogonal family. For instance,

family	D	$w(u)$
Tchebichev 1st kind (T_n)	$(-1, 1)$	$1/\sqrt{1-u^2}$
Tchebichev 2nd kind (U_n)	$(-1, 1)$	$\sqrt{1-u^2}$
Legendre (P_n)	$(-1, 1)$	1
Laguerre (L_n)	$(0, +\infty)$	e^{-u}
Hermite probabilistic (H_n)	$(-\infty, +\infty)$	$e^{-u^2/2}$
Trigonometrical $(1, \sin(n\pi u), \cos(n\pi u))$	$(-1, +1)$	1

As a result, an *a priori* condition for orthogonality is the compatibility between the distribution of U and the weight. For instance, the use of an orthogonal family H_n supposes a Gaussian density of probability. In practice, we may overcome this limitation by using *truncated series*, which corresponds to an approximation using a finite-dimensional subspace: *the family is used to generate a basis and the orthogonality is unnecessary*. As in the previous situation, we have:

$$PX = \lim P_k X \text{ and } \|X - PX\| = \lim \|X - P_k X\|.$$

3.1.6.3. *Case of a Hilbert basis*

A third situation of interest is the one where S possesses a *Hilbert basis* F, i.e., a family F that is both *countable and orthonormal*:

$$F = \{\varphi_n\}_{n \in \mathbb{N}}; \ (\varphi_m, \varphi_n) = \begin{cases} 1, \text{ if } m = n, \\ 0, \text{ if } m \neq n. \end{cases}$$

such that any element of S may be *uniquely represented by a series* (see [DE 08]):

$$S = \left\{ s \in L^2(\Omega) : s = \sum_{i=0}^{+\infty} s_i \varphi_i(U); \ s_i \in \mathbb{R}, \ \forall i \in \mathbb{N} \right\}.$$

In this case:

$$PX = \sum_{j=0}^{+\infty} x_j \varphi_j(U) \in S$$

and

$$\left(\sum_{j=0}^{+\infty} x_j \varphi_j(U), \varphi_i(U) \right) = (X, \varphi_i(U)), \ \forall i \in \mathbb{N}.$$

Since the family is orthonormal, we have:

$$x_i = (X, \varphi_i(U)), \ \forall i \in \mathbb{N}.$$

Let us observe that any countable family $G = \{\psi_n\}_{n \in \mathbb{N}}$ such that

$$S = \left\{ s \in L^2(\Omega) : s = \sum_{i=0}^{+\infty} s_i \psi_i(U); \ s_i \in \mathbb{R}, \ \forall i \in \mathbb{N} \right\},$$

(i.e. such that any element of S may be uniquely represented by a series) may be transformed in a Hilbert basis (i.e. may be orthonormalized) by using Gram–Schmidt's procedure:

$$\phi_0 = \psi_0 \; ; \; \varphi_0 = \phi_0/\|\phi_0\| \, ;$$

$$n > 0 : \phi_n = \psi_n - \sum_{i=0}^{n-1} (\psi_n, \varphi_i)\, \varphi_i \; ; \; \varphi_n = \phi_n/\|\phi_n\|.$$

In particular, any of the orthogonal families previously presented may be transformed in an orthonormal one. When using a Hilbert basis, we also have:

$$PX = \lim P_k X \quad \text{and} \quad \|X - PX\| = \lim \|X - P_k X\|.$$

EXAMPLE 3.7.– Let us consider $\Omega = (0,1)$ and $P((a,b)) = b - a$ (i.e. $P(d\omega) = \ell(d\omega)$ where ℓ is the Lebesgue's measure). Let $U(\omega) = \sqrt{\omega}$ and $X(\omega) = \omega^2$. Let us consider the family of the Legendre polynomials. We are looking for:

$$P_k X = \sum_{j=0}^{k} x_j \varphi_j(U) \quad (\varphi_j = \text{Legendre polynomial of order } j).$$

For $k = 6$, we have:

$$\mathbf{A}_k = \begin{pmatrix} 1 & \frac{2}{3} & \frac{1}{4} & 0 & -\frac{1}{24} & 0 & \frac{1}{64} \\ \frac{2}{3} & \frac{1}{2} & \frac{4}{15} & \frac{1}{12} & 0 & -\frac{1}{96} & 0 \\ \frac{1}{4} & \frac{4}{15} & \frac{1}{4} & \frac{6}{35} & \frac{13}{192} & 0 & -\frac{1}{80} \\ 0 & \frac{1}{12} & \frac{6}{35} & \frac{3}{16} & \frac{8}{63} & \frac{3}{64} & 0 \\ -\frac{1}{24} & 0 & \frac{13}{192} & \frac{8}{63} & \frac{9}{64} & \frac{10}{99} & \frac{31}{768} \\ 0 & -\frac{1}{96} & 0 & \frac{3}{64} & \frac{99}{10} & \frac{99}{15} & \frac{768}{12} \\ \frac{1}{64} & 0 & -\frac{1}{80} & 0 & \frac{31}{768} & \frac{128}{143} & \frac{143}{256} \end{pmatrix},$$

$$\mathbf{B}_k = \begin{pmatrix} \frac{1}{3} \\ \frac{1}{2} \\ \frac{2}{7} \\ \frac{5}{24} \\ \frac{8}{63} \\ \frac{1}{16} \\ \frac{16}{693} \\ \frac{1}{192} \end{pmatrix},$$

and the solution is:

$$\mathbf{X}_k = \begin{pmatrix} \frac{1}{5} \\ 0 \\ \frac{4}{7} \\ 0 \\ \frac{8}{35} \\ 0 \\ 0 \end{pmatrix},$$

which corresponds to:

$$P_k X = \frac{1}{5} + \frac{2}{7}\left(-1 + 3U^2\right) + \frac{1}{35}\left(3 - 30U^2 + 35U^4\right),$$

i.e.:

$$P_k X = U^4,$$

what is the exact representation of X as a function of U. We may also consider Laguerre polynomials: in this case:

$$P_k X = \sum_{j=0}^{k} x_j \varphi_j (U) \ (\varphi_j = \text{Laguerre polynomial of order } j).$$

For $k = 6$, we have:

$$\mathbf{A}_k = \begin{pmatrix}
1 & \frac{1}{3} & -\frac{1}{12} & -\frac{19}{60} & -\frac{151}{360} & -\frac{1091}{2520} & -\frac{7841}{20160} \\
\frac{1}{3} & \frac{1}{6} & \frac{1}{20} & -\frac{1}{36} & -\frac{191}{2520} & -\frac{341}{3360} & -\frac{20117}{181440} \\
-\frac{1}{12} & \frac{1}{20} & \frac{7}{60} & \frac{173}{1260} & \frac{859}{6720} & \frac{6067}{60480} & \frac{14449}{226800} \\
-\frac{19}{60} & -\frac{1}{36} & \frac{173}{1260} & \frac{1079}{5040} & \frac{8359}{36288} & \frac{26891}{129600} & \frac{17929}{110880} \\
-\frac{151}{360} & -\frac{191}{2520} & \frac{859}{6720} & \frac{8359}{36288} & \frac{47611}{181440} & \frac{824141}{3326400} & \frac{907189}{4435200} \\
-\frac{1091}{2520} & -\frac{341}{3360} & \frac{6067}{60480} & \frac{26891}{129600} & \frac{824141}{3326400} & \frac{4845121}{19958400} & \frac{574273}{2745600} \\
-\frac{7841}{20160} & -\frac{20117}{181440} & \frac{14449}{226800} & \frac{17929}{110880} & \frac{907189}{4435200} & \frac{574273}{2745600} & \frac{97926401}{518918400}
\end{pmatrix},$$

$$\mathbf{B}_k = \begin{pmatrix} \frac{1}{3} \\ \frac{1}{21} \\ -\frac{19}{168} \\ -\frac{281}{1512} \\ -\frac{1507}{7560} \\ -\frac{14591}{83160} \\ -\frac{43427}{332640} \end{pmatrix},$$

and the solution is:

$$\mathbf{X}_k = \begin{pmatrix} 24 \\ -96 \\ 144 \\ -96 \\ 24 \\ 0 \\ 0 \end{pmatrix},$$

which corresponds to:

$$P_k X = 48 - 96(1 - U) - 96U + 72U^2 - 16U^3 + U^4 \\ + 72\left(2 - 4U + U^2\right) - 16\left(6 - 18U + 9U^2 - U^3\right),$$

i.e.

$$P_k X = U^4,$$

that corresponds to the exact representation. When using Hermite polynomials:

$$P_k X = \sum_{j=0}^{k} x_j \varphi_j(U) \quad (\varphi_j = \text{Hermite polynomial of order } j)$$

and

$$\mathbf{A}_k = \begin{pmatrix} 1 & \frac{4}{3} & 0 & -\frac{24}{5} & -\frac{20}{3} & \frac{176}{7} & 96 \\ \frac{4}{3} & 2 & \frac{8}{15} & -\frac{20}{3} & -\frac{464}{35} & \frac{88}{3} & \frac{10720}{63} \\ 0 & \frac{8}{15} & \frac{4}{3} & -\frac{16}{35} & -\frac{32}{3} & -\frac{800}{63} & \frac{496}{5} \\ -\frac{24}{5} & -\frac{20}{3} & -\frac{16}{35} & 24 & \frac{11552}{315} & -\frac{624}{5} & -\frac{40000}{77} \\ -\frac{20}{3} & -\frac{464}{35} & -\frac{32}{3} & \frac{11552}{315} & \frac{656}{5} & -\frac{53056}{693} & -\frac{22016}{15} \\ \frac{176}{7} & \frac{88}{3} & -\frac{800}{63} & -\frac{624}{5} & -\frac{53056}{693} & \frac{2528}{3} & \frac{5181056}{3003} \\ 96 & \frac{10720}{63} & \frac{496}{5} & -\frac{40000}{77} & -\frac{22016}{15} & \frac{5181056}{3003} & \frac{121920}{7} \end{pmatrix},$$

$$\mathbf{B}_k = \begin{pmatrix} \frac{1}{3} \\ \frac{4}{7} \\ \frac{1}{3} \\ -\frac{104}{63} \\ -\frac{24}{5} \\ \frac{3152}{693} \\ \frac{164}{3} \end{pmatrix},$$

so that the solution is:

$$\mathbf{X}_k = \begin{pmatrix} \frac{3}{4} \\ 0 \\ \frac{3}{4} \\ 0 \\ \frac{1}{16} \\ 0 \\ 0 \end{pmatrix},$$

which corresponds to:

$$P_k X = \frac{3}{4} + \frac{3}{4}\left(-2 + 4U^2\right) + \frac{1}{16}\left(12 - 48U^2 + 16U^4\right),$$

i.e.:

$$P_k X = U^4,$$

and we obtain the exact representation. For Tchebichev polynomials of first kind, we have:

$$\mathbf{A}_k = \begin{pmatrix} 1 & \frac{2}{3} & 0 & -\frac{2}{5} & -\frac{1}{3} & -\frac{2}{21} & 0 \\ \frac{2}{3} & \frac{1}{2} & \frac{2}{15} & -\frac{1}{6} & -\frac{26}{105} & -\frac{1}{6} & -\frac{22}{315} \\ 0 & \frac{2}{15} & \frac{1}{3} & \frac{2}{7} & 0 & -\frac{2}{9} & -\frac{1}{5} \\ -\frac{2}{5} & -\frac{1}{6} & \frac{2}{7} & \frac{1}{2} & \frac{14}{45} & -\frac{1}{30} & -\frac{82}{385} \\ -\frac{1}{3} & -\frac{26}{105} & 0 & \frac{14}{45} & \frac{7}{15} & \frac{74}{231} & 0 \\ -\frac{2}{21} & -\frac{1}{6} & -\frac{2}{9} & -\frac{1}{30} & \frac{74}{231} & \frac{1}{2} & \frac{38}{117} \\ 0 & -\frac{22}{315} & -\frac{1}{5} & -\frac{82}{385} & 0 & \frac{38}{117} & \frac{17}{35} \end{pmatrix},$$

$$\mathbf{B}_k = \begin{pmatrix} \frac{1}{3} \\ \frac{2}{7} \\ \frac{1}{6} \\ \frac{2}{63} \\ -\frac{1}{15} \\ -\frac{74}{693} \\ -\frac{1}{10} \end{pmatrix},$$

and the solution is:

$$\mathbf{X}_k = \begin{pmatrix} \frac{3}{8} \\ 0 \\ \frac{1}{2} \\ 0 \\ \frac{1}{8} \\ 0 \\ 0 \end{pmatrix},$$

which corresponds to:

$$P_k X = \frac{3}{8} + \frac{1}{2}\left(-1 + 2U^2\right) + \frac{1}{8}\left(1 - 8U^2 + 8U^4\right),$$

i.e.

$$P_k X = U^4.$$

For Tchebichev polynomials of second kind, we have:

$$\mathbf{A}_k = \begin{pmatrix} 1 & \frac{4}{3} & 1 & \frac{8}{15} & \frac{1}{3} & \frac{12}{35} & \frac{1}{3} \\ \frac{4}{3} & 2 & \frac{28}{15} & \frac{4}{3} & \frac{92}{105} & \frac{2}{3} & \frac{188}{315} \\ 1 & \frac{28}{15} & \frac{7}{3} & \frac{232}{105} & \frac{5}{3} & \frac{356}{315} & \frac{13}{15} \\ \frac{8}{15} & \frac{4}{3} & \frac{232}{105} & \frac{8}{3} & \frac{776}{315} & \frac{28}{15} & \frac{4616}{3465} \\ \frac{1}{3} & \frac{92}{105} & \frac{5}{3} & \frac{776}{315} & \frac{43}{15} & \frac{9236}{3465} & \frac{31}{15} \\ \frac{12}{35} & \frac{2}{3} & \frac{356}{315} & \frac{28}{15} & \frac{9236}{3465} & \frac{46}{15} & \frac{127628}{45045} \\ \frac{1}{3} & \frac{188}{315} & \frac{13}{15} & \frac{4616}{3465} & \frac{31}{15} & \frac{127628}{45045} & \frac{337}{105} \end{pmatrix},$$

$$\mathbf{B}_k = \begin{pmatrix} \frac{1}{3} \\ \frac{4}{7} \\ \frac{2}{3} \\ \frac{40}{63} \\ \frac{8}{15} \\ \frac{292}{693} \\ \frac{1}{3} \end{pmatrix},$$

so that:

$$\mathbf{X}_k = \begin{pmatrix} \frac{1}{8} \\ 0 \\ \frac{3}{16} \\ 0 \\ \frac{1}{16} \\ 0 \\ 0 \end{pmatrix},$$

and:

$$P_k X = \frac{1}{8} + \frac{3}{16}\left(-1 + 4U^2\right) + \frac{1}{16}\left(1 - 12U^2 + 16U^4\right),$$

i.e. the exact representation:

$$P_k X = U^4. \blacksquare$$

EXAMPLE 3.8.– The popular families of finite elements correspond to increasing finite-dimensional subspaces. For instance, let us consider $k > 0$, $h = 1/k$, $x_i = ih$, $0 \leq i \leq k$ and a family of $P1$ finite elements:

$$\varphi_j(s) = \begin{cases} 1 - \frac{|s-x_i|}{h}, & \text{if } |s - x_i| \leq h \\ 0, & \text{otherwise} \end{cases}.$$

Let us consider again $\Omega = (0,1)$, $P((a,b)) = b - a$ (i.e. $P(d\omega) = \ell(d\omega)$, where ℓ is the Lebesgue's measure), $U(\omega) = \sqrt{\omega}$ and $X(\omega) = \omega^2$. We are looking for:

$$P_k X = \sum_{j=0}^{k} x_j \varphi_j(U).$$

For instance, when $k = 10$, we have:

$$\mathbf{A}_k = \begin{pmatrix} \frac{1}{600} & \frac{1}{600} & 0 & 0 & 0 & 0 & 0 & 0 & 0 & 0 & 0 \\ \frac{1}{600} & \frac{1}{75} & \frac{1}{200} & 0 & 0 & 0 & 0 & 0 & 0 & 0 & 0 \\ 0 & \frac{1}{200} & \frac{2}{75} & \frac{1}{120} & 0 & 0 & 0 & 0 & 0 & 0 & 0 \\ 0 & 0 & \frac{1}{120} & \frac{1}{25} & \frac{7}{600} & 0 & 0 & 0 & 0 & 0 & 0 \\ 0 & 0 & 0 & \frac{7}{600} & \frac{4}{75} & \frac{3}{200} & 0 & 0 & 0 & 0 & 0 \\ 0 & 0 & 0 & 0 & \frac{3}{200} & \frac{1}{15} & \frac{11}{600} & 0 & 0 & 0 & 0 \\ 0 & 0 & 0 & 0 & 0 & \frac{11}{600} & \frac{2}{25} & \frac{13}{600} & 0 & 0 & 0 \\ 0 & 0 & 0 & 0 & 0 & 0 & \frac{13}{600} & \frac{7}{75} & \frac{1}{40} & 0 & 0 \\ 0 & 0 & 0 & 0 & 0 & 0 & 0 & \frac{1}{40} & \frac{8}{75} & \frac{17}{600} & 0 \\ 0 & 0 & 0 & 0 & 0 & 0 & 0 & 0 & \frac{17}{600} & \frac{3}{25} & \frac{19}{600} \\ 0 & 0 & 0 & 0 & 0 & 0 & 0 & 0 & 0 & \frac{19}{600} & \frac{13}{200} \end{pmatrix}$$

$$\mathbf{B}_k = \begin{pmatrix} \frac{1}{21000000} \\ \frac{3}{500000} \\ \frac{23}{250000} \\ \frac{289}{500000} \\ \frac{283}{125000} \\ \frac{667}{100000} \\ \frac{4069}{250000} \\ \frac{17381}{500000} \\ \frac{4203}{62500} \\ \frac{60267}{500000} \\ \frac{594323}{7000000} \end{pmatrix}.$$

The results obtained with $k = 50$ are shown in Figure 3.4.

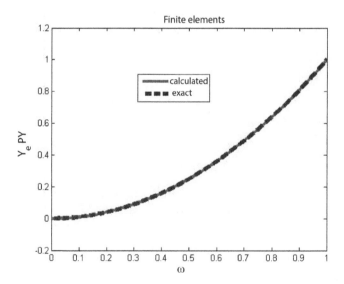

Figure 3.4. *Finite element approximation. For a color version of the figure, see www.iste.co.uk/souzadecursi/quantification.zip*

All the examples may be implemented in Matlab by using the programs given in section 3.1.5.

3.2. Approximations based on statistical properties (moment matching method)

The preceding approximations involve the evaluation of expressions that depend on the joint distribution of the pair (U, X). For instance, the evaluation of quantities such as $f(x \mid U = u)$, or:

$$(U, X) = E(UX) = \int uxf(u, x)\, dudx$$

or, more generally,

$$(X, \varphi_i(U)) = E(X\varphi_i(U)) = \int y\varphi_i(u) f(u, x)\, dudx\ ,$$

where f is the joint density probability of the pair (U, X).

When the distribution of the pair is not known *a priori*, it has to be evaluated from samples. This may involve practical difficulties since, on the one hand, the generation of samples is mandatory – with a cost – and, on the other hand, the samples must be used for the estimation of these quantities – with an error. In practice, the usual situation involves quantities having unknown or difficult to determinate distributions and we have only a sample of the pair.

This remark suggests that in some situations, it may be useful to use a procedure that *does not involve the evaluation of quantities depending upon the joint distribution of the pair* and *estimates all the necessary statistical properties directly from a sample*. For instance, we may determine an element $PX \in S$ such that some statistics of X coincide with those of PX.

An example of such an approach is provided by the *moment matching method*, which looks for a finite expansion such that chosen moments of X and PX coincide: $M_i(X) = E(X^i) = E\left((PX)^i\right) = M_i(PX)$ for some chosen values of i (for instance $1 \leq i \leq n$). This kind of approximation is connected with *Lévy's theorem* (see section 1.12.3, [DE 92]) and the *convergence in law* or *convergence in distribution* (see section 1.12.5): let us recall that the characteristic function of X is (see [1.10]):

$$\varphi(t) = E(\exp(itX))$$

and that a sequence $\{X_n\}_{n \in \mathbb{N}}$ converges to X in law if and only if $\varphi_n(t) = E(\exp(itX_n))$ converges pointwise to $\varphi(t)$ (i.e., $\varphi_n(t) \longrightarrow \varphi(t)$ a.e.). The characteristic function is closely connected to the moments. As previously mentioned (see section 1.10):

– if $M_p(X) < \infty$, then $\varphi^{(p)}$ (derivative of order p of φ) satisfies $\varphi^{(p)}(t) = i^p E\left(X^p e^{itX}\right)$. In particular, $\varphi^{(p)}(0) = i^p M_p(Y)$;

– if $M_p(X) < \infty$, $\forall\, p \in \mathbb{N}$ and the series $S(t) = \sum_{n \in \mathbb{N}} \dfrac{(it)^n}{n!} M_n(X)$ has a strictly positive convergence radius, then $\varphi(t) = S(t)$.

These two properties suggest that, on the one hand, it is possible to represent φ – and so, the law of X – by using the moments of X and, on the other hand, to approximate the law of X by using an expansion based on its first moments: a variable having the same moments as X is expected to have a distribution that closely approaches the distribution of X. The first idea (representation of the distribution of X by a series $S(t)$) of moments is connected to the *problem of the moments* (see, for instance, [CHO 62]) and the second idea has been exploited in the literature (see, for instance, [GAV 03, GAV 08, GAV 09]).

From the numerical standpoint, we consider:

$$PX = \sum_{j=1}^{k} x_j \varphi_j(U) \in S \;;\; \mathbf{X} = (x_1, ..., x_k)^t \;;$$

and we determine $\mathbf{X}$ such that:

$$M_p(X) = M_p(PX) \text{ for } 1 \leq p \leq n.$$

When $k = n$, these equations define a set of n nonlinear algebraical equations that may be solved by standard methods. However, in practice, we may consider a larger number of moments ($k \neq n$), and so, more equations than unknowns. In this case, it is convenient to look for solutions based on a minimization procedure. For instance, let us consider a map $T : \mathbb{R}^k \longrightarrow S$ given by:

$$\mathbf{v} = (v_1, ..., v_k)^t \longrightarrow T(\mathbf{v}) = \sum_{j=1}^{k} v_j \varphi_j (U) \in S$$

and a map $M : \mathbb{R}^k \longrightarrow \mathbb{R}^n$ given by:

$$\mathbf{v} = (v_1, ..., v_k)^t \longrightarrow \mathbf{M}(\mathbf{v}) = (M_1(T(\mathbf{v})), ..., M_n(T(\mathbf{v})))^t \in \mathbb{R}^n.$$

Taking:

$$\mathbf{M}_X = (M_1(X), ..., M_n(X))^t \in \mathbb{R}^n,$$

we may introduce $J : \mathbb{R}^k \longrightarrow \mathbb{R}$ given by:

$$J(\mathbf{v}) = dist(\mathbf{M}(\mathbf{v}), \mathbf{M}_X),$$

where $dist$ is a measure of distance: for instance, we may consider a norm $\|\bullet\|_{\mathbb{R}^n}$ and $\mathbb{R}^n$ and define $dist(\mathbf{u}, \mathbf{v}) = \|\mathbf{u} - \mathbf{v}\|_{\mathbb{R}^n}$. In this case, we look for:

$$\mathbf{x} = \arg \min_{\mathbb{R}^k} J.$$

The main difficulty remains in the non-convexity of J: the quality of the approximation depends on the quality of the optimization – if the numerically determined point is far from a global optimum of J, the approximation has a poor quality. As a result, it becomes mandatory to use adapted optimization procedures, able to solve non-convex continuous problems.

In addition, we observe that Levy's theorem ensures the convergence in distribution, which is a weak convergence involving approximation of the cumulative distribution function of the variable, *but not the approximation of the variables themselves*: the moment matching method is not expected to provide a good approximation of the variables, but only of their distributions. In practice, we may obtain a good representation of the variables if an excellent quality is obtained in the numerical solution of the nonlinear system of algebraical equations (or in the global optimization problem), while the approximation of the cumulative distribution function requests less computational effort.

EXAMPLE 3.9.– Let us consider again $\Omega = (0,1)$, $P((a,b)) = b - a$ (i.e. $P(d\omega) = \ell(d\omega)$ where ℓ is Lebesgue's measure), $U(\omega) = \sqrt{\omega}$ and $X(\omega) = \omega^2$. If we look for an approximation of X by a polynomial function of degree 5 of U, but the joint distribution of the pair is unknown, we must estimate the values of $B_i = E(XU^i)$ from the sample, what introduces an error. For instance, here are the results obtained by different values of ns:

ns	**X**
10	$(-1.4309, 28.4332, -169.5479, 421.2838, -459.5838, 183.1038)$
100	$(-0.1089, 2.1753, -13.0411, 32.5765, -34.8053, 14.3105)$
1000	$(-0.0105, 0.2108, -1.2644, 3.1608, -2.4766, 1.3905)$

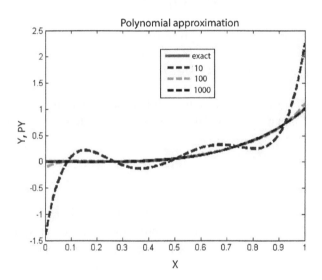

Figure 3.5. *Polynomial approximation with **B** estimated. For a color version of the figure, see www.iste.co.uk/souzadecursi/quantification.zip*

The results are presented in Figure 3.5: it shows that the approximation is poor until we use about 100 points. The situation may become more complex if the distribution of U is also unknown. In this case, the matrix **A** has to be estimated from the given data: the results for such a situation are shown in the table given below.

ns	X
10	(0.5230, − 0.0748, − 14.9273, 46.0987, − 49.9426, 19.3635)
100	(0.4373, − 0.0170, − 15.6027, 51.8225, − 59.8655, 24.2747)
1,000	(0.4294, − 0.0135, − 15.4512, 51.4904, − 59.6683, 24.2615)

As we can see, results are poor even for $1,000$ data points: An alternative approach consists, as mentioned, of using an approximation for which the n first empirical moments of X coincide with those of the approximation. The results for $n = 5$, minimizing the sum of the squares of the relative errors (analogously to a χ^2 measure of distance), are shown in Figure 3.7. The final distance between the moments is $1E-13$ and the quality of the approximation is good – even the density of probability is relatively correct. ∎

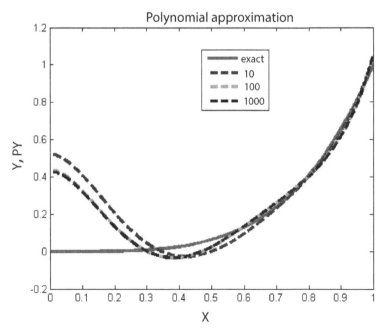

Figure 3.6. *Polynomial approximation with **A** and **B** estimated. For a color version of the figure, see www.iste.co.uk/souzadecursi/quantification.zip*

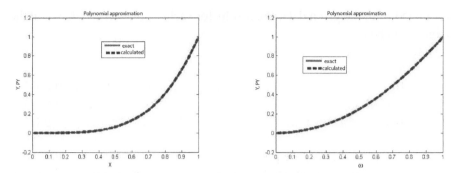

Figure 3.7. *Approximation by moment matching. For a color version of the figure, see www.iste.co.uk/souzadecursi/quantification.zip*

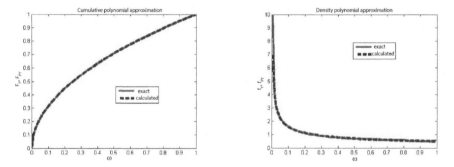

Figure 3.8. *Approximation by moment matching. For a color version of the figure, see www.iste.co.uk/souzadecursi/quantification.zip*

All the examples may be implemented in Matlab by using the programs below. They assume that, on the one hand, $\varphi_k(U)$ is provided by a subprogram phi(k,U) while, on the other hand, a sample $\{(U_i, X_i) : 1 \leq i \leq ns\}$ of the pair (U, X) is available. In addition, a subprogram dist(Y,Z) provides the distance between data vectors **Y** and **Z** of same length. The program uses fminsearch in order to minimize the objective function. Of course, the reader may replace it by his own optimization program. Analogously, centered moments may be used instead simple ones by modifying the program empirical_moments.

Listing 3.5. *Moment Matching*

```
function x = expcoef(phi, N_X, Xs, Us, n, dist, maxit,maxf,
    tol_x ,tol_f )
%
% determines the coefficients of the expansion
% using a sample (U_i, X_i) from (X,U) and
% moment matching
%
% IN:
% phi : basis function phi(k,U) − type anonymous function
% N_X : order of the expansion − type integer
% Xs : table of values of X − type array of double
% Us : table of values of U − type array of double
% (Us(i),Xs(i)) is a variate from (U,X)
% n: number of moments to be considered
% dist: function avaluating the distance between data vectors −
      type anonymous function
% maxit : max of minimization iterations − type integer
% maxf : max of objective function evaluations − type integer
% tol_x :precision requested on the coefficients − type double
% tol_f : precision requested on the minimal value of the
      objective − type
% double
%
% OUT:
% x: table 1 x N_X of coefficients − type array of double
%
M_X = empirical_moments(Xs,n);
fobj = @(c) objective_m3(c,Us,phi,M_X,dist);
x0 = randn(1, N_X);
options = optimset('MaxIter',maxit,'MaxFunEvals',maxf,'TolX',
    tol_x ,'TolFun',tol_f );
x = fminsearch(fobj, x0,options);
return;
end

function v = objective_m3(c,Us,phi,M_X,dist)
%
% determines the objective function to be minimized
%
% IN:
% c : table 1 x N_X of the coefficients of the expansion −
      type array of double
% Us : table ns x k of the sample points − type array of double
% (Us(i),Xs(i)) is a variate from (U,X)
% phi : basis function phi(j,U) − type anonymous function
```

```
% M_X : table of moments to be approached — type array of
    double
% dist: function avaluating the distance between data vectors —
    type anonymous function
%
% OUT:
% A: table ns x N_X of values — type array of double
%     A(i,j) = phi(j,U_i)
%
PXs = projection(c,Us,phi);
Ms = empirical_moments(PXs,length(M_X));
v = dist(Ms, M_X);
return;
end

function m = empirical_moments(Y,n)
%
% determines the empirical moments
%
% IN:
% Y : vector of data — type array of double
% n : number of moments — type integer
%
% OUT:
% m = vector of moments — type array of double
%
m = zeros(n,1);
for i = 1: n
    m(i) = mean( Y.^i );
end;
return;
end
```

3.3. Interpolation-based approximations (collocation)

An alternative use of a sample $\{(U_i, X_i) : i = 1, ..., ns\}$ consists of determining:

$$PX(U) = \sum_{j=1}^{k} x_j \varphi_j(U) \in S$$

such that:

$$PX(U_i) = X_i, i = 1, ..., ns.$$

In this case, $\mathbf{X} = (x_1, ..., x_k)^t$ is the solution of a linear system:

$$\mathbf{AX} = \mathbf{B},\ A_{ij} = \varphi_j(U_i),\ B_i = X_i\ (1 \leq i \leq ns, 1 \leq j \leq k).$$

This linear system involves ns equations for k unknowns. For $ns > k$, the number of equations is higher than the number of unknowns and the system is overdetermined: we look for generalized solutions, such as, for instance, minimum square ones – in this case, the solution may be interpreted as a discrete version of the Hilbertian approximations introduced in the case of finite-dimensional subspaces.

When such an approach is adopted, other interpolation techniques may be used, such as, for instance, spline approximations or collocation by intervals – this last situation may be interpreted as a discrete version of the approximations based on increasing sequences of finite-dimensional subspaces.

The wide variety of interpolation techniques avoids any tentative of a rapid summary, but the reader may refer to the literature in order to obtain more information about interpolation. Matlab implementation may be performed as follows: assume that, on the one hand, $\varphi_k(U)$ is provided by a subprogram phi(k,U) while, on the other hand, a sample $\{(U_i, X_i) : 1 \leq i \leq ns\}$ of the pair (U, X) is available. Then, we may use the code:

Listing 3.6. *Collocation*

```
function x = expcoef(phi, N_X, Xs, Us)
%
%   determines the coefficients of the expansion
%   using a sample (U_i, X_i) from (X,U)
%
% IN:
% phi : basis function phi(k,U) - type anonymous function
% N_X : order of the expansion - type integer
% Xs  : table of values of X - type array of double
% Us  : table of values of U - type array of double
% (Us(i),Xs(i)) is a variate from (U,X)
%
% OUT:
% x: table 1 x N_X of coefficients - type array of double
%
A = collocation_matrix(phi, N_X, Us);
x = A \ Xs;
x = x'; % transpose to 1 x N_x vector
return;
```

end

```
function A = collocation_matrix(phi, N_X, Us)
%
% creates the matrix of the linear system by using a sample
%
% IN:
% phi : basis function phi(k,U) - type anonymous function
% N_X : order of the expansion - type integer
% Us  : table of values of U - type array of double
% (Us(i),Xs(i)) is a variate from (U,X)
%
% OUT:
% A: table ns x N_X of values - type array of double
%     A(i,j) = phi(j,U_i)
%
A = zeros(ns, N_X);
for i = 1: ns
    for j = 1:N_X
        A(i,j) = phi(j,Us(i,:));
    end;
end;
return;
end
```

EXAMPLE 3.10.– Let us consider again $\Omega = (0,1)$, $P((a,b)) = b - a$ (i.e. $P(d\omega) = \ell(d\omega)$ where ℓ is Lebesgue's measure), $U(\omega) = \sqrt{\omega}$ and $X(\omega) = \omega^2$. If we look for the approximation of X by a polynomial function of degree 5 of U, a simple idea consists of generating a sample (U_i, X_i) for *conveniently chosen values* of U. Then, we may interpolate the values of X. For instance, let us consider a uniform grid on Ω: $\omega_i = ih$, $0 \leq i \leq np$, $h = 1/np$ and the corresponding values $U_i = U(\omega^i)$ and $X_i = X(\omega^i)$. For a polynomial function of degree $d = 5$ of U, we have:

$$\mathbf{X} = (-0.0000, 0.0000, -0.0000, 0.0000, 1.0000, 0.0000)$$

The results are shown in Figures 3.9 and 3.10. ■

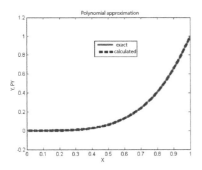 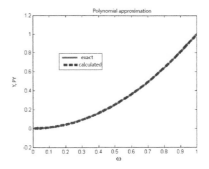

Figure 3.9. *Approximation by interpolation. For a color version of the figure, see www.iste.co.uk/souzadecursi/quantification.zip*

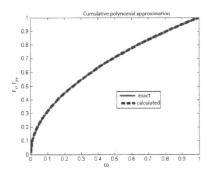 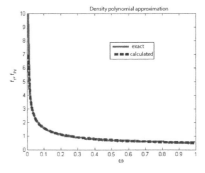

Figure 3.10. *Approximation by interpolation. For a color version of the figure, see www.iste.co.uk/souzadecursi/quantification.zip*

4

Linear Algebraic Equations Under Uncertainty

In this chapter, we consider a given square $n \times n$ matrix, $\mathbf{A} = (A_{ij}) \in \mathcal{M}(n,n)$ and classical associated problems such as:

– linear equations:

$$\mathbf{AX} = \mathbf{B},$$

where $\mathbf{B} = (B_i) \in \mathcal{M}(n,1)$ is a known $n \times 1$ vector and $\mathbf{X} = (X_i) \in \mathcal{M}(n,1)$ is an unknown $n \times 1$ vector (situations where $\mathbf{A}$ is not square are also examined – see below);

– eigenvalue or eigenvector problems:

$$\mathbf{X} \neq \mathbf{0} \text{ and } \mathbf{AX} = \lambda \mathbf{X},$$

where λ is an unknown real number and $\mathbf{X}$ is an unknown $n \times 1$ vector.

We will examine two basically different situations:

– In the first situation, we consider that $\mathbf{A}$ or $\mathbf{B}$ may contain uncertainty generated by a random vector $\mathbf{v}$ – i.e. $\mathbf{A} = \mathbf{A}(\mathbf{v})$, $\mathbf{B} = \mathbf{B}(\mathbf{v})$, where $\mathbf{v}$ is a random vector – and we are interested in the determination of the distribution of probability or statistics of the solution $\mathbf{X}$.

– In the second situation, $\mathbf{A}$ and $\mathbf{B}$ are deterministic, but we are interested in stochastic methods for the determination of the unknowns.

Let us recall that there are classical methods for dealing with the first situation, based on the *condition number* of $\mathbf{A}$, but these methods do not produce information about the distribution of probability of $\mathbf{X}$. The condition number is defined as (see, for instance, [DAT 09])

$$cond(\mathbf{A}) = \|\mathbf{A}\| \, \|\mathbf{A}^{-1}\|$$

and verifies (see, for instance, [DAT 09])

$$\frac{\|\Delta \mathbf{X}\|}{\|\mathbf{X}\|} \leq cond(\mathbf{A}) \frac{\|\Delta \mathbf{B}\|}{\|\mathbf{B}\|} \text{ if } \mathbf{A}(\Delta \mathbf{X}) = \Delta \mathbf{B}.$$

This inequality gives an estimative of the error $\Delta \mathbf{X}$ in $\mathbf{X}$ when an error $\Delta \mathbf{B}$ arises in $\mathbf{B}$. As observed, it does not carry an information about the distribution of $\mathbf{X}$. Analogously, we have, for infinitesimal variations of $\mathbf{A}$ (see, for instance, [DAT 09]):

$$\|d\lambda\| \leq cond(\mathbf{A}) \, \|d\mathbf{A}\|.$$

In the same way as previously, this inequality does not provide information about the distribution of λ.

4.1. Representation of the solution of uncertain linear systems

In this section, we consider the linear system:

$$\mathbf{AX} = \mathbf{B},$$

where $\mathbf{A} = (A_{ij}) \in \mathcal{M}(n,n)$ is a square matrix $n \times n$, $\mathbf{B} = (B_i) \in \mathcal{M}(n,1)$ is a vector of dimension $n \times 1$ and $\mathbf{X} = (X_i) \in \mathcal{M}(n,1)$ is the vector of the unknowns. We assume that there exists a random vector $\mathbf{v} = (v_1, ..., v_{nr})$ such that $\mathbf{A} = \mathbf{A}(\mathbf{v})$, $\mathbf{B} = \mathbf{B}(\mathbf{v})$. Our objective is the numerical determination of the probability distribution of $\mathbf{X}$ in three basic situations:

– when the distribution of $\mathbf{v}$ is known and, in addition, the functions $\mathbf{v} \longrightarrow \mathbf{A}(\mathbf{v})$ and $\mathbf{v} \longrightarrow \mathbf{B}(\mathbf{v})$ are known;

– when a sample of $\mathbf{v}$ is given, but the functions $\mathbf{v} \longrightarrow \mathbf{A}(\mathbf{v})$ and $\mathbf{v} \longrightarrow \mathbf{B}(\mathbf{v})$ are known;

– when a sample of the pair $(\mathbf{A}, \mathbf{B})$ is given.

In all these situations, this objective may be attained by the construction of an approximation $\mathbf{PX}$ of $\mathbf{X}$ in a convenient subspace of random variables. For instance, if the components v_i take their values on an interval $\Omega = (a, b)$, then we may use the procedure introduced in Chapter 3: let us consider a *total family* $F = \{\varphi_k\}_{k \subset \mathbb{N}}$ from the functional space $L^2(\Omega)$ and an approximation $\mathbf{PX}$ belonging to the finite-dimensional subspace

$$S = [\{\varphi_1(\xi), ..., \varphi_{N_X}(\xi)\}]^n$$
$$= \left\{ \sum_{k=1}^{N_X} \mathbf{D}_k \varphi_k(\xi) : \mathbf{D}_k \in \mathbb{R}^n, \ 1 \leq k \leq N_X \right\}, \qquad [4.1]$$

where $N_X \in \mathbb{N}^*$ and ξ is conveniently chosen. Let us consider $\varphi(\xi) = (\varphi_1(\xi), ..., \varphi_{N_X}(\xi))^t \in \mathcal{M}(N_X, 1)$. Then:

$$\mathbf{Y} \in \mathbf{S} \iff \mathbf{Y} = \mathbf{D}\varphi(\xi), \mathbf{D} = (D_{ij}) \in \mathcal{M}(n, N_X).$$

Thus

$$\mathbf{PX} = \chi \varphi(\xi), \qquad [4.2]$$

where $\chi = (\chi_{ij}) \in \mathcal{M}(n, N_X)$ are unknown coefficients to be determined. Once χ is calculated, we may generate

$$\mathbf{PX} = \sum_{k=1}^{N_X} \chi_k \varphi_k(\xi) \quad \left(\text{i.e. } (\mathbf{PX})_j = \sum_{k=0}^{N_X} \chi_{jk} \varphi_k(\mathbf{v}) \right). \qquad [4.3]$$

4.1.1. *Case where the distributions are known*

In this section, we consider the situation where the distribution of $\mathbf{v}$ and the functions $\mathbf{v} \longrightarrow \mathbf{A}(\mathbf{v})$ and $\mathbf{v} \longrightarrow \mathbf{B}(\mathbf{v})$ are given.

As previously observed (see Chapter 3), a convenient random variable is $\xi = \mathbf{v}$. Let us assume that $\mathbf{v}$ takes its values on $\Omega \subset \mathbb{R}^{nr}$ and $\mathbf{X}(\mathbf{v}) \in V = [L^2(\Omega)]^n$. In this case, let $F = \{\varphi_k\}_{k \subset \mathbb{N}}$ be a *total family* of the functional space $L^2(\Omega)$. We have

$$(\mathbf{Y}, \mathbf{AX}) = E(\mathbf{Y}^t \mathbf{AX}) = E(\mathbf{Y}^t \mathbf{B}) = (\mathbf{Y}, \mathbf{B}), \forall \mathbf{Y} \in V$$

and **X** is the solution of the variational equation:

$$\mathbf{X} \in V \text{ and } E\left(\mathbf{Y}^t \mathbf{A} \mathbf{X}\right) = E\left(\mathbf{Y}^t \mathbf{B}\right), \forall \, \mathbf{Y} \in V. \tag{4.4}$$

This equation is approximated as

$$\mathbf{P}\mathbf{X} \in S \text{ and } E\left(\mathbf{Y}^t \mathbf{A} \left(\mathbf{P}\mathbf{X}\right)\right) = E\left(\mathbf{Y}^t \mathbf{B}\right), \forall \, \mathbf{Y} \in S. \tag{4.5}$$

Since (from equation [4.2]),

$$\mathbf{A}\left(\mathbf{P}\mathbf{X}\right) = \mathbf{A}\mathbf{C}\varphi\left(\mathbf{v}\right),$$

we have

$$E\left(\varphi\left(\mathbf{v}\right)^t \mathbf{D}^t \mathbf{A} \chi \varphi\left(\mathbf{v}\right)\right) = E\left(\varphi\left(\mathbf{v}\right)^t \mathbf{D}^t \mathbf{B}\right), \forall \, \mathbf{D} \in \mathcal{M}(n,p). \tag{4.6}$$

Thus

$$E\left(\varphi\left(\mathbf{v}\right)^t \mathbf{D}^t \mathbf{A} \chi \varphi\left(\mathbf{v}\right)\right) = \sum_{i,j=1}^{n} \sum_{k,m=1}^{N_X} E\left(\varphi_m\left(\mathbf{v}\right) D_{im} A_{ij} \chi_{jk} \varphi_k\left(\mathbf{v}\right)\right)$$

and

$$E\left(\varphi\left(\mathbf{v}\right)^t \mathbf{D}^t \mathbf{B}\right) = \sum_{i=1}^{n} \sum_{m=1}^{N_X} E\left(\varphi_m\left(\mathbf{v}\right) D_{im} B_i\right).$$

So, by taking $D_{im} = \delta_{ir}\delta_{ms}$ in equation [4.6], we obtain

$$\sum_{j=1}^{n} \sum_{k=1}^{N_X} E\left(\varphi_s\left(\mathbf{v}\right) A_{rj} \chi_{jk} \varphi_k\left(\mathbf{v}\right)\right) = E\left(\varphi_s\left(\mathbf{v}\right) B_r\right),$$

$$1 \leq r \leq n, 1 \leq s \leq N_X.$$

Using the notation

$$\mathcal{A}_{rsjk} = E\left(\varphi_s\left(\mathbf{v}\right) A_{rj} \varphi_k\left(\mathbf{v}\right)\right), \; \mathcal{B}_{rs} = E\left(\varphi_s\left(\mathbf{v}\right) B_r\right), \tag{4.7}$$

we have

$$\sum_{j=1}^{n}\sum_{k=1}^{N_X} \mathcal{A}_{rsjk}\chi_{jk} = \mathcal{B}_{rs}, 1 \leq r \leq n, 1 \leq s \leq p.$$

These equations form a linear system: let us consider

$$ind(j,k) = (k-1)n + j$$

and the matrices $\mathbf{M} = (M_{\alpha\beta}) \in \mathcal{M}(nN_X, nN_X)$, $\mathbf{C} = (C_\beta) \in \mathcal{M}(nN_X, 1)$, $\mathbf{N} = (N_\alpha) \in \mathcal{M}(nN_X, 1)$ by

$$M_{\alpha\beta} = \mathcal{A}_{rsjk}, N_\alpha = \mathcal{B}_{rs}, C_\beta = \chi_{jk}, \alpha = ind(r,s), \beta = ind(j,k).$$

Then, we have

$$\mathbf{MC} = \mathbf{N}. \qquad [4.8]$$

The solution of this linear system determines $\mathbf{C}$ and, as a result, χ. This approach is implemented under Matlab by using the programs phi(k,U) and scalprod(Y,Z) previously introduced. Let us recall that the first program evaluates $\varphi_k(U)$, while the second program evaluates the scalar product of two variables Y and Z: for instance, if $Y = Y(\omega), Z = Z(\omega)$ and the density of ω is $f_\omega(\omega)$, defined on (a,b), then scalprod(Y,Z) returns the value of $\int_a^b Y(\omega)Z(\omega)f_\omega(\omega)d\omega$. In this case

Listing 4.1. *UQ of a Linear System by integration*

```
function  chi = expcoef(A,B,phi,n,N_X,U,f_omega,a,b)
%
%  determines the coefficients of the expansion.
%     by integrating the variables
%
% IN:
% A : n x n matrix of the linear system - type anonymous
      function
% B : n x 1 second member of the linear system - type anonymous
      function
% n : number of unknowns  -  type integer
% N_X : order of the expansion  -  type integer
% U : the random variable argument of the expansion  -  type
      anonymous function
```

```
% f_omega : density of omega - type anonymous function
% a,b : bounds of integration - type double
%
% OUT:
% chi : n x N_X matrix of the coefficients - type array of
%     double
%
[M,N] = variational_matrices(A,B,phi,n,N_X,U,f_omega,a,b);
C = M \ N;
chi = zeros(n,N_X);
for r = 1: n
    for s = 1: N_X
        alfa = index_map(r,s,n,N_X);
        chi(r,s) = C(alfa);
    end;
end;
return;
end

function [M,N] = variational_matrices(A,B,phi,n,N_X,U,f_omega,a
    ,b)
%
% determines matrices M, N such that
% M(alpha, beta) = E( phi_s A_rj phi_k ) , N(alpha) = E(B_r
%    phi_s )
%    alpha = index_map(r,s) , beta = index_map(j,k)
%
% IN:
% A : n x n matrix of the linear system - type anonymous
%     function
% B : n x 1 second member of the linear system - type anonymous
%     function
% n : number of unknowns - type integer
% N_X : order of the expansion - type integer
% U : the random variable argument of the expansion - type
%     anonymous function
% f_omega : density of omega - type anonymous function
% a,b : bounds of integration - type double
%
% OUT:
% M : nN_X x nN_X matrix of the linear system
% B : nN_X x 1 second member of the linear system - type
%     anonymous function
%
M = zeros(n*N_X,n*N_X);
N = zeros(n*N_X,1);
for r = 1: n
```

```
            B_r = @(om) select_index([r],B,U(om));
            for s = 1: N_X
                phi_s = @(om) phi(s,U(om));
                alfa = index_map(r,s,n,N_X);
                N(alfa) = scalprod(B_r,phi_s,f_omega,a,b);
                for j = 1: n
                    A_rj = @(om) select_index([r,j],A,U(om));
                    for k = 1: N_X
                        phi_k = @(om) phi(k,U(om))*phi_s(om);
                        betta = index_map(j,k,n,N_X);
                        M(alfa,betta) = scalprod(A_rj, phi_k);
                    end;
                end;
            end;
        end;
        return;
        end

function val = select_index(index, M, U)
%
%   selects the element M_index(U) from matrix M(U)
%
% IN:
% index - the vector of index - type array of integer
% M - the matrix as function of U - type anonymous function
% U - the random variable - type double
%
% OUT:
% val - the value of M_index(U) - type double
%
mmm = M(U);
if length(index) == 1
    val = mmm(index(1));
elseif length(index) == 2
    val = mmm(index(1), index(2));
end;
return;
end
```

EXAMPLE 4.1.– Let us consider the case where $\mathbf{v} = (v_1)$ is uniformly distributed on $(-1, 1)$: its density is

$$f(v) = \begin{cases} 1/2, \text{if } -1 < v < 1 \\ 0, \text{otherwise.} \end{cases}$$

Let us consider

$$A(v) = \begin{pmatrix} 2v_1 + 3 & v_1 + 3 \\ v_1 + 1 & v_1 + 2 \end{pmatrix} , \quad B(v) = \begin{pmatrix} v_1 \\ 1 \end{pmatrix}.$$

In this case,

$$\det(A) = v_1^2 + 3v_1 + 3 \geq 1$$

and

$$X = \frac{1}{v_1^2 + 3v_1 + 3} \begin{pmatrix} v_1^2 + v_1 - 3 \\ -v_1^2 + v_1 + 3 \end{pmatrix}.$$

The method exposed is applied with a polynomial basis:

$$\varphi_k(v) = \left(\frac{v+1}{2}\right)^{k-1}.$$

The results obtained with $p = 5$ are exposed in Figure 4.1. The relative mean quadratic error is of 0.5%.

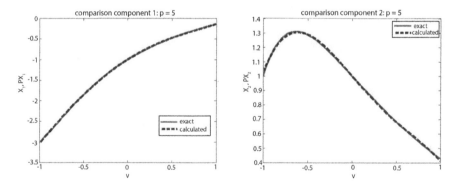

Figure 4.1. *Results obtained in example 4.1. For a color version of the figure, see www.iste.co.uk/souzadecursi/quantification.zip*

EXAMPLE 4.2.– Let us consider a bar, i.e. a structure modeled as a unidimensional continuous medium having the geometry of a segment of line and having only longitudinal ones as admissible displacement. A bar of length $\ell > 0$ is described by a variable $s \in (0, \ell)$ and its field of displacements is a function $x : (0, \ell) \longrightarrow \mathbb{R}$. The mechanical behavior of the bar is characterized by its elasticity modulus E, density ρ and right section S. The equilibrium of a bar under the loads of the gravity g and a force F applied to one of its extremities is described by the following equation:

$$\frac{d}{ds}\left(ES\frac{dx}{ds}\right) + \rho S g = 0 \text{ on } (0, \ell), \, ES\frac{dx}{ds}(\ell) = F, \, x(0) = 0.$$

In real structures, all these parameters are affected by variations: E, S, ρ, F may be considered as random variables taking their values on given intervals. In addition, internal and/or external variability may occur: different values may be observed at the different points of the same bar and/or different values may be obtained for two structures expected to be identical. For instance, wood structures present internal variability when considering layers corresponding to different years and external variability when considering different trees.

When the equilibrium equations of the bar are solved by a finite element method, the interval $(0, \ell)$ is discretized in n subintervals of length $h = \ell/n$ corresponding to the nodes $s_i = ih$, $0 \leq i \leq n$. The unknowns are the approximated values $X_i \approx x(s_i)$, for $1 \leq i \leq n$, which verify the linear system $\mathbf{AX} = \mathbf{B}$, where the stiffness matrix $\mathbf{A}$ is obtained by assembling the elementary stiffness matrices issued from each element I_i, while $\mathbf{B}$ is constructed by assembling elementary mass and force matrices from each element I_i.

In order to study the effects of an external variability, we may consider F as deterministic and E, S, ρ as a constant for each structure, but varying among the structures. In this case,

$$\mathbf{A} = \frac{ES}{h}\begin{pmatrix} 2 & -1 & 0 & \ldots & \ldots & 0 \\ -1 & 2 & -1 & 0 & \ldots & \ldots \\ 0 & -1 & \ldots & \ldots & 0 & 0 \\ \ldots & 0 & \ldots & 0 & 2 & -1 \\ 0 & \ldots & \ldots & 0 & -1 & 1 \end{pmatrix}, \, \mathbf{B} = \frac{\rho S g h}{2}\begin{pmatrix} 2 \\ 2 \\ \ldots \\ 2 \\ 1 \end{pmatrix} - \begin{pmatrix} 0 \\ 0 \\ \ldots \\ 0 \\ F \end{pmatrix},$$

where E, S and ρ are random variables. Taking $\mathbf{v} = (v_1, v_2)$, $v_1 = ES$, $v_2 = \rho S$, this situation corresponds to the case under consideration in this section. For instance, let us assume that each v_i is an independent random variable uniformly distributed on an interval (a_i, b_i). Let us consider two strictly positive integers $n_1 > 0$, $n_2 > 0$ and the functions given by ($0 \leq r \leq n_1, 0 \leq s \leq n_2$)

$$\varphi_k(\mathbf{v}) = \left(\frac{v_1 - a_1}{b_1 - a_1}\right)^r \left(\frac{v_2 - a_2}{b_2 - a_2}\right)^s, \quad k = sn_1 + r \ . \qquad [4.9]$$

The procedure introduced in this section may be applied. For instance, let us consider the situation where $a_1 = 2.2 \, MN$, $b_1 = 2.7 \, MN$, $a_2 = 0.11 \, kN/m$, $b_2 = 0.14 \, kN/m$, $F = 1 \, kN$ and $\ell = 5 \, m$. Using $n_1 = n_2 = 3$, $n = 10$, we obtain the results shown in Figure 4.2. The mean quadratic error is less than 0.02%.

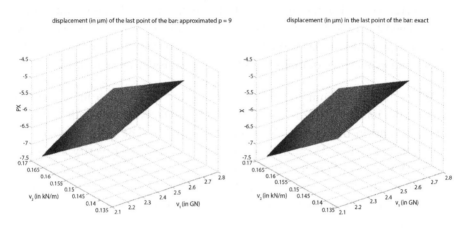

Figure 4.2. *Results obtained in example 4.2 (independent variables). For a color version of the figure, see www.iste.co.uk/souzadecursi/quantification.zip*

In practice, the variables E and ρ are not independent. In real situations, analysis of samples of wood (often of a given type from the same forest) shows a significant correlation between these two parameters and we may consider approaching E by a function of ρ. When a sample from the pair (E, ρ) is available, such data may be used in order to generate a function $E = E(\rho)$ (see Chapter 3). For instance, let us assume that $E \approx (1,840 + 15\rho) \times 10^6$ in SI units. In this case, the single random variable to be considered is $\mathbf{v} = (v_1)$, $v_1 = \rho S$. Let us assume that the variable ρS is uniformly distributed on (a, b),

$a = 0.11 \ kN/m$, $b = 0.14 \ kN/m$, $F = 1 \ kN$, $\ell = 5 \ m$. We may apply the approach presented in this section with a polynomial basis:

$$\varphi_k(v) = \left(\frac{v+a}{b-a}\right)^{k-1}.$$

The results obtained using $n = 10$, $p = 5$ may be found in Figure 4.3. The mean quadratic error is lower than $1E - 5\%$.

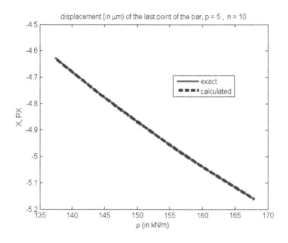

Figure 4.3. *Results obtained in example 4.2 (E as a function of ρ). For a color version of the figure, see www.iste.co.uk/souzadecursi/quantification.zip*

Internal variability may be studied in an analogous manner.

4.1.2. *Case where the distributions are unknown*

When considering variables having unknown densities, the means appearing in equations defining the matrices **M** and **N** of the linear system which determines the coefficients χ of the projection **PX** cannot be evaluated directly, but they have to be estimated from a *sample*: the means involved in equation [4.7] are approximated by using empirical means obtained from the sample in order to generate the linear system [4.8]. For instance, these unknown values may be estimated by using a sample

$\left(\left(\mathbf{v}^1, \mathbf{A}^1, \mathbf{B}^1\right), ..., \left(\mathbf{v}^{ns}, \mathbf{A}^{ns}, \mathbf{B}^{ns}\right)\right)$ of ns variates from the triplet $(\mathbf{v}, \mathbf{A}, \mathbf{B})$:

$$\mathcal{A}_{rsjk} \approx \frac{1}{ns} \sum_{i=1}^{ns} \varphi_s\left(\mathbf{v}^i\right) A_{rj}^i \varphi_k\left(\mathbf{v}^i\right), \quad \mathcal{B}_{rs} \approx \frac{1}{ns} \sum_{i=1}^{ns} \varphi_s\left(\mathbf{v}^i\right) B_r^i.$$

If the functions $\mathbf{v} \longrightarrow \mathbf{A}(\mathbf{v})$ and $\mathbf{v} \longrightarrow \mathbf{B}(\mathbf{v})$ are explicitly known, all the quantities may be approximated by using the sample $\left(\mathbf{v}^1, ..., \mathbf{v}^{ns}\right)$ of ns variates from $\mathbf{v}$:

$$\mathcal{A}_{rsjk} \approx \frac{1}{ns} \sum_{i=1}^{ns} \varphi_s\left(\mathbf{v}^i\right) A_{rj}\left(\mathbf{v}^i\right) \varphi_k\left(\mathbf{v}^i\right), \quad \mathcal{B}_{rs} \approx \frac{1}{ns} \sum_{i=1}^{ns} \varphi_s\left(\mathbf{v}^i\right) B_r\left(\mathbf{v}^i\right).$$

Once these values have been evaluated, the procedure is the same as in the preceding section: the solution of the linear system [4.8] determines $\mathbf{C}$ and, as a result, χ. This approach is implemented under Matlab by using the program scalprod(Y,Z) evaluating the scalar product by using a sample. In this case

Listing 4.2. *UQ of a Linear System by using a sample*

```
function  chi = expcoef(A,B,phi,n,N_X,Us)
%
%   determines the coefficients of the expansion.
%     by using a sample
%
% IN:
% A : n x n matrix of the linear system - type anonymous
       function
% B : n x 1 second member of the linear system - type anonymous
       function
% n : number of unknowns - type integer
% N_X : order of the expansion - type integer
% Us : table of values of U - type array of double
% Us(:,i) is a variate from U
%
% OUT:
% chi : n x N_X matrix of the coefficients - type array of
       double
%
[M,N] = variational_matrices(A,B,phi,n,N_X,Us);
C = M \ N;
chi = zeros(n,N_X);
for r = 1: n
```

```
        for s = 1: N_X
            alfa = index_map(r,s,n,N_X);
            chi(r,s) = C(alfa);
        end;
    end;
return;
end

function [M,N] = variational_matrices(A,B,phi, N_X, Us)
%
% determines matrices M, N such that
% M(alpha, beta) = E( phi_s A_rj phi_k) , N(alpha) = E(B_r
   phi_s)
%   alpha = index_map(r,s) , beta = index_map(j,k)
%
% IN:
% A : n x n matrix of the linear system − type anonymous
     function
% B : n x 1 second member of the linear system − type anonymous
     function
% n : number of unknowns − type integer
% N_X : order of the expansion − type integer
% Us : table of values of U − type array of double
% Us(:,i) is a variate from U
%
% OUT:
% M : nN_X x nN_X matrix of the linear system
% B : nN_X x 1 second member of the linear system − type
     anonymous function
%
M = zeros(n*N_X,n*N_X);
N = zeros(n*N_X,1);
for r = 1: n
    B_r = @(U) select_index([r],B,U);
    for s = 1: N_X
        phi_s = @(U) phi(s,U);
        alfa = index_map(r,s,n,N_X);
        f1 = @(U) phi_s(U)*B_r(U);
        Y = map(f1, Us,1);
        N(alfa) = mean(Y);
        for j = 1: n
            A_rj = @(om) select_index([r,j],A,U(om));
            for k = 1: N_X
                phi_sk = @(U) phi(k,U)*phi_s(U);
                f1 = @(U) phi_sk(U)*A_rj(U);
                Y = map(f1, Us,1);
                betta = index_map(j,k,n,N_X);
```

```
            M( alfa , betta ) = mean(Y);
         end;
      end;
   end;
end;
return;
end
```

EXAMPLE 4.3.– Let us consider again the situation studied in example 4.1: here, we assume that the distribution of **v** is unknown, but that a sample of ns variates from **v** has been furnished – in this example, such a sample is generated by using the intrinsic function `rand` of Matlab: the instruction `v = a + (b-a){*}rand(ns,1);` generates a vector of ns real numbers from the uniform distribution on (a, b). The results obtained with $p = 5, ns = 25$ are shown in Figure 4.4. The relative mean quadratic error between data and the approximation is less than 0.5% and a comparison with the exact values leads to an analogous result.

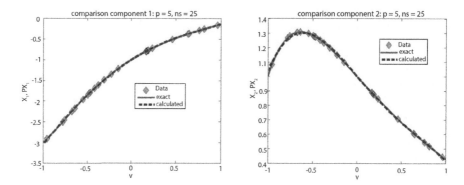

Figure 4.4. *Results obtained in example 4.3 (sample of 25 random points). For a color version of the figure, see www.iste.co.uk/souzadecursi/quantification.zip*

Figure 4.5 shows the results obtained with $p = 5, ns = 100$. The relative mean quadratic error between data and the approximation is less than 0.5% and a comparison with the exact values leads to an analogous result.

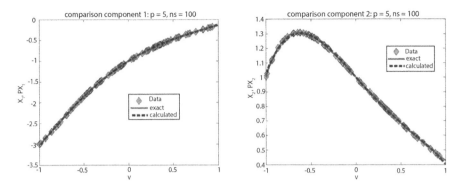

Figure 4.5. *Results obtained in example 4.3 (sample of 100 random points). For a color version of the figure, see www.iste.co.uk/souzadecursi/quantification.zip*

As previously observed, we may also consider the use of a uniform grid, analogously to the use of a grid of uniformly distributed collocation points: for instance, when using $ns = 11$ equally spaced "<collocation">" points, we obtain the results shown in Figure 4.6. The relative mean quadratic error is less than 0.6% when the approximation is compared with the data or the exact solution.

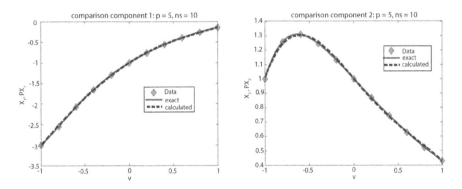

Figure 4.6. *Results obtained in example 4.3 (11 equidistributed points). For a color version of the figure, see www.iste.co.uk/souzadecursi/quantification.zip*

EXAMPLE 4.4.– Let us consider the situation presented in example 4.2: analogously to example 4.3, we assume that the distribution of **v** is unknown, but a sample formed by ns variates from the variable **v** is available – in this example, such a sample is generated in the same way as in the preceding example. The results obtained in the case where $a_1 = 2.2\ MN$, $b_1 = 2.7\ MN$, $a_2 = 0.11\ kN/m$, $b_2 = 0.14\ kN/m$, $F = 1\ kN$, $\ell = 5\ m$, $n_1 = n_2 = 3$, $n = 10$, $ns = 100$ are shown in Figure 4.7. The relative mean quadratic error is less than 0.02%.

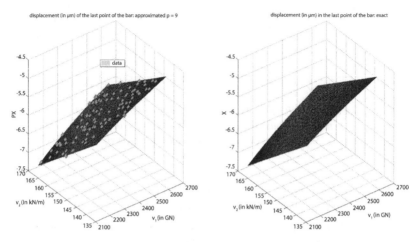

Figure 4.7. *Results obtained in example 4.4 (sample of 100 random points). For a color version of the figure, see www.iste.co.uk/souzadecursi/quantification.zip*

When using a uniform grid of 36 points, the relative mean quadratic error remains lower than 0.02%. The results are shown in Figure 4.8.

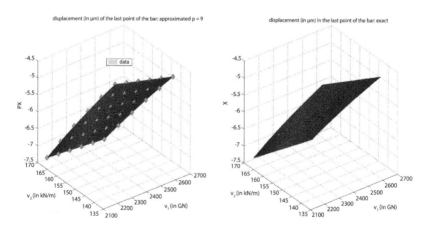

Figure 4.8. *Results obtained in example 4.4 (uniform grid of 36 points). For a color version of the figure, see www.iste.co.uk/souzadecursi/quantification.zip*

The situation is analogous when E is considered as a function of ρ. With the same parameters given in example 4.2, a uniform grid of 11 equally spaced values of ρ furnishes the results in Figure 4.9, with a relative mean quadratic error of $1E-5\%$ for the comparison between the approximation and the exact solution.

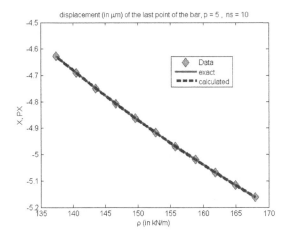

Figure 4.9. *Results obtained in example 4.4 (11 equidistributed points). For a color version of the figure, see www.iste.co.uk/souzadecursi/quantification.zip*

As in the preceding situation, internal variability may be studied in an analogous manner.

4.2. Representation of eigenvalues and eigenvectors of uncertain matrices

In this section, we consider the determination of the pairs $(\lambda_i, \mathbf{X}_i)$, where λ_i is an eigenvalue and $\mathbf{X}_i$ is an eigenvector of the square matrix $\mathbf{A} = (A_{ij}) \in \mathcal{M}(n, n)$:

$$\mathbf{A}\mathbf{X}_i = \lambda_i \mathbf{X}_i \, , \, \mathbf{X}_i \neq \mathbf{0} \, .$$

In order to simplify the notation, we do not write the index i and we use the notation $\mathbf{X}$ instead of writing $\mathbf{X}_i$, λ in place of λ_i: the readers must keep in mind that the method constructed in the following considers eigenvalues and eigenvectors. In addition, we denote $\|\mathbf{y}\| = \sqrt{\mathbf{y}^t . \mathbf{y}}$ for a vector $\mathbf{y} \in \mathbb{R}^n$.

We assume that the matrix **A** depends on a random vector $\mathbf{v} = (v_1, ..., v_{nr})$ (i.e. $\mathbf{A} = \mathbf{A}(\mathbf{v})$, where $\mathbf{v}$ is random).

Eigenvalues and eigenvectors of random matrices have been studied in the literature by using various procedures and approaches. Examples include analysis of the eigenvalues and eigenvectors of random matrices and their distributions in various situations (see, for instance, [FOR 09, CHO 09, HAN 78]), construction of estimators and determination of statistics (see, for instance, [BHA 10, TAO 09, TAO 10]), polynomial decomposition for the determination of the coefficients of polynomial chaos expansions associated with random variables (see, for instance, [RAH 06, RAH 07, RAH 09, RAH 11]) and analysis of continuous dynamical systems (see, for instance, [MAN 93b, MAN 93a, CRO 10]). In the following, we consider the classical methods of representation, based on the approaches by collocation, moment fitting or projection; we also consider the adaptation of classical numerical methods for the determination of eigenvalues and eigenvectors, such as iterated powers, subspace iteration and Krylov iterations. In a coherent approach with the presentation given in Chapter 3, the eigenvalues and eigenvectors studied are considered as unknown functions of the random variables involved – we have $\lambda = \lambda(\mathbf{v})$ e $\mathbf{X} = \mathbf{X}(\mathbf{v})$ – and a formal representation in terms of a series of these random variables is introduced in order to generate an approximation of these functions and reduce the problem to the determination of the coefficients of the terms of the series. Analogously to the preceding sections, we also consider the situation where only one sample is available.

Let us assume, without loss of generality, that the eigenvalues and eigenvectors under consideration are all real, i.e. their imaginary parts are null (the method presented in the following extends straightly to complex eigenvalues and eigenvectors, with a similar implementation given by the replacement of $\mathbb{R}$ by $\mathbb{C}$). As previously mentioned, we may consider an approximation **PX** corresponding to the truncation of an infinite series giving the representation of the solution on a Hilbert basis (see equations [4.1]–[4.3]):

$$\mathbf{PX} \in S_X = [\{\varphi_1(\boldsymbol{\xi}), ..., \varphi_{N_X}(\boldsymbol{\xi})\}]^n$$
$$= \left\{ \sum_{k=1}^{N_X} \mathbf{D}_k \varphi_k(\boldsymbol{\xi}) : \mathbf{D}_k \in \mathbb{R}^n,\ 1 \leq k \leq N_X \right\},$$

In an analogous manner, we consider $P\lambda$ such that

$$P\lambda \in S_\lambda = [\{\psi_1(\boldsymbol{\xi}), ..., \psi_q(\boldsymbol{\xi})\}]$$

$$= \left\{ \sum_{k=1}^{N_\lambda} D_k \psi_k(\boldsymbol{\xi}) : D_k \in \mathbb{R}, 1 \leq k \leq N_\lambda \right\},$$

$$\mathbf{PX} = \sum_{k=1}^{N_X} \boldsymbol{\chi}_k \varphi_k(\boldsymbol{\xi}), P\lambda = \sum_{k=1}^{N_\lambda} \ell_k \psi_k(\boldsymbol{\xi}) \qquad [4.10]$$

Taking $\varphi(\boldsymbol{\xi}) = (\varphi_1(\boldsymbol{\xi}), ..., \varphi_{N_X}(\boldsymbol{\xi}))^t \in \mathcal{M}(N_X, 1)$ and $\psi(\boldsymbol{\xi}) = (\psi_1(\boldsymbol{\xi}), ..., \psi_{N_\lambda}(\boldsymbol{\xi}))^t \in \mathcal{M}(N_\lambda, 1)$, we have:

$$\mathbf{PX} = \boldsymbol{\chi}\varphi(\boldsymbol{\xi}) \text{ and } P\lambda = \ell\psi(\boldsymbol{\xi}), \qquad [4.11]$$

where $\boldsymbol{\chi} = (\chi_{ij}) \in \mathcal{M}(n, N_X)$ and $\ell = (\ell_i) \in \mathcal{M}(1, N_\lambda)$ are the unknowns to be determined.

In practice, the same family and the same degree of approximation may be used for both the elements $\mathbf{X}$ and λ.

4.2.1. *Determination of the distribution of eigenvalues and eigenvectors by collocation*

The simplest way to determine an approximation of the distribution of an eigenvalue or eigenvector consists of the use of a sample in order to generate a projection. For instance, let us assume that we are interested in the dominant eigenvalue: if we may obtain a sample $\mathbf{S} = (\boldsymbol{\xi}_1, ..., \boldsymbol{\xi}_{ns})$ of ns variates of $\boldsymbol{\xi}$ (or if such a sample may be generated), we may use it for the generation of a sample of the dominant eigenvalue $\overline{\lambda}$ and, then, for the determination of the values of ℓ corresponding to the best approximation having the form given in equation [4.11], with respect to the data furnished by the sample $(\overline{\lambda}_1, ..., \overline{\lambda}_{ns})$. For instance, we may determine the coefficients of the expansion by solving the linear system

$$\overline{\lambda}_i = P\lambda(\boldsymbol{\xi}_i) = \ell\psi(\boldsymbol{\xi}_i) \quad (i = 1, ..., ns).$$

This linear system is expressed as

$$\mathbf{L}\ell = \overline{\lambda}, \quad L_{ij} = \psi_j(\boldsymbol{\xi}_i)$$

and it is, in general, overdetermined (more equations than unknowns). We may look for a solution furnished by minimum squares, with a Tikhonov regularization parameter $\varepsilon > 0$ – in this case, we have

$$\left(\mathbf{L}^t\mathbf{L} + \varepsilon Id\right)\ell = \mathbf{L}^t\overline{\lambda}.$$

This approach may be implemented in Matlab by using the programs given in section 3.3.

EXAMPLE 4.5.– Let us consider a single random variable $v = u^2 \in \mathbb{R}$, u uniformly distributed on $(0, 2)$ and the random matrix

$$\mathbf{A}(v) = \begin{pmatrix} \cos\left(\frac{\pi}{2}v\right) & -\sin\left(\frac{\pi}{2}v\right) \\ \sin\left(\frac{\pi}{2}v\right) & \cos\left(\frac{\pi}{2}v\right) \end{pmatrix} \begin{pmatrix} (2-v) & 0 \\ 0 & v \end{pmatrix} \begin{pmatrix} \cos\left(\frac{\pi}{2}v\right) & \sin\left(\frac{\pi}{2}v\right) \\ -\sin\left(\frac{\pi}{2}v\right) & \cos\left(\frac{\pi}{2}v\right) \end{pmatrix}.$$

[4.12]

The eigenvalues of $\mathbf{A}$ are $2 - v$ and v. Thus

$$\overline{\lambda} = \begin{cases} 2-v, & \text{if } 0 < v < 1; \\ v, & \text{if } 1 < v < 4. \end{cases} \quad ; \quad \underline{\lambda} = \begin{cases} v, & \text{if } 0 < v < 1; \\ 2-v, & \text{if } 1 < v < 2. \end{cases} \quad [4.13]$$

In this case, the exact cumulative distribution function of the dominating eigenvalue is

$$F_{\overline{\lambda}}(\lambda) = \begin{cases} \frac{1}{2}\left(\sqrt{\lambda} - \sqrt{2-\lambda}\right), & \text{se } 1 < \lambda < 2; \\ \frac{1}{2}\sqrt{\lambda}, & \text{se } 2 < v < 4. \end{cases} \quad [4.14]$$

Let $(v_1, ..., v_{ns})$ be a sample of ns variates from v, which leads to the dominating eigenvalues $(\overline{\lambda}_1, ..., \overline{\lambda}_{ns})_i$. These values may be used in order to generate and solve the linear system

$$\lambda_i = P\lambda(v_i) = \ell\psi(v_i).$$

As previously stated, this linear system is, in general, overdetermined and has more equations than unknowns. The results obtained by using a sample of

$ns = 21$ realizations are given in Figures 4.10 (results corresponding to a Tikhonov regularizing parameter equal to zero) and 4.11 (results corresponding to a Tikhonov regularizing parameter of $1E - 3$). The figure on the left compares the variable λ to the projection $P\lambda$, while the figure on the right compares the cumulative distribution function of the approximation to the exact one. We use an approximation by a polynomial of degree 8.

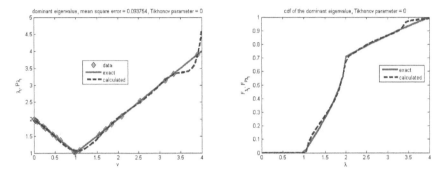

Figure 4.10. *Results obtained in example 4.5 (random sample of 21 points). For a color version of the figure, see www.iste.co.uk/souzadecursi/quantification.zip*

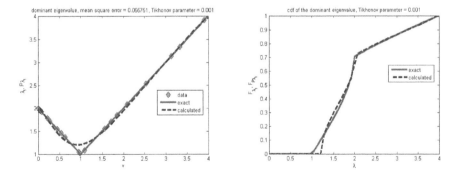

Figure 4.11. *Results obtained in example 4.5 (sample of 21 random points). For a color version of the figure, see www.iste.co.uk/souzadecursi/quantification.zip*

The results given by a sample $v_i = u_i^2$ generated from the equally spaced points $(u_1, ..., u_{ns})$ are shown in Figure 4.12.

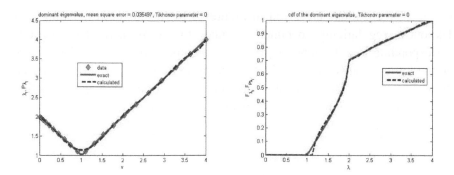

Figure 4.12. *Results obtained in example 4.5 (21 equidistributed points). For a color version of the figure, see www.iste.co.uk/souzadecursi/quantification.zip*

If we are interested in the eigenvalues, we may consider a sample of pairs $(\overline{\lambda}_i, \overline{\mathbf{X}}_i)$ such that $\|\overline{\mathbf{X}}_i\| = 1$ and $\max\{\overline{\mathbf{X}}_i.\overline{\mathbf{e}}_1, \overline{\mathbf{X}}_i.\overline{\mathbf{e}}\} > 0$. Then, we may use an analogous procedure and approximate the eigenvectors by using the representation given in equation [4.3]. An example of results is given in Figure 4.13 – they are obtained by using $ns = 51$ equally spaced values of u and a trigonometrical basis of $N_X = 23$ elements.

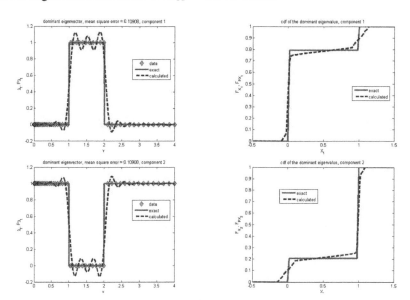

Figure 4.13. *Results obtained in example 4.5 (51 equidistributed points). For a color version of the figure, see www.iste.co.uk/souzadecursi/quantification.zip*

4.2.2. Determination of the distribution of eigenvalues and eigenvectors by moment fitting

A variation of the method presented in the previous section consists of using a sample $\Lambda = (\lambda_1, ..., \lambda_{ns})$ in order to generate a representation by the *moment matching method*: a sample provides the empirical moments $\mathbf{M^e} = (M_1^e, ..., M_q^e)$, $M_i^e = \frac{1}{ns}\sum_{j=1}^{ns}(\lambda_j^i)$ and the approximated values $\mathbf{M^a}(\boldsymbol{\ell}) = (M_1^a, ..., M_q^a)$, $M_i^a = \frac{1}{ns}\sum_{j=1}^{ns}(\lambda(\mathbf{v_j}^i))$, generated by using the representation $\boldsymbol{\ell}$. These values may be used in order to determine the coefficients of the expansion that fit values $\mathbf{M^a}(\boldsymbol{\ell})$ to $\mathbf{M^e}$: we may solve the nonlinear algebraical system $\mathbf{M^a}(\boldsymbol{\ell}) = \mathbf{M^e}$ or, as an alternative, minimize an objective function $d(\mathbf{M^a}(\boldsymbol{\ell}), \mathbf{M^e})$, which measures a pseudo-distance between them. This approach may be implemented in Matlab by using the programs given in section 3.2.

EXAMPLE 4.6.– Let us consider again the situation described in example 4.5. We compare the four different methods for moment fitting: three methods are based on the alternative minimization of a pseudo-distance (minimization of the relative error, minimization of the sum of relative errors and minimization of the absolute error) and the fourth method is the numerical solution of the nonlinear algebraical system introduced above (see also section 3.2). The results corresponding to a sample of $ns = 21$ points are shown in Figure 4.14: we observe that the eigenvalue itself is not correctly approximated, but the results are good for its cumulative density function – as observed in section 3.2. The results given by a sample of $ns = 21$ equally spaced values of u are shown in Figure 4.15. Finally, the results obtained by using 51 equidistributed values of u are shown in Figure 4.16. In all these situations, better results are obtained when solving the nonlinear algebraical system of equations.

4.2.3. Representation of extreme eigenvalues by optimization techniques

Let $F : \mathbb{R}^n \longrightarrow \mathbb{R}$ be given by

$$F(\mathbf{X}) = \mathbf{X}^t.\mathbf{A}.\mathbf{X} \Big/ \mathbf{X}^t.\mathbf{X} \ . \qquad [4.15]$$

F is Rayleigh's quotient associated with the matrix $\mathbf{A}$. Its minimum value coincides with the minimal eigenvalue $\underline{\lambda}$ of $\mathbf{A}$, and it is attained when $\mathbf{X}$ coincides with an eigenvector associated with $\underline{\lambda}$ (see, for instance, [GRI 02]).

In an analogous manner, its maximum value coincides with the dominating eigenvalue $\overline{\lambda}$ of the matrix $\mathbf{A}$ and it is attained when $\mathbf{X}$ is one of the associated eigenvectors.

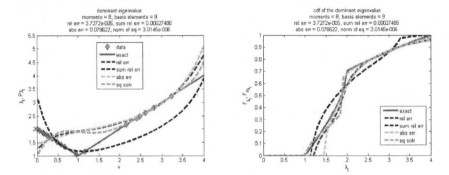

Figure 4.14. *Results obtained in example 4.6 (21 random points). For a color version of the figure, see www.iste.co.uk/souzadecursi/quantification.zip*

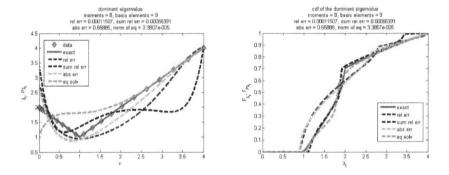

Figure 4.15. *Results obtained in example 4.6 (21 equidistributed points). For a color version of the figure, see www.iste.co.uk/souzadecursi/quantification.zip*

Rayleigh's coefficient provides an alternative approach for the numerical determination of the distribution of extreme eigenvalues by minimizing or maximizing the mean $E\left(F(\mathbf{PX})\right)$. For instance, we may use the representation given in equations [4.10]–[4.11] in order to introduce

$$f(\chi) = E\left(F\left(\chi\varphi(\xi)\right)\right) \ . \qquad [4.16]$$

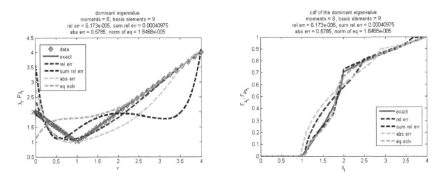

Figure 4.16. *Results obtained in example 4.6 (51 equidistributed points). For a color version of the figure, see www.iste.co.uk/souzadecursi/quantification.zip*

Then, we may numerically determine χ such that f is minimal or maximal: the solution defines **PX** – the numerical optimization may be carried out by using, for instance, the methods presented in [LOP 11]. In addition, on the one hand, the minimum may be restrained to a ball $\chi_i^t \cdot \chi_i \leq r$, $(r > 0)$ and, on the other hand, singular value decompositions (SVD) may be used: if $\mathbf{A}(\mathbf{v}) = \mathbf{W}(\mathbf{v})\mathbf{S}(\mathbf{v})\mathbf{U}(\mathbf{v})^t$, with $\mathbf{W}(\mathbf{v})^t\mathbf{W}(\mathbf{v}) = \mathbf{U}(\mathbf{v})^t\mathbf{U}(\mathbf{v}) = Id$, then, for $\mathbf{B}(\mathbf{v}) = \mathbf{U}(\mathbf{v})^t\mathbf{W}(\mathbf{v})\mathbf{S}(\mathbf{v})$ and $Y = \mathbf{U}(\mathbf{v})^t X$,

$$F(\mathbf{X}) = G(\mathbf{Y}) = \mathbf{Y}^t.\mathbf{B}.\mathbf{Y} \Big/ \mathbf{Y}^t.\mathbf{Y} \qquad [4.17]$$

and **B** may be used instead of **A**. The algorithm corresponding to this method is algorithm 4.1.

Algorithm 4.1. Rayleigh's quotient optimization

Require: $N_X > 0$, $kmax > 0$, $precmin > 0$, $\chi^{(0)} \in \mathcal{M}(n, N_X)$;
Require: a sample $\mathbf{v}_s$ of ns variates from $\mathbf{v}$;
Require: a method for the optimization of f;
  generate : χ **minimizing** f (equation [4.16]);
  generate : **PX** associated to χ;
  $\lambda := \mathbf{PX}^t.\mathbf{A}.\mathbf{PX} \big/ \mathbf{PX}^t.\mathbf{PX}$;
  return $\mathbf{PX}, \lambda$

An alternative use of Rayleigh's quotient consists of its use for the generation of a sample of $\underline{\lambda}$ or $\overline{\lambda}$ (by minimization or maximization): then, the sample may be used for collocation, as in section 3.3. For instance:

– *either* minimize $f_i(\chi) = F(\chi \varphi(\mathbf{v_i}))$ for each i: each minimization produces a variate λ_i of λ;

– *or* minimize $F_i(\mathbf{X}) = \mathbf{X}^t.\mathbf{A}(\mathbf{v_i}).\mathbf{X} \Big/ \mathbf{X}^t.\mathbf{X}$ for each i: in this case too, each minimization generates a variate from λ_i of λ.

EXAMPLE 4.7.– Let us consider again the situation described in example 4.5. Here, we are interested in determining the extreme (maximal and minimal) eigenvalues, which correspond to the maximization and minimization of the Rayleigh's quotient (equation [4.17]. We use the approach introduced in [LOP 11], with a sample of $ns = 21$ equally spaced values of u. The results obtained are shown in Figures 4.17 and 4.18, respectively.

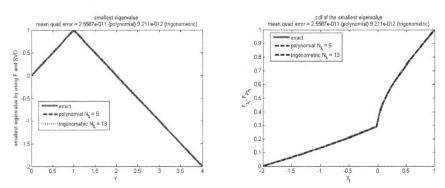

Figure 4.17. *Minimal eigenvalue in example 4.7 (21 equidistant points). For a color version of the figure, see www.iste.co.uk/souzadecursi/quantification.zip*

4.2.4. *Power iterations*

A second classical method for the determination of the dominating eigenvalue when $\mathbf{A}$ possesses n different eigenvalues is the method of *power iterations*, which consists of iterations starting from an initial vector $\mathbf{X}^{(0)}$:

$$\mathbf{X}^{(k+1)} = \mathbf{T}^{(k+1)} \Big/ \left\| \mathbf{T}^{(k+1)} \right\|, \quad \mathbf{T}^{(k+1)} = \mathbf{A}.\mathbf{X}^{(k)}. \quad [4.18]$$

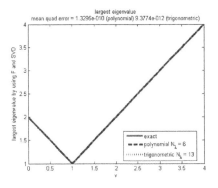 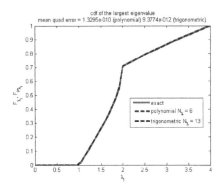

Figure 4.18. *Maximal eigenvalue in example 4.7 (21 equidistant points). For a color version of the figure, see www.iste.co.uk/souzadecursi/quantification.zip*

Let us consider the representation introduced in equations [4.10]–[4.11]: we define $\mathbf{X}^{(k+1)}(v) = \chi^{(k+1)}\varphi(\mathbf{v})$ and, in an analogous manner, $\mathbf{T}^{(k+1)} = \tau^{(k+1)}\varphi(\mathbf{v})$. Then, we have:

$$\sum_{j=1}^{N_X} \tau_{pj}^{(k+1)} \varphi_j(\mathbf{v}) = \sum_{j=1}^{N_X} \sum_{q=1}^{n} A_{pq}(\mathbf{v}) \chi_{qj}^{(k)} \varphi_j(\mathbf{U}). \qquad [4.19]$$

Multiplying both sides by $\varphi_i(\mathbf{v})$, we obtain:

$$\sum_{j=1}^{N_X} \tau_{pj}^{(k+1)} \varphi_j(\mathbf{v})\varphi_i(\mathbf{v}) = \sum_{j=1}^{N_X} \sum_{q=1}^{n} A_{pq}(\mathbf{v}) \chi_{qj}^{(k)} \varphi_j(\mathbf{v})\varphi_i(\mathbf{v}). \qquad [4.20]$$

and, by taking the mean of both sides, it results that

$$\sum_{j=1}^{N_X} \sum_{q=1}^{n} \mathcal{C}_{piqj} \tau_{qj}^{(k+1)} = \sum_{j=1}^{N_X} \sum_{q=1}^{n} \mathcal{D}_{piqj}(\mathbf{v}) \chi_{qj}^{(k)}, \qquad [4.21]$$

where

$$\mathcal{C}_{piqj} = E\left(\delta_{pq}\varphi_i(\mathbf{v})\varphi_j(\mathbf{v})\right), \quad \mathcal{D}_{piqj} = E\left(A_{pq}(\mathbf{v})\varphi_i(\mathbf{v})\varphi_j(\mathbf{v})\right). \qquad [4.22]$$

Let us introduce the transformation $ind(a,b) = a + (b-1)N_X$: we denote, for $r = ind(p,i)$ and $s = ind(q,j)$,

$$C_{rs} = \mathcal{C}_{piqj}, \; D_{rs} = \mathcal{D}_{piqj}, \; t_s^{(k+1)} = \tau_{qj}^{(k+1)}, \; x_s^{(k)} = \chi_{qj}^{(k)}. \qquad [4.23]$$

and equation [4.22] expressed as

$$\mathbf{C}\mathbf{t}^{(k+1)} = \mathbf{D}\mathbf{x}^{(k)}. \qquad [4.24]$$

Once $\mathbf{t}^{(k+1)}$ is evaluated, it may be normalized in order to generate

$$\mathbf{x}^{(k+1)} = \mathbf{t}^{(k+1)} \Big/ \left\|\mathbf{t}^{(k+1)}\right\|, \; \chi_{qj}^{(k+1)} = x_s^{(k+1)}. \qquad [4.25]$$

Equations [4.24]–[4.25] define an iterative procedure: $\mathbf{X}^{(k)}$ is given, which determines $\left(\chi^{(k)}\right)$ and, as a result, $\mathbf{x}^{(k)}$ and $\mathbf{t}^{(k+1)}$ are given by equation [4.24]). Thus, $\mathbf{x}^{(k+1)}$ and $\left(\chi_1^{(k+1)}, \ldots, \chi_{N_X}^{(k+1)}\right)$ are given by equation [4.25], which furnishes $\mathbf{X}^{(k+1)}$. At the end of the iterations, the eigenvalue is approximated by the Rayleigh's quotient defined in equation [4.15]: $\lambda \approx F(\mathbf{X})$. Here yet, SVD may be used: $\mathbf{A}(\mathbf{v})$ may be replaced by $\mathbf{B}(\mathbf{v})$.

A deflating approach may also be used in order to generate other eigenvalues: once $\mathbf{X}$ and λ are determined, we may consider $\tilde{\mathbf{A}} = \mathbf{A} - \lambda \mathbf{X}.\mathbf{X}^t \big/ \mathbf{X}^t.\mathbf{X}$. Then, the dominating eigenvalue of $\tilde{\mathbf{A}}$ is the second larger modulus eigenvalue of $\mathbf{A}$, and the procedure presented may be applied to $\tilde{\mathbf{A}}$ in order to generate a representation for the second eigenvalue.

At least, we observe that the determination of the minimal eigenvalue may be carried out by using the inverse $\mathbf{A}^{-1}$ of $\mathbf{A}$ instead of this last one. In practice, the determination of the inverse matrix may be avoided by using inverse iterations: $\mathbf{A}.\mathbf{T}^{(k+1)} = \mathbf{X}^{(k)}$, and we have $\mathbf{D}\mathbf{t}^{(k+1)} = \mathbf{C}\mathbf{x}^{(k)}$ instead of equation [4.24].

The algorithm corresponding to this method is shown in algorithm 4.2.

As in the preceding situation, an alternative consists of using power iterations as a generator of variates: a sample is obtained in this way and a collocation or moment matching method may be used (see sections 3.3 and 3.2).

Algorithm 4.2. Power Iterations

Require: $N_X > 0, kmax > 0, precmin > 0, \chi^{(0)} \in \mathcal{M}(n, N_X)$;
  local : k, $prec$, t, xnew;
  generate : $\mathcal{C}$ and $\mathcal{D}$;
  generate : $\mathbf{x}^{(0)}$ associated to $\chi^{(0)}$;
  $k := 0$;
  $\mathbf{x} := \mathbf{x}^{(0)}$;
  $prec := precmin + 1$;
  while $k < kmax$ **and** $prec > precmin$ **do**
    determine t : Ct := D.x (or Dt := C.x for the inverse iterations);
    xnew := t $/ \|\mathbf{t}\|$;
    $prec := \|\mathbf{xnew} - \mathbf{x}\|$;
    x := xnew;
    $k := k + 1$;
  end while
  generate : χ associated to x;
  generate : PX associated to χ;
  $\lambda := \mathbf{PX}^t.\mathbf{A}.\mathbf{PX} \,/\, \mathbf{PX}^t.\mathbf{PX}$;
  return X, λ

EXAMPLE 4.8.– Let us consider again the situation described in example 4.5. The results obtained for the minimal and maximal eigenvalues are shown in Figures 4.19–4.22. In Figures 4.19 and 4.20, the means have been evaluated by numerical integration using the intrinsic function *quad* of Matlab. In Figures 4.21 and 4.22, these means have been estimated by using a sample of equally spaced values of u.

4.2.5. *Subspace iterations and Krylov iterations*

A simple extension of power iterations is obtained by the simultaneous use of a set of initial vectors (in the place of a single vector): we do not consider a solitary initial vector, but a set $S^{(0)} = \left\{ \mathbf{X}_1^{(0)}, ..., \mathbf{X}_{kd}^{(0)} \right\}$ of kd initial vectors. $S^{(0)}$ may be generated in order to form an orthonormal set by applying Gram–Schmidt's procedure to a given set $\mathcal{S}^{(0)} = \left\{ \mathcal{X}_1^{(0)}, ..., \mathcal{X}_{kd}^{(0)} \right\}$

of linearly independent vectors provided by the user. Some usual methods for the generation of $\mathcal{S}^{(0)}$ are given in the following:

– random choice;

– eigenvectors of an auxiliary matrix;

– Krylov vectors: $\mathcal{X}_1^{(0)}$ is given and $\mathcal{X}_{i+1}^{(0)} = \mathbf{A}.\mathcal{X}_i^{(0)}$.

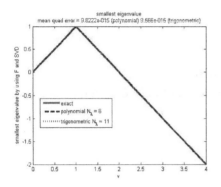
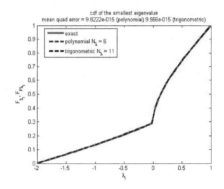

Figure 4.19. *Minimal eigenvalue in example 4.8 (numerical integration). For a color version of the figure, see www.iste.co.uk/ souzadecursi/quantification.zip*

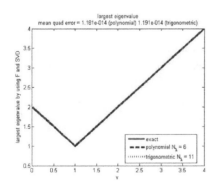
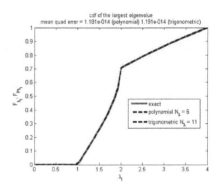

Figure 4.20. *Maximal eigenvalue in example 4.8 (numerical integration). For a color version of the figure, see www.iste.co.uk/souzadecursi/quantification.zip*

Linear Algebraic Equations Under Uncertainty 257

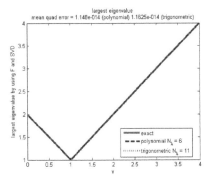 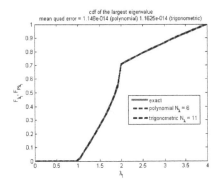

Figure 4.21. *Minimal Eigenvalue in example 4.8 (21 equidistant points). For a color version of the figure, see www.iste.co.uk/souzadecursi/quantification.zip*

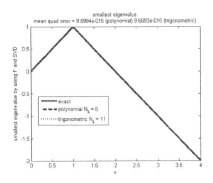 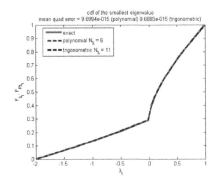

Figure 4.22. *Maximal eigenvalue in example 4.8 (21 equidistant points). For a color version of the figure, see www.iste.co.uk/souzadecursi/quantification.zip*

At step k, $S^{(k)} = \left\{ \mathbf{X}_1^{(k)}, ..., \mathbf{X}_{kd}^{(k)} \right\}$ is given and $S^{(k+1)} = \left\{ \mathbf{X}_1^{(k+1)}, ..., \mathbf{X}_{kd}^{(k+1)} \right\}$ is obtained by applying Gram–Schmidt's orthonormalization procedure to the set $\mathcal{S}^{(k+1)} = \left\{ \mathcal{X}_1^{(k+1)}, ..., \mathcal{X}_{kd}^{(k+1)} \right\}$, generated by one iteration of the power method applied to the elements of $S^{(k)}$: $\mathcal{X}_i^{(k+1)}$ is the result of the application of the procedure corresponding to equations [4.23]–[4.24] to $\mathbf{X}_i^{(k)}$. The algorithm corresponding to this method is algorithm 4.3.

Algorithm 4.3. Subspace Iterations

Require: $kmax > 0, precmin > 0, \chi^{(0)} \in \mathcal{M}(n, N_X)$;
Require: $N_X > 0, kmax > 0, precmin > 0, np > 0$;
  local : k, $prec$, **t**, **xnew**;
  generate : $\mathcal{C}$ e $\mathcal{D}$;
  generate $T^0 = \{\chi_1^{(0)}, \ldots, \chi_{np}^{(0)}\} \subset \mathcal{M}(n, N_X)$;
  for $i := 1$ to np **do**
    generate : $\mathbf{t}_i^{(0)}$ associated to $\chi_i^{(0)}$;
  end for
  generate $S^0 = \{\mathbf{x}_1^{(0)}, \ldots, \mathbf{x}_{np}^{(0)}\}$ **by Gram-Schmidt's orthonormalization of** $\{\mathbf{t}_1^{(0)}, \ldots, \mathbf{t}_{np}^{(0)}\}$;
  $k := 0$;
  $prec := precmin + 1$;
  while $k < kmax$ **and** $prec > precmin$ **do**
    for $i := 1$ a np **do**
      generate : x associated to $\chi_i^{(k)}$;
      determine t : $\mathbf{C}\mathbf{t} := \mathbf{D}.\mathbf{x}$;
      generate : χ associated to t;
      $\eta_i^{(k)} := \eta$;
    end for
    generate $T^k = \{\eta_1^{(k)}, \ldots, \eta_{np}^{(k)}\}$;
    generate $S^{(k+1)} = \{\chi_1^{(k+1)}, \ldots, \chi_{np}^{(k+1)}\}$ **by Gram-Schmidt's orthonormalization of** T^k;
    $prec := 0$;
    for $i := 1$ a np **do**
      $prec := prec + \left\|\chi_i^{(k+1)} - \chi_i^{(k)}\right\|$;
    end for;
    $k := k + 1$;
  end while
  for $i := 1$ to np **do**
    generate : $\mathbf{PX}_i$ associated to $\chi_i^{(k+1)}$;
    $\lambda_i := \mathbf{PX_i}^t.\mathbf{A}.\mathbf{PX_i} / \mathbf{PX_i}^t.\mathbf{PX_i}$;
  end for
  return $\mathbf{X}_1, \ldots, \mathbf{X}_{np}, \lambda_1, \ldots, \lambda_{np}$

As in the preceding situations, the SVD may be used and we may replace $(\mathbf{A}(\mathbf{v}))$ by $\mathbf{B}(\mathbf{v})$) previously defined. In addition, the procedure may also be used in order to generate samples of vectors and eigenvectors to be used in collocation or moment matching approaches (sections 3.3 and 3.2).

EXAMPLE 4.9.– Let us consider again the situation presented in example 4.5. We want to determine simultaneously the maximal and minimal eigenvalues by using the starting set suggested by Krylov. Results obtained are shown in Figures 4.23 and 4.26, respectively. In Figures 4.23 and 4.24, the means have been evaluated by numerical integration using the intrinsic function *quad* of Matlab. In Figures 4.25 and 4.26, these means have been estimated by using a sample of equally spaced values of u.

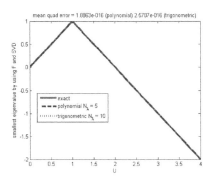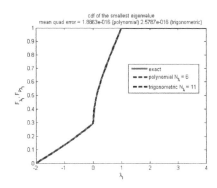

Figure 4.23. *Minimal eigenvalue in example 4.9 (numerical integration). For a color version of the figure, see www.iste.co.uk/souzadecursi/quantification.zip*

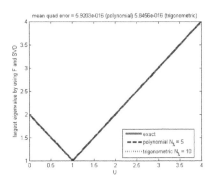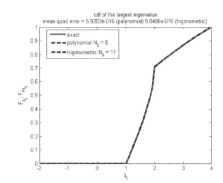

Figure 4.24. *Maximal eigenvalue in example 4.9 (numerical integration). For a color version of the figure, see www.iste.co.uk/ souzadecursi/quantification.zip*

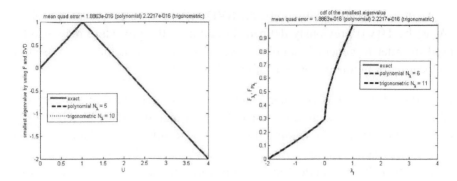

Figure 4.25. *Minimal eigenvalue in example 4.9 (21 equidistributed points). For a color version of the figure, see www.iste.co.uk/ souzadecursi/quantification.zip*

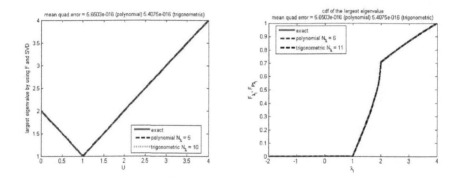

Figure 4.26. *Maximal eigenvalue in example 4.9 (21 equidistributed points). For a color version of the figure, see www.iste.co.uk/ souzadecursi/quantification.zip*

4.3. Stochastic methods for deterministic linear systems

Deterministic linear systems have been extensively and deeply studied in the literature. Efficient methods of numerical solution are widely spread and may easily be found. However, the existing methods generally request the complete solution of the system, i.e. the determination of all the components of the solution, even if we are interested in a few ones. In some special situations, the direct determination of a small subset of components of the solution may be interesting and rare approaches consider this question. One of them is furnished by stochastic methods tending to generate an independent determination of each component of the solution. Eventually, a single component may be directly determined by using a stochastic approach. The model situation is the following one: let us consider the linear system

$$\mathbf{AX} = \mathbf{B}, \qquad [4.26]$$

where $\mathbf{X} = (X_i) \in \mathcal{M}(n,1)$ is the unknown; $\mathbf{B} = (B_i) \in \mathcal{M}(n,1)$ is a vector and $\mathbf{A} = (A_{ij}) \in \mathcal{M}(n,n)$ verifies

$$\mathbf{A} = \alpha \mathbf{Id} - \mathbf{\Pi}, \qquad [4.27]$$

where $\alpha > 1$ is a real number and $\mathbf{\Pi} = (\Pi_{ij}) \in \mathcal{M}(n,n)$ is a matrix such that

$$\pi_{ij} \geq 0, \forall\, i,\, j\,;\; \sum_{j=1}^{n} \pi_{ij} = 1, \forall\, i\; (1 \leq i,\, j \leq n). \qquad [4.28]$$

Let us consider $\Omega = \{1, ..., n\}$; a function $f : \Omega \longrightarrow \mathbb{R}$, given by

$$f(i) = B_i\,,\, i = 1, ..., n \qquad [4.29]$$

and a sequence of random variables $\{U_k\}_{k \in \mathbb{N}} \subset \Omega$ such that

$$P(U_{k+1} = j \mid U_k = i) = \Pi_{ij}\,, \forall\, k \in \mathbb{N}\,;\; U_0 \in \Omega\,; \qquad [4.30]$$

and

$$\forall\, k \in \mathbb{N} : P(U_{k+1} = j \mid U_k = i, U_{k-1} = i_{k-1}, ..., U_0 = i_0) = \\ \Pi_{ij}\,, \forall\, (j, i, i_{k-1}, ..., i_0) \in \Omega^{k+2}\,; \qquad [4.31]$$

Let

$$\mathbf{P}_0 = (P(U_0 = i)) \in \mathcal{M}(n,1).$$

Our objective is to establish that:

THEOREM 4.1.– Assume that conditions [4.26]–[4.31] are satisfied. Then

$$X_i = \sum_{k=0}^{+\infty} \frac{1}{\alpha^k} E\left(f\left(U_k\right) \mid U_0 = i\right). \blacksquare$$

PROOF.–

1) Let $i \in \Omega$: if $P(U_0 = i) = 0$, then

2) Let us show that

$$P\left(U_{k+1} = i_{k+1}, U_k = i_k, U_{k-1} = i_{k-1}, ..., U_0 = i_0\right) =$$
$$\Pi_{i_0 i_1} \Pi_{i_1 i_2} ... \Pi_{i_{k+1} i_k} P(U_0 = i_0), \quad [4.32]$$
$$\forall \left(i_{k+1}, i_k, i_{k-1}, ..., i_0\right) \in \Omega^{k+2}$$

The proof is carried out by recurrence: let $k = 0$ and consider the events $E = "U_1 = i_1"$; $F = "U_0 = i_0"$. We have

$$P\left(E \mid F\right) = \Pi_{i_0 i_1} \; ; P\left(F\right) = P\left(U_0 = i_0\right).$$

Since

$$P\left(E \cap F\right) = P\left(E \mid F\right).P\left(F\right),$$

we have

$$P\left(U_1 = i_1, U_0 = i_0\right) = \Pi_{i_0 i_1} P(U_0 = i_0).$$

So, the first domino is placed. For recurrence, let us assume that equation [4.32] is verified for a given $k > 0$. Taking $E = "U_{k+2} = i_{k+2}"$; $F = "U_{k+1} = i_k, U_k = i_k, ..., U_0 = i_0"$, we obtain, from equation [4.31],

$$P\left(E \mid F\right) = \Pi_{i_{k+1} i_{k+2}} \; ; P\left(F\right) = P\left(U_{k+1} = i_{k+1}, U_k = i_k, ..., U_0 = i_0\right)$$

and, using that $P\left(E \cap F\right) = P\left(E \mid F\right).P\left(F\right)$, we have

$$P\left(U_{k+2} = i_{k+2}, U_{k+1} = i_{k+1}, U_k = i_k, ..., U_0 = i_0\right) =$$
$$\Pi_{i_{k+1} i_{k+2}} P\left(U_{k+1} = i_{k+1}, U_k = i_k, ..., U_0 = i_0\right).$$

Thus, by using the recurrence assumption for k (equation [4.32])

$$P(U_{k+2} = i_{k+2}, U_{k+1} = i_{k+1}, U_k = i_k, ..., U_0 = i_0) = \Pi_{i_0 i_1} \Pi_{i_1 i_2} ... \Pi_{i_k i_{k+1}} \Pi_{i_{k+1} i_{k+2}} P(U_0 = i_0)$$

and the relation given in equation [4.32] is verified for the value $k+1$, which completes the proof by recurrence.

3) Since

$$P(U_{k+1} = j, U_0 = i) = \sum_{i_k, i_{k-1}, ..., i_1 = 1}^{n} P(U_{k+1} = j, U_k = i_k, U_{k-1} = i_{k-1}, ..., U_0 = i),$$

we have

$$P(U_{k+1} = j, U_0 = i) = P(U_0 = i) \sum_{i_k, i_{k-1}, ..., i_1 = 1}^{n} \Pi_{i i_1} \Pi_{i_1 i_2} ... \Pi_{i_k j},$$

that is

$$P(U_{k+1} = j, U_0 = i) = \left[\mathbf{\Pi}^{k+1}\right]_{ij} P(U_0 = i).$$

4) Let us establish that

$$\forall\, k \in \mathbb{N} : P(U_{k+1} = j \mid U_0 = i) = \left[\mathbf{\Pi}^{k+1}\right]_{ij}, \forall\, i, j\ (1 \leq i, j \leq n)$$

Let us assume that $P(F) \neq 0$: we consider $E = "U_{k+1} = j"$; $F = "U_0 = i"$. Then

$$P(U_{k+1} = j \mid U_0 = i) = \frac{P(E \cap F)}{P(F)} = \left[\mathbf{\Pi}^{k+1}\right]_{ij}.$$

Now, let us assume that $P(F) = 0$: in this case

$$\Pi_{ij} = P(U_1 = j \mid U_0 = i) = 0, \forall\, j\ (1 \leq j \leq n)$$

and

$$\left[\mathbf{\Pi}^{k+1}\right]_{ij} = \sum_{i_k,i_{k-1},\ldots,i_1=1}^{n} \Pi_{ii_1} \Pi_{i_1 i_2} \ldots \Pi_{i_k j} = 0$$
$$= P(U_1 = j \mid U_0 = i), \forall j \ (1 \leq j \leq n),$$

and the equality remains valid.

5) Thus

$$E(f(U_k) \mid U_0 = i) = \sum_{j=1}^{n} f(j) P(U_k = j \mid U_0 = i) =$$
$$\sum_{j=1}^{n} \left[\mathbf{\Pi}^k\right]_{ij} B_j = \left[\mathbf{\Pi}^k \mathbf{B}\right]_i$$

6) In addition

$$\left(\mathrm{Id} - \frac{1}{\alpha}\mathbf{\Pi}\right)^{-1} = \sum_{k=0}^{+\infty} \frac{1}{\alpha^k} \mathbf{\Pi}^k$$

So

$$\sum_{k=0}^{+\infty} \frac{1}{\alpha^k} E(f(U_k) \mid U_0 = i) = \sum_{k=0}^{+\infty} \frac{1}{\alpha^k} \left[\mathbf{\Pi}^k \mathbf{B}\right]_i =$$
$$\left[\left(\sum_{k=0}^{+\infty} \frac{1}{\alpha^k} \mathbf{\Pi}^k\right) \mathbf{B}\right]_i = \left[\left(\mathrm{Id} - \frac{1}{\alpha}\mathbf{\Pi}\right)^{-1} \mathbf{B}\right]_i = X_i.$$

and the proof of the theorem is complete. ■

5

Nonlinear Algebraic Equations Involving Random Parameters

5.1. Nonlinear systems of algebraic equations

In this section, we consider systems of algebraic equations:

$$\mathbf{F}(\mathbf{X}, \mathbf{v}) = \mathbf{0}, \qquad [5.1]$$

where $\mathbf{F}: \mathcal{M}(n,1) \times \mathcal{M}(nr,1) \longrightarrow \mathcal{M}(n,1)$ is a regular function, $\mathbf{X} = (X_i) \in \mathcal{M}(n,1)$ is the vector of the unknowns, $\mathbf{v} = (v_1, ..., v_{nr})$ is a random vector. In this case, $\mathbf{X}$ is an implicit function of $\mathbf{v}$ and, so, it is also a random vector, which may be considered as a function of $\mathbf{v}$: $\mathbf{X} = \mathbf{X}(\mathbf{v})$. Our objective is the numerical determination of the variable $\mathbf{X}(\mathbf{v})$ by using expansions, as previously introduced (Chapter 3): let us consider an approximation $\mathbf{PX}$ of $\mathbf{X}$ on a convenient subspace of random variables, such that, for instance,

$$\mathbf{PX} = \sum_{k=1}^{N_X} \chi_k \varphi_k(\boldsymbol{\xi}) \quad \left(\text{i.e. } (\mathbf{PX})_j = \sum_{k=0}^{N_X} \chi_{jk} \varphi_k(\mathbf{v}) \right). \qquad [5.2]$$

In this expression, the unknown to be determined is $\chi = (\chi_{ij}) \in \mathcal{M}(n, N_X)$. $F = \{\varphi_k\}_{k \in \mathbb{N}}$ is a family conveniently chosen, such as, for instance, a *total family* of a functional space – if $\mathbf{v}$ takes its values in $\Omega \subset \mathbb{R}^{nr}$, a convenient choice is $L^2(\Omega)$. $\boldsymbol{\xi}$ is a random variable conveniently chosen and $N_X \in \mathbb{N}^*$ is the dimension of the subspace where the approximation is to be determined.

Taking $\boldsymbol{\varphi}(\boldsymbol{\xi}) = (\varphi_1(\boldsymbol{\xi}), ..., \varphi_{N_X}(\boldsymbol{\xi}))^t \in \mathcal{M}(N_X, 1)$, we have:

$$\mathbf{PX} = \chi \boldsymbol{\varphi}(\boldsymbol{\xi}). \qquad [5.3]$$

In the follwing, we apply the methods presented in the Chapter 3 and, in addition, we present also other approaches, particularly adapted to equation [5.1].

5.1.1. *Collocation*

When a sample $\mathcal{X} = (\mathbf{X}_1, \ldots, \mathbf{X}_{ns})$ of ns variates from $\mathcal{X}$ is available, we may consider the linear system:

$$\mathbf{PX}(\boldsymbol{\xi}_i) = \mathbf{X}_i, \quad i = 1, \ldots, ns. \qquad [5.4]$$

The numerical solution of these equations furnishes the coefficients χ – as previously observed, this linear system is overdetermined and an adapted method must be used, such as, for instance, a solution by minimum squares.

If, in addition, a sample $\mathcal{V} = (\mathbf{v}_1, \ldots, \mathbf{v}_{ns})$ of ns variates from $\mathbf{v}$, the natural choice is $\mathbf{X}_i = \mathbf{X}(\mathbf{v}_i)$ and $\boldsymbol{\xi}_i = \mathbf{v}_i$, for $i = 1, \ldots, ns$. The situation is analogous if the values of $\mathbf{v}_i$ may be determined from those of $\mathbf{X}_i$. However, if we have only a direct sample from $\mathbf{X}$, i.e. if the values of $\mathbf{X}_i$ are given, but the corresponding values of $\mathbf{v}_i$ cannot be determined, we may introduce an artificial random vector $\mathbf{a}$ and take $\boldsymbol{\xi} = \mathbf{a}$. As observed in Chapter 3, the results may be poor if the variables are independent. In this case, it is mandatory to create some form of dependence between the variables $\mathbf{a}$ and $\mathbf{X}$, what may be obtained, for instance, by reordering the samples in an increasing order. This approach is particularly useful namely in the situation where multiple solutions exist (see examples mentioned afterwards). It may be implemented in Matlab by using the programs of section 3.3.

EXAMPLE 5.1.– Let us consider the second degree equation:

$$F(X, v) = X^2 - 2X + v = 0,$$

where v is uniformly distributed on $(0, 1)$. This equation has as solutions:

$$X_1 = 1 - \sqrt{1 - v}, \quad X_2 = 1 + \sqrt{1 - v}.$$

Their cumulative distributions are, respectively,

$$F_1(x) = \begin{cases} 0, \text{ if } x \leq 0 \\ 1 - (x-1)^2, \text{ if } 0 < x < 1 \\ 1, \text{ otherwise.} \end{cases} ;$$

$$F_2(x) = \begin{cases} 0, \text{ if } x \leq 1 \\ (x-1)^2, \text{ if } 1 < x < 2 \\ 1, \text{ otherwise.} \end{cases} .$$

Let us assume that a sample of $ns = 11$ variates from v is given. In this case, we may generate the corresponding samples from X_1 and X_2, by taking $X_{1i} = X_1(v_i)$ and $X_{2i} = X_2(v_i)$. If such a sample is not available (for instance, if the values of v are unknown), but only the values of X_1 and X_2 are given, we may consider a random variable a uniformly distributed on $(-1, 1)$ and consider that $X_1 = X_1(a)$, $X_2 = X_2(a)$. In order to generate dependence between these variables, we may, for instance, increasingly order X_{1i} and X_{2i} and use an increasingly ordered sample a. A uniform grid $a_i = -1 + 2(i-1)/ns$ may also be used – these points are ordered in an increasing way.

The results furnished by a polynomial basis with $ns = 11$ points and a solution of equation [5.4] by minimum squares are shown in Figures 5.1 and 5.2.

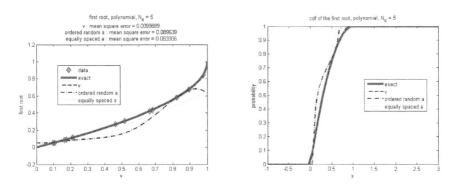

Figure 5.1. *Results for X_1 in Example 5.1 (11 random points). For a color version of the figure, see www.iste.co.uk/souzadecursi/quantification.zip*

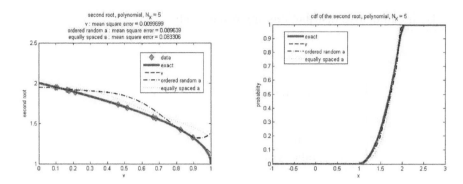

Figure 5.2. *Results for X_2 in Example 5.1 (11 random points). For a color version of the figure, see www.iste.co.uk/souzadecursi/quantification.zip*

The results may be improved if additional points – thus, additional information – are available. For instance, the results obtained for $ns = 21$ points are shown in Figures 5.3 and 5.4.

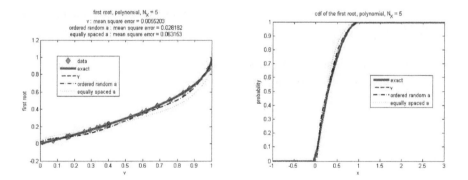

Figure 5.3. *Results for X_1 in Example 5.1 (21 random points). For a color version of the figure, see www.iste.co.uk/souzadecursi/quantification.zip*

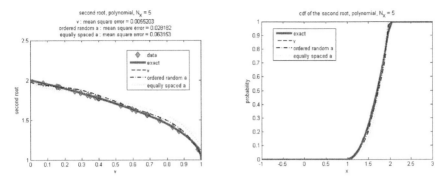

Figure 5.4. *Results for X_2 in Example 5.1 (21 random points). For a color version of the figure, see www.iste.co.uk/souzadecursi/quantification.zip*

It must be noted that the results furnished by a sample of ns equally distributed values of v_i are identical to those furnished by ns equally distributed values of a_i – this result is not surprising, since, in such a case, there exists a connection between these values: $a_i = 2v_i - 1$ and the variables a and v are each one a function of the other one. The results corresponding to this situation are exhibited in Figures 5.5 and 5.6.

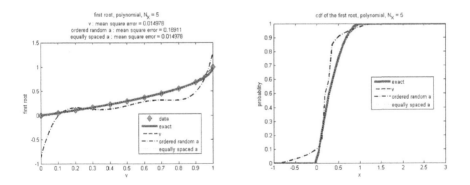

Figure 5.5. *Results for X_1 in Example 5.1 (11 equidistant points). For a color version of the figure, see www.iste.co.uk/souzadecursi/quantification.zip*

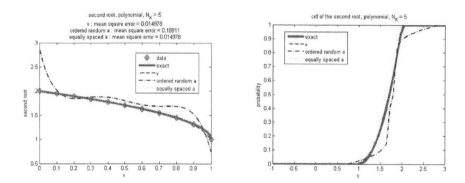

Figure 5.6. *Results for X_2 in Example 5.1 (11 equidistant points). For a color version of the figure, see www.iste.co.uk/souzadecursi/quantification.zip*

A situation having a practical interest is these where values of both X_1 and X_2 are mixed in a sample and cannot be distinguished: the sample contains values generated by both the roots and we do not have supplementary information that makes possible a separation in two samples formed by a single variable. For instance, let us consider a sample formed by 40 values, from which 20 corresponds to X_1 and 20 others to X_2. Let us consider 40 equally spaced artificial values a_i, such that $a_1 = -1$ and $a_{40} = 1$. The results furnished by the method described are shown in Figure 5.7.

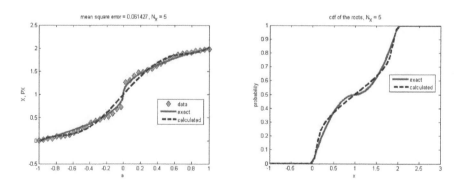

Figure 5.7. *Results for a sample where the roots are mixed – Example 5.1 (20 random points of each root). For a color version of the figure, see www.iste.co.uk/souzadecursi/quantification.zip*

5.1.2. Moment fitting

A second traditional method for the numerical exploitation of samples is the use of the *moment matching method*, presented in Chapter 3: the coefficients of the expansion are determined in such a way that the first moments of the approximation $\mathbf{M^a(PX)} = \left(M_1^a, ..., M_q^a\right)$ coincide with $\mathbf{M^e} = \left(M_1^e, ..., M_q^e\right)$, the empirical moments on the data. This may be achieved by solving the nonlinear system of equations $\mathbf{M^a(PX)} = \mathbf{M^e}$ or minimizing an objective function corresponding to a pseudo-distance $d\left(\mathbf{M^a(Px)}, \mathbf{M^e}\right)$ measuring the distance between the vectors of moments. It may be implemented in Matlab by using the programs of section 3.2.

EXAMPLE 5.2.– Let us consider the situation described in Example 5.1. Assume that it is given a sample of $ns = 21$ variates from v. Then, analogous to Example 5.1, we may generate samples from X_1 and X_2, by taking $X_{1i} = X_1(v_i)$ and $X_{2i} = X_2(v_i)$, which permits the application of the moment matching method. The results furnished by a polynomial basis are presented in Figures 5.8 and 5.9.

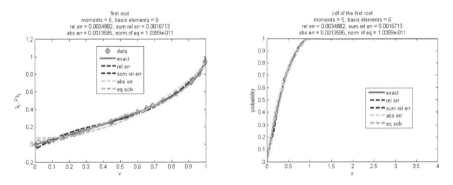

Figure 5.8. *Results for X_1 in Example 5.2 (21 random points). For a color version of the figure, see www.iste.co.uk/souzadecursi/quantification.zip*

272 Uncertainty Quantification and Stochastic Modeling with Matlab®

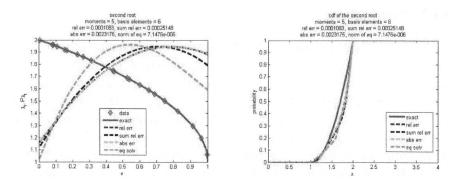

Figure 5.9. *Results for X_2 in Example 5.2 (21 random points). For a color version of the figure, see www.iste.co.uk/souzadecursi/quantification.zip*

The results obtained with a sample of equidistant points $v_i = (i-1)/ns$ are exhibited in Figures 5.10 and 5.11.

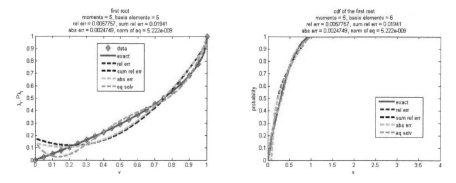

Figure 5.10. *Results for X_1 in Example 5.2 (21 equidistant points). For a color version of the figure, see www.iste.co.uk/souzadecursi/quantification.zip*

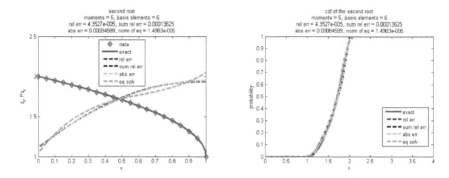

Figure 5.11. *Results for X_2 in Example 5.2 (21 equidistant points). For a color version of the figure, see www.iste.co.uk/souzadecursi/quantification.zip*

When the values of v_i are unknown, we may consider – as in Example 5.1 – an artificial variable a. For instance, let us consider a sample of 20 variates from X_1 and 20 variates from X_2, as a sample of 40 equally spaced points a_i such that $a_1 = -1$ e $a_{40} = 1$. The results are shown in Figure 5.12.

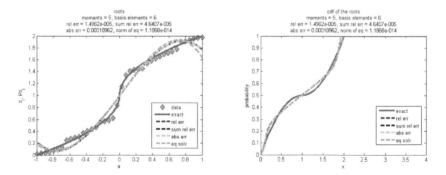

Figure 5.12. *Results for a sample of mixed roots in Example 5.1 (20 random points of each root). For a color version of the figure, see www.iste.co.uk/souzadecursi/quantification.zip*

5.1.3. *Variational approximation*

From the variational standpoint, equation [5.1] is approximated by:

$$\mathbf{F}\left(\mathbf{PX}, \mathbf{v}\right) = \mathbf{0}, \qquad [5.5]$$

Since (from equation [5.3]),

$$\mathbf{F}(\mathbf{PX}, \boldsymbol{\xi}) = \mathbf{F}(\chi\varphi(\boldsymbol{\xi}), \mathbf{v}),$$

we have:

$$\varphi(\boldsymbol{\xi})^t \mathbf{D}^t \mathbf{F}(\chi\varphi(\boldsymbol{\xi}), \mathbf{v}) = 0, \forall \mathbf{D} \in \mathcal{M}(n, N_X). \quad [5.6]$$

Since:

$$\varphi(\boldsymbol{\xi})^t \mathbf{D}^t \mathbf{F}(\chi\varphi(\boldsymbol{\xi}), \mathbf{v}) = \sum_{i=1}^{n} \sum_{m=1}^{N_X} \varphi_m(\boldsymbol{\xi}) D_{im} F_i \left(\sum_{k=1}^{N_X} \chi_k \varphi_k(\boldsymbol{\xi}), \mathbf{v} \right), \forall \mathbf{D} \in \mathcal{M}(n, N_X),$$

we have:

$$\sum_{i=1}^{n} \sum_{m=1}^{N_X} D_{im} E\left(\varphi_m(\boldsymbol{\xi}) F_i \left(\sum_{k=1}^{N_X} \chi_k \varphi_k(\boldsymbol{\xi}), \mathbf{v} \right) \right) = 0, \forall \mathbf{D} \in \mathcal{M}(n, N_X).$$

Taking $D_{im} = \delta_{ir}\delta_{ms}$, we obtain:

$$E\left(\varphi_s(\boldsymbol{\xi}) F_r \left(\sum_{k=1}^{N_X} \chi_k \varphi_k(\boldsymbol{\xi}), \mathbf{v} \right) \right) = 0, 1 \le r \le n, 1 \le s \le N_X. \quad [5.7]$$

Equation [5.7] form a system of $n \times N_X$ nonlinear equations for the $n \times N_X$ unknowns $\chi = (\chi_{ij}) \in \mathcal{M}(n, N_X)$. It must be solved by an adequate method in order to furnish the coefficients χ. It must be noticed that the construction of equation [5.7] requests the knowledge of the joint distribution of the pair $(\boldsymbol{\xi}, \mathbf{v})$ – theoretically or empirically, by means of a sample of the pair. In addition, better results are obtained by using $\boldsymbol{\xi} = \mathbf{v}$.

This approach is implemented in Matlab by a modification of the method introduced in section 3.1.5. For instance, assume that $\mathbf{F}(\mathbf{PX}, \mathbf{v})$ is furnished by the subprogram F(X,v), while the value of $\varphi_k(\mathbf{v})$ is furnished by the subprogram phi(k,v). Moreover, assume that a program eqsolver(G) is available and furnishes the solution for the nonlinear system of equations $\mathbf{G}(\mathbf{C}) = \mathbf{0}$. In practice, eqsolver(G) may involve additional parameters, such as, for instance, a starting point, a maximum number of iterations etc. In this case, the reader must adapt the code below in order to include the additional parameters. An example of code is given as following:

Listing 5.1. *UQ of nonlinear equations by variational approach*

```matlab
function chi = expcoef(F,vs,phi,eqsolver)
%
% determines the coefficients of the expansion.
%  by using a sample
%
% IN:
% F : vector of equations  - type anonymous function
% vs : table of values of v - type array of double
% vs(:,i) is a variate from X
% phi : basis function phi(k,v) - type anonymous function
% eqsolver: subprogram solving nonlinear system FC(C) = 0 -
%   type anonymous
% function
%
% OUT:
% chi : n x N_X matrix of the coefficients - type array of
%   double
%
FC = @(C) equations_for_coefficients(F,C,vs,phi,n,N_X);
Csol = eqsolver(FC);
chi = zeros(n, N_X);
for r = 1: n
    for s = 1: N_X
        alffa = index_map(r,s,n,N_X);
        chi(r,s) = Csol(alffa);
    end;
end;
return;
end
%
function w = function_proj(f,chi,vs,phi)
%
% maps f( PX(v) , v ) on the sample vs from v
% f is assumed to have the same dimension as the
% number of lines of chi
%
% IN:
% f : the function to be evaluated - type anonymous function
% chi : n x N_X matrix of the coefficients - type array of
%   double
% vs : table of values of v - type array of double
% vs(:,i) is a variate from v
% phi : basis function phi(k,v) - type anonymous function
%
% OUT:
% w : n x ns table of values of f - type array of double
```

```
%
ns = size(vs,1);
n = size(chi,1);
PXs = projection(chi,vs,phi);
w = zeros(n,ns);
for i = 1: ns
    w(:,i) = f(PXs(:,i),vs(:,i));
end;
return;
end

%
function N = iteration_variational_matrix(phi, f, chi, vs,n,N_X
    )
%
% generates the matrix N such that
%    N_rs = E(phi_s(v)f_r(X(v)))
% assumes that f furnishes a vector of length n
% and the number of lines of Xs is also n
%
% IN:
% phi : basis function phi(k,v) - type anonymous function
% f : Iteration function    - type anonymous function
% chi : n x N_X matrix of the coefficients - type array of
%     double
% vs : table of values of v - type array of double
% vs(:,i) is a variate from X
% n: number of unknowns (length of X) - type integer
% N_X : order of the expansion - type integer
%
% OUT:
% N: N_X x n table of scalar products - type array of double
%
N = zeros(N_X, n);
w = function_proj(f,chi,vs,phi);
for r = 1: n
    Y = w(r,:);
    for s = 1:N_X
        f1 = @(U) phi(s,U);
        Z = map(f1, vs, 1);
        aux = scalprod(Y,Z);
        N(r,s) = aux;
    end;
end;
return;
end
%
```

```
function B = equations_for_coefficients(F,C,vs,phi,n,N_X)
%
% evaluates the equations for the coefficients of the expansion
%
% IN:
% F : vector of equations    - type anonymous function
% C : nN_X x 1 vector of the coefficients - type array of
%     double
% vs : table of values of v - type array of double
% vs(:,i) is a variate from X
% phi : basis function phi(k,v) - type anonymous function
% n: number of unknowns  - type integer
% N_X : order of the expansion - type integer
%
% OUT:
% B : nN_X x 1 mvector of the equations - type array of double
%
chi = zeros(n, N_X);
for r = 1: n
    for s = 1: N_X
        alffa = index_map(r,s,n,N_X);
        chi(r,s) = C(alffa);
    end;
end;
N = iteration_variational_matrix(phi, F, chi, vs,n,N_X);
B = zeros(size(C));
for r = 1: n
    for s = 1: N_X
        alffa = index_map(r,s,n,N_X);
        B(alffa) = N(r,s);
    end;
end;
return;
end
%
function v = index_map(i,j,n,N_X)
v = i + (j-1)*n;
return;
end
```

EXAMPLE 5.3.– Let us consider the situation presented in Example 5.1. We may determine an approximation under the form of a polynomial function with $\xi = v$ and numerically solve equation [5.6] by the Newton–Raphson iterations. The results obtained are shown in Figures 5.13–5.15. In the first figure, the means have been evaluated by numerical integration. In the second figure, we consider 21 equally spaced values of v. The last figure corresponds to a sample of 21 random realizations of v.

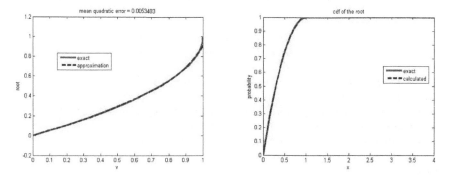

Figure 5.13. *Results obtained in Example 5.3 (numerical integration). For a color version of the figure, see www.iste.co.uk/souzadecursi/quantification.zip*

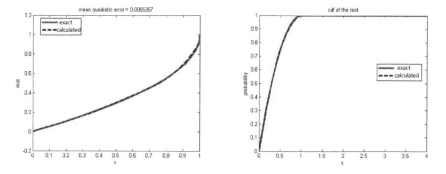

Figure 5.14. *Results obtained in Example 5.3 (equidistant points). For a color version of the figure, see www.iste.co.uk/souzadecursi/quantification.zip*

The use of a variable $\xi \neq v$ is illustrated in Figures 5.16 and 5.17: we consider, on the one hand, a sample of $ns = 21$ variates from v and 21 points a_i, equally spaced on $(-1, 1)$, such that $a_1 = -1$ and $a_{21} = 1$. The results obtained by using $\xi = a$ and a polynomial function of degree 5 are shown in Figure 5.16, while the results for a polynomial function of degree 8 are shown in Figure 5.17.

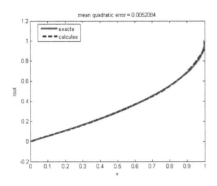 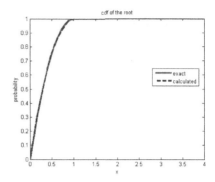

Figure 5.15. *Results obtained in Example 5.3 (random points). For a color version of the figure, see www.iste.co.uk/souzadecursi/quantification.zip*

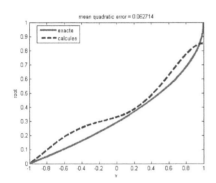 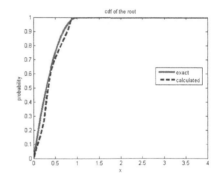

Figure 5.16. *Results obtained in Example 5.3 ($\xi \neq v$, sample of 21 variates). For a color version of the figure, see www.iste.co.uk/souzadecursi/quantification.zip*

5.1.4. *Adaptation of iterative methods*

Let us consider an iterative method for the numerical solution of equation [5.1], having as iterating function Ψ: the method generates a sequence $\left\{ \mathbf{X}^{(p)} \right\}_{p \geq 0}$, starting from the initial point $\mathbf{X}^{(0)}$ and verifying:

$$\mathbf{X}^{(p+1)} = \Psi\left(\mathbf{X}^{(p)}\right). \tag{5.8}$$

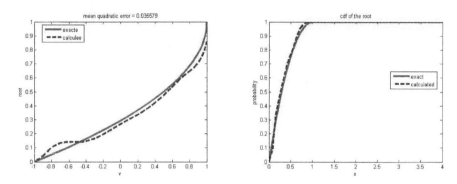

Figure 5.17. *Results obtained in Example 5.3 ($\xi \neq v$, sample of 21 points). For a color version of the figure, see www.iste.co.uk/souzadecursi/quantification.zip*

Such a method may be adapted to the determination of **PX**. Let us introduce:

$$\mathbf{PX}^{(p)} = \sum_{k=1}^{N_X} \chi_k^{(p)} \varphi_k(\boldsymbol{\xi}) \quad \left(\text{i.e. } \left(\mathbf{PX}^{(p)}\right)_j = \sum_{k=0}^{N_X} \chi_{jk}^{(p)} \varphi_k(\mathbf{v}) \right). \quad [5.9]$$

Then, the iterations may be approximated as follows:

$$\mathbf{PX}^{(p+1)} = \boldsymbol{\Psi}\left(\mathbf{PX}^{(p)}\right). \quad [5.10]$$

let us adopt the variational standpoint; we have:

$$\varphi(\boldsymbol{\xi})^t \mathbf{D}^t \mathbf{PX}^{(p+1)} = \varphi(\boldsymbol{\xi})^t \mathbf{D}^t \boldsymbol{\Psi}\left(\mathbf{PX}^{(p)}\right), \forall \mathbf{D} \in \mathcal{M}(n, N_X).$$

and

$$\sum_{k,m=1}^{N_X} \sum_{i=1}^{n} D_{im} \varphi_m(\boldsymbol{\xi}) \chi_{ik}^{(p+1)} \varphi_k(\boldsymbol{\xi}) =$$
$$\sum_{m=1}^{N_X} \sum_{i=1}^{n} D_{im} \varphi_m(\boldsymbol{\xi}) \Psi_i \left(\sum_{k=1}^{N_X} \chi_k^{(p)} \varphi_k(\boldsymbol{\xi}) \right),$$
$$1 \leq i \leq n, 1 \leq m \leq N_X.$$

So,

$$\sum_{k,m=1}^{N_X}\sum_{i=1}^{n} D_{im} E\left(\varphi_m(\boldsymbol{\xi})\chi_{ik}^{(p+1)}\varphi_k(\boldsymbol{\xi})\right) =$$
$$\sum_{m=1}^{N_X}\sum_{i=1}^{n} D_{im} E\left(\varphi_m(\boldsymbol{\xi})\Psi_i\left(\sum_{k=1}^{N_X}\chi_k^{(p)}\varphi_k(\boldsymbol{\xi})\right)\right),$$
$$1 \leq i \leq n, 1 \leq m \leq N_X.$$

Taking $D_{im} = \delta_{ir}\delta_{ms}$ in this equality, we have, for, $1 \leq r \leq n, 1 \leq s \leq N_X$

$$\sum_{k}^{N_X} E\left(\varphi_s(\boldsymbol{\xi})\varphi_k(\boldsymbol{\xi})\right)\chi_{rk}^{(p+1)} = E\left(\varphi_s(\boldsymbol{\xi})\Psi_r\left(\sum_{k=1}^{N_X}\chi_k^{(p)}\varphi_k(\boldsymbol{\xi})\right)\right). \quad [5.11]$$

These equations form a linear system for the unknowns $\chi_{rk}^{(p+1)}$; let us denote:

$$\mathcal{A}_{rsjk} = \delta_{jr} E\left(\varphi_s(\boldsymbol{\xi})\varphi_k(\boldsymbol{\xi})\right), \ \mathcal{B}_{rs}$$
$$= E\left(\varphi_s(\boldsymbol{\xi})\Psi_r\left(\sum_{k=1}^{N_X}\chi_k^{(p)}\varphi_k(\boldsymbol{\xi})\right)\right) \quad [5.12]$$

Then,

$$\sum_{j=1}^{n}\sum_{k=1}^{N_X}\mathcal{A}_{rsjk}\chi_{jk}^{(p+1)} = \mathcal{B}_{rs}, 1 \leq r \leq n, 1 \leq s \leq N_X.$$

Let us redefine the indexes by:

$$ind(j,k) = (k-1)n + j$$

and set $\mathbf{A} = (A_{\alpha\beta}) \in \mathcal{M}(nN_X, nN_X)$, $\mathbf{C} = (C_\beta) \in \mathcal{M}(nN_X, 1)$, $\mathbf{B} = (B_\alpha) \in \mathcal{M}(nN_X, 1)$ given by:

$$A_{\alpha\beta} = \mathcal{A}_{rsjk}, B_\alpha = \mathcal{B}_{rs}, C_\beta = \chi_{jk}^{(p+1)},$$
$$\alpha = ind(r,s), \beta = ind(j,k). \quad [5.13]$$

then:

$$\mathbf{AC} = \mathbf{B}. \tag{5.14}$$

The solution of this linear system determines $\mathbf{C}$ and, so, $\chi^{(p+1)}$. In practice, the solution may be determined by solving n linear systems for N_X unknowns and a fixed matrix $M \in \mathcal{M}(N_X, N_X)$: let $\mathbf{N^r} \in \mathcal{M}(N_X, 1)$ and $\mathbf{C^r} \in \mathcal{M}(N_X, 1)$ be such that:

$$M_{sk} = E\left(\varphi_s\left(\boldsymbol{\xi}\right)\varphi_k\left(\boldsymbol{\xi}\right)\right),$$
$$N_s^r = E\left(\varphi_s\left(\boldsymbol{\xi}\right)\Psi_r\left(\sum_{k=1}^{N_X}\chi_k^{(p)}\varphi_k\left(\boldsymbol{\xi}\right)\right)\right), C_k^r = \chi_{rk}^{(p+1)} \tag{5.15}$$

then,

$$\mathbf{MC^r} = \mathbf{N^r}, 1 \leq r \leq n. \tag{5.16}$$

The solution of equation [5.16] for a given r furnishes the values of χ_{rk}, for $1 \leq k \leq N_X$.

In many practical situations, the iteration function reads as:

$$\boldsymbol{\Psi}\left(\mathbf{X}\right) = \mathbf{X} + \boldsymbol{\Phi}\left(\mathbf{X}\right).$$

In this case, the iterations [5.8] take the form:

$$\mathbf{X}^{(p+1)} = \mathbf{X}^{(p)} + \boldsymbol{\Delta}\mathbf{X}^{(p)} ; \boldsymbol{\Delta}\mathbf{X}^{(p)} = \boldsymbol{\Phi}\left(\mathbf{X}^{(p)}\right). \tag{5.17}$$

and we have:

$$\chi^{(p+1)} = \chi^{(p)} + \boldsymbol{\Delta}\chi^{(p)}, \tag{5.18}$$

where $\boldsymbol{\Delta}\chi^{(p)}$ is determined by solving the linear system $\mathbf{A}\boldsymbol{\Delta}\mathbf{C} = \boldsymbol{\Delta}\mathbf{B}$, analogous to [5.14], with Φ replacing Ψ in [5.12] (recall that $\beta = ind(j,k)$):

$$\Delta\mathcal{B}_{rs} = E\left(\varphi_s\left(\boldsymbol{\xi}\right)\Phi_r\left(\sum_{k=1}^{N_X}\chi_k^{(p)}\varphi_k\left(\boldsymbol{\xi}\right)\right)\right), \Delta B_\alpha$$
$$= \Delta\mathcal{B}_{rs}, \Delta C_\beta = \Delta\chi_{jk}^{(p+1)} \tag{5.19}$$

Analogous to [5.16], we may determine $\Delta \chi^{(p)}$ by solving n linear systems $\mathbf{M}\Delta\mathbf{C}^r = \Delta\mathbf{N}^r$, analogous to [5.16], with Φ replacing Ψ in [5.15]:

$$\Delta N_s^r = E\left(\varphi_s(\boldsymbol{\xi})\,\Phi_r\left(\sum_{k=1}^{N_X}\chi_k^{(p)}\varphi_k(\boldsymbol{\xi})\right)\right),\ \Delta C_k^r = \Delta\chi_{rk}^{(p+1)}. \quad [5.20]$$

This approach is implemented as follows: assume that, on the one hand, the value of $\varphi_k(\mathbf{v})$ is furnished by the subprogram phi(k,v) while, on the other hand, the iteration function $\Phi(\mathbf{X},\mathbf{v})$ is evaluated by a subprogram Phiiter (X,v). The first step is the evaluation of the means involved in the equations presented. Such a means correspond to scalar products and may be evaluated by integration or using a sample. Here, we assume that a sample of ns variates from $\mathbf{v}$ is available. Then:

Listing 5.2. *Adaptation of an iterative method*

```
function chi = expcoef(f_iter,chi_ini,vs,phi,nitmax,errmax)
%
%   determines the coefficients of the expansion.
%     by using a sample
%
% IN:
% f_iter : Iteration function    - type anonymous function
% chi_ini : initial n x N_X matrix of the coefficients - type
          array of double
% vs : table of values of v - type array of double
% vs(:,i) is a variate from X
% phi : basis function phi(k,v) - type anonymous function
% nitmax = max iteration number - type integer
% errmax = max precision - type double
%
% OUT:
% chi : n x N_X matrix of the coefficients - type array of
       double
%
M = fixed_variational_matrix(phi, vs);
A = tab4(M,n,N_X);
chi = chi_ini;
nit = 0;
not_stop = 1;
while not_stop
    nit = nit + 1;
    delta_chi = iteration_sample(f_iter,chi,vs,phi,A);
    chi = chi + delta_chi;
    err = norm(delta_chi);
```

```
        not_stop = nit < nitmax && err > errmax;
end;
return;
end
%
function w = function_proj(f,chi,vs,phi)
%
%  maps f( PX(v) , v ) on the sample vs from v
%  f is assumed to have the same dimension as the
%  number of lines of chi
%
% IN:
% f : the function to be evaluated - type anonymous function
% chi : n x N_X matrix of the coefficients - type array of
        double
% vs : table of values of v - type array of double
% vs(:,i) is a variate from v
% phi : basis function phi(k,v) - type anonymous function
%
% OUT:
% w : n x ns table of values of f - type array of double
%
ns = size(vs,1);
n = size(chi,1);
PXs = projection(chi,vs,phi);
w = zeros(n,ns);
for i = 1: ns
    w(:,i) = f(PXs(:,i),vs(:,i));
end;
return;
end
%
function A = tab4(M,n,N_X)
%
%  generates the table A from the table of
%  scalar products of the basis functions
%    M_ij = E( phi_i(v)phi_j(v) )
%
% IN:
% M: N_X x N_X table of scalar products - type array of double
% n: number of unknowns (length of X) - type integer
% N_X : order of the expansion - type integer
%
% OUT:
% A = nN_X x nN_X table - type array of double
% contains A(alpha, beta)
%
```

```
aaaa = zeros(n,N_X,n,N_X);
for r = 1: n
    for s = 1: N_X
        for j = 1: n
            for k=s:N_X
                if r == j
                    aux = M(s,k);
                    aaaa(r,s,j,k) = aux;
                    aaaa(j,k,r,s) = aux;
                end;
            end;
        end;
    end;
end;
nn = n*N_X;
A = zeros(nn, nn);
for r = 1: n
    for s = 1: N_X
        alffa = index_map(r,s);
        for j = 1: n
            for k = 1: N_X
                betta = index_map(j,k,n,N_X);
                A(alffa,betta) = aaaa(r,s,j,k);
            end;
        end;
    end;
end;
return;
end

%
function B = tab2(N,n,N_X)
%
% generates the table B from table N
%    N_rs = E(phi_s(v)f_r(PX(v)))
%
% IN:
% N: N_X x n table of scalar products — type array of double
% n: number of unknowns (length of X) — type integer
% N_X : order of the expansion — type integer
%
% OUT:
% B = nN_X x 1 table — type array of double
% contains B(alpha)
%
nn = n*N_X;
B = zeros(nn, 1);
```

```
        for r = 1: n
            for s = 1: N_X
                alffa = index_map(r,s,n,N_X);
                B(alffa) = N(r,s);
            end;
        end;
        return;
    end
%
function M = fixed_variational_matrix(phi, vs)
%
% generates the matrix M such that
%    M_ij = E(phi_i(v)phi_j(v))
%
% IN:
% phi : basis function phi(k,v) - type anonymous function
% vs : table of values of v - type array of double
% vs(:,i) is a variate from X
%
% OUT:
% M: N_X x N_X table of scalar products - type array of double
%
    M = zeros(N_X, N_X);
    for i = 1: N_X
        f1 = @(U) phi(i,U);
        Y = map(f1, vs, 1);
        A(i,i) = scalprod(Y,Y);
        for j = i+1:N_X
            f1 = @(U) phi(j,U);
            Z = map(f1, vs, 1);
            aux = scalprod(Y,Z);
            M(i,j) = aux;
            M(j,i) = aux;
        end;
    end;
    return;
end
%
function N = iteration_variational_matrix(phi, f, chi, vs,n,N_X
    )
%
% generates the matrix N such that
%    N_rs = E(phi_s(v)f_r(X(v)))
% assumes that f furnishes a vector of length n
% and the number of lines of Xs is also n
%
% IN:
```

```
% phi : basis function phi(k,v) − type anonymous function
% f : Iteration function − type anonymous function
% chi : n x N_X matrix of the coefficients − type array of
    double
% vs : table of values of v − type array of double
% vs(:,i) is a variate from X
% n: number of unknowns (length of X) − type integer
% N_X : order of the expansion − type integer
%
% OUT:
% N: N_X x n table of scalar products − type array of double
%
N = zeros(N_X, n);
w = function_proj(f,chi,vs,phi);
for r = 1: n
    Y = w(r,:);
    for s = 1:N_X
        f1 = @(U) phi(s,U);
        Z = map(f1, vs, 1);
        aux = scalprod(Y,Z);
        N(r,s) = aux;
    end;
end;
return;
end
%
function delta_chi = iteration_sample(f_iter,chi_old,vs,phi,A)
%
% evaluates the variation of the coefficients
%
% IN:
% f_iter : Iteration function − type anonymous function
% chi_old : n x N_X matrix of the coefficients − type array of
    double
% vs : table of values of v − type array of double
% vs(:,i) is a variate from X
% phi : basis function phi(k,v) − type anonymous function
% A = nN_X x nN_X table − type array of double
% contains A(alpha, beta)
%
% OUT:
% delta_chi : n x N_X matrix of the coefficients − type array
    of double
% chi = chi_old + delta_chi
%
n = size(chi_old,1);
N_X = size(chi_old,2);
```

```
N = iteration_variational_matrix(phi, f_iter, chi_old, vs,n,N_X
    );
B = tab2(N,n,N_X);
delta_C = A \B;
delta_chi = zeros(size(chi_old));
for r = 1: n
    for s = 1: N_X
        alffa = index_map(r,s,n,N_X);
        delta_chi(r,s) = delta_C(alffa);
    end;
end;
return;
end
%
function v = index_map(i,j,n,N_X)
v = i + (j-1)*n;
return;
end
```

EXAMPLE 5.4.– Let us consider again the situation presented in Example 5.1. When using Newtons iterations for the solution of the second degree equation, the iteration function is:

$$\Phi(X, v) = \frac{X^2 - 2X + v}{2X - 2},$$

We show in Figure 5.18 the results obtained after 100 iterations, starting from $\chi_{jk}^{(0)} = 1$, $\forall j, k$. In these calculations, we use an approximation by a polynomial function of degree 5 and the means are estimated by using $ns = 21$ equally spaced points v_i.

In Figure 5.18, we exhibit the results furnished by 100 iterations starting from $\chi_{jk}^{(0)} = 0$, $\forall j, k$. In these calculations, the means have been evaluated by numerical integration using the internal function *quad* of Matlab.

5.2. Numerical solution of noisy deterministic systems of nonlinear equations

In this section, we consider the solution of deterministic algebraic systems of nonlinear equations:

$$\mathbf{F}(\mathbf{x}) = \mathbf{0},$$

where $\mathbf{F}\colon \mathcal{M}(n,1) \longrightarrow \mathbb{R}^n$ is function corresponding to the vector of equations and $\mathbf{x} = (x_i) \in \mathcal{M}(n,1)$ is the vector of unknowns. We assume that these equations are deterministic and do not contain random variables: $\mathbf{x}$ is a vector of real numbers.

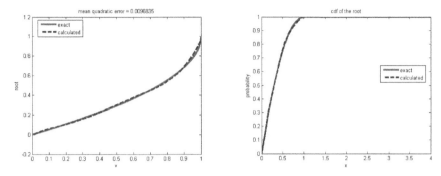

Figure 5.18. *Results obtained in Example 5.4 ($ns = 21$ equidistributed points v_i). For a color version of the figure, see www.iste.co.uk/ souzadecursi/quantification.zip*

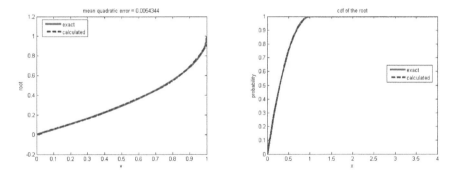

Figure 5.19. *Results obtained in Example 5.4 (numerical integration). For a color version of the figure, see www.iste.co.uk/souzadecursi/quantification.zip*

We are mainly interested in the situations where the evaluation of $\mathbf{F}(\mathbf{x})$ involves errors or noise: for instance, when only approximations of $\mathbf{F}(\mathbf{x})$ may be constructed. In such a situation, standard methods, for instance Newton–Raphson, may fail, since the errors in the evaluation of $\mathbf{F}(\mathbf{x})$ and its derivatives may introduce spurious oscillations or divergence. In the framework of stochastic methods, an alternative is the use of the algorithm of

Robbins–Monro, which may be considered as an extension of the simple fixed point iterations

$$\mathbf{x}^{(k+1)} = \mathbf{x}^{(k)} - \theta \mathbf{F}(\mathbf{x}^{(k)}), \theta > 0, \mathbf{x}^{(0)} \text{ given}.$$

Such a simple procedure of relaxation may be generalized by the introduction of a different θ at each iteration k:

$$\mathbf{x}^{(k+1)} = \mathbf{x}^{(k)} - \theta_k \mathbf{F}(\mathbf{x}^{(k)}), \theta_k > 0, \mathbf{x}^{(0)} \text{ prescribed}.$$

When the sequence $\{\theta_k\}_{k \in \mathbb{N}}$ verifies:

$$\theta_k > 0, \forall\, k \in \mathbb{N}\,;\; \sum_{i=0}^{+\infty} \theta_i = +\infty\,;\; \sum_{i=0}^{+\infty} \theta_i^2 < +\infty, \qquad [5.21]$$

the method is referred to as being the algorithm of *Robbins–Monro*. These conditions may be satisfied simply using:

$$\theta_k = \frac{\theta}{a + b(k+1)}\,, \; a, b, \theta > 0.$$

This algorithm has been initially used in order to solve equations involving random variables. Let us consider the iterations:

$$\mathbf{x}^{(k+1)} = \mathbf{x}^{(k)} - \theta_k \mathbf{Y}^{(k)},$$

where $\mathbf{Y}^{(k)}$ is an approximation of $\mathbf{F}(\mathbf{x}^{(k)})$. By assuming that the error $\varepsilon^{(k)} = \mathbf{F}(\mathbf{x}^{(k)}) - \mathbf{Y}^{(k)}$ is a centered random variable, independent from $\{\mathbf{x}^{(0)}, ..., \mathbf{x}^{(k-1)}, \mathbf{Y}^{(0)}, ..., \mathbf{Y}^{(k-1)}\}$, the algorithm converges to x.

The convergence is established by the following theorem:

THEOREM 5.1.– let $(,)$ be the usual scalar product of $\mathbb{R}^n$. Assume that:

1) $\mathbf{F} : \mathbb{R}^n \longrightarrow \mathbb{R}^n$ is continuous and bounded:

$$\sup\{\|\mathbf{F}(\mathbf{x})\| : x \in \mathbb{R}^n\} = M < \infty\,;$$

2) There exists $\mathbf{x} \in \mathbb{R}^n$ such that $\mathbf{F}(\mathbf{x}) = 0$;

3) Let, for $\delta > 0$, $A(\mathbf{x}, \delta) = \{\mathbf{y} \in \mathbb{R} : \|\mathbf{y} - \mathbf{x}\| \geq \delta\}$. Assume that there exists a function $m : \mathbb{R} \longrightarrow \mathbb{R}$, such that:

4) The sequence $\{\theta_k\}_{k \in \mathbb{N}}$ verifies [5.21].

Under these assumptions, the sequence $\{\mathbf{x}^{(k)}\}_{k \in \mathbb{N}}$ converges to $\mathbf{x} : \mathbf{x}^{(k)} \longrightarrow \mathbf{x}$ for $k \longrightarrow +\infty$. ∎

We observe that the condition (iii) implies the uniqueness of the solution, since $\mathbf{F}(\mathbf{y}) \neq \mathbf{F}(\mathbf{x})$ for $\mathbf{y} \neq \mathbf{x}$.

PROOF.–

1) Let $\mathbf{T}^{(k)} = \mathbf{x}^{(k)} - \mathbf{x}$. Then:

$$\mathbf{T}^{(k+1)} = \mathbf{T}^{(k)} - \theta_k \left(\mathbf{F}\left(\mathbf{x}^{(k)}\right) - \mathbf{F}(\mathbf{x}) \right)$$

and

$$\left\| \mathbf{T}^{(k+1)} \right\|^2 = \left\| \mathbf{T}^{(k)} \right\|^2 + \theta_k^2 \left\| \mathbf{F}\left(\mathbf{x}^{(k)}\right) \right\|^2 - 2\theta_k \left(\mathbf{T}^{(k)}, \mathbf{F}\left(\mathbf{x}^{(k)}\right) \right).$$

Let:

$$S = \sum_{k=0}^{+\infty} \theta_k^2$$

and

$$z_k = \left\| \mathbf{T}^{(k)} \right\|^2 + M^2 S - \sum_{i=0}^{k-1} \theta_i^2 \left\| \mathbf{F}\left(\mathbf{x}^{(i)}\right) \right\|^2$$

The assumptions on f show that:

$$0 < \sum_{i=0}^{k-1} \theta_i^2 \left\| \mathbf{F}\left(\mathbf{x}^{(i)}\right) \right\|^2 \leq M^2 \sum_{i=0}^{k-1} \theta_i^2 \leq M^2 S,$$

which implies the existence of a real number $A > 0$ such that:

$$0 < A \leq M^2 S \text{ and } \sum_{i=0}^{+\infty} \theta_i^2 \left\| \mathbf{F}\left(\mathbf{x}^{(i)}\right) \right\|^2 = A.$$

In addition,

$$z_k \geq 0, \forall\, k \in \mathbb{N}.$$

2) We also have:

$$z_{k+1} = \left\| \mathbf{T}^{(k+1)} \right\|^2 + M^2 S - \sum_{i=0}^{k} \theta_i^2 \left\| \mathbf{F}\left(\mathbf{x}^{(i)}\right) \right\|^2.$$

So,

$$z_{k+1} - z_k = \left\| \mathbf{T}^{(k+1)} \right\|^2 - \left\| \mathbf{T}^{(k)} \right\|^2 - \theta_k^2 \left\| \mathbf{F}\left(\mathbf{x}^{(k)}\right) \right\|^2$$
$$= -2\theta_k \left(\mathbf{T}^{(k)}, \mathbf{F}\left(\mathbf{x}^{(k)}\right) \right)$$

and we have:

$$z_{k+1} = z_k - 2\theta_k \left(\mathbf{T}^{(k)}, \mathbf{F}\left(\mathbf{x}^{(k)}\right) \right).$$

Since,

$$\left(\mathbf{T}^{(k)}, \mathbf{F}\left(\mathbf{x}^{(k)}\right) \right) = \left(\mathbf{T}^{(k)}, \mathbf{F}\left(\mathbf{x}^{(k)}\right) - \mathbf{F}(\mathbf{x}) \right)$$
$$= \left(\mathbf{x}^{(k)} - \mathbf{x}, \mathbf{F}\left(\mathbf{x}^{(k)}\right) - \mathbf{F}(\mathbf{x}) \right) \geq m\left(\left\| \mathbf{x}^{(k)} - \mathbf{x} \right\| \right) \geq 0,$$

we obtain:

$$z_{k+1} \leq z_k, \forall\, k \in \mathbb{N}.$$

3) Thus, $\{z^{(k)}\}_{k \in \mathbb{N}}$ is decreasing and bounded from below. There exists a real number $z \geq 0$ such that:

$$z^{(k)} \longrightarrow z \text{ for } k \longrightarrow +\infty.$$

4) Consequently,

$$\left\|\mathbf{T}^{(k)}\right\|^2 = z_k - M^2 S + \sum_{i=0}^{k-1} \theta_i^2 \left\|\mathbf{F}\left(\mathbf{x}^{(i)}\right)\right\|^2 \longrightarrow B = z - M^2 S + A.$$

Since $\left|T^{(k)}\right|^2 \geq 0, \forall\, k \in \mathbb{N}$, we have $B \geq 0$.

5) Assume that $B > 0$. Then, there exists $k_0 > 0$ such that:

$$k \geq k_0 \Longrightarrow \left\|\mathbf{T}^{(k)}\right\|^2 \geq \frac{B}{2} \Longrightarrow \left\|\mathbf{T}^{(k)}\right\| \geq \sqrt{\frac{B}{2}} > 0.$$

So,

$$k \geq k_0 \Longrightarrow \left(\mathbf{T}^{(k)}, \mathbf{F}\left(\mathbf{x}^{(k)}\right)\right) = \left(\mathbf{T}^{(k)}, \mathbf{F}\left(\mathbf{x}^{(k)}\right) - \mathbf{F}(\mathbf{x})\right)$$
$$= \left(\mathbf{x}^{(k)} - \mathbf{x}, \mathbf{F}\left(\mathbf{x}^{(k)}\right) - \mathbf{F}(\mathbf{x})\right) \geq m\left(\sqrt{\frac{B}{2}}\right) > 0$$

and, by taking

$$\lambda = 2m\left(\sqrt{\frac{B}{2}}\right) > 0,$$

we have:

$$k \geq k_0 \Longrightarrow z_{k+1} = z_k - 2\theta^{(k)}\left(\mathbf{T}^{(k)}, \mathbf{F}\left(\mathbf{x}^{(k)}\right)\right) \leq z_k - \lambda \theta_k.$$

This inequality shows that:

$$\forall\, k > k_0 : z_k - z_{k_0} = \sum_{i=k_0}^{k-1} (z_{i+1} - z_i) \leq -\lambda \sum_{i=k_0}^{k-1} \theta_i.$$

Since,

$$z_{k+1} \leq z_k \leq z_0, \forall\, k \in \mathbb{N},$$

this last inequality implies that:

$$\forall\, k > k_0 : 0 \leq \sum_{i=k_0}^{k-1} \theta_i \leq \frac{z_{k_0} - z_k}{\lambda} \leq \frac{z_0 - z_k}{\lambda} \leq \frac{z_0 - z}{\lambda}.$$

6) But in this case:

$$\sum_{i=0}^{+\infty} \theta_i < +\infty,$$

which is in contradiction with the assumptions. So, $B = 0$ and the proof is complete. ∎

Matlab implementation is performed as follows.

Listing 5.3. *Robbins-Monro*

```
function X = Robbins_Monro_solution(f_iter,X_ini,theta_0,a,b,
    nitmax,errmax)
%
%   determines X by Robbins-Monro iteraations of function f_iter
%   assumes that f_iter and X_ini have same dimension
%
% IN:
% f_iter : Iteration function   - type anonymous function
% X_ini : initial vector of the coefficients - type array of
    double
% a,b,theta_ini : coefficients - type double
% nitmax = max iteration number - type integer
% errmax = max precision - type double
%
% OUT:
% X : vector containing the solution - type array of double
%
X = X_ini;
nit = 0;
not_stop = 1;
while not_stop
    nit = nit + 1;
    theta_k = theta_fading(nit,a,b,theta_0);
    f_k = f_iter(X);
    X = X + theta_k*f_k;
    err = norm(f_k);
    not_stop = nit < nitmax && err > errmax;
```

```
end;
return;
end

function theta = theta_fading(k,a,b,theta_0)
%
%   determines the coefficient of Robbins-Monro iteration
%
% IN:
% k : Iteration number    - type integer
% a,b,theta_0 : coefficients - type double
%
% OUT:
% theta : coefficient - type double
%
theta = theta_0/(a + b*(k+1));
return;
end
```

6

Differential Equations Under Uncertainty

From the mathematical standpoint, differential equations are essential equations. The methods presented in Chapters 4 and 5 may be used for uncertainty quantification in differential equations. The main difficulty lies in the number of unknowns to be determined.

From the formal standpoint, we must keep in mind that the solutions of differential equations are not elements from a finite-dimensional space, but vectors from an infinite-dimensional one: while the solution of the linear system $\mathbf{AX} = \mathbf{B}$ (with $\mathbf{A} \in \mathcal{M}(n,n)$ and $\mathbf{B} \in \mathcal{M}(n,1)$) is a vector $\mathbf{X} \in \mathcal{M}(n,1)$, the solution of the differential equation $x' = ax$ on $(0,T)$, $x(0) = x_0$ is a function $x(t) = x_0 \exp(at)$. Likewise, while the determination of $\mathbf{X}$ consists of the determination of n real numbers, the determination of x expresses the determination of $x(t)$ for each $t \in (0,T)$ – i.e. the determination of infinitely many real numbers. This specificity has a significant impact on the complexity of the calculations connected to the problem of uncertainty quantification: in the case of a finite-dimensional linear system $\mathbf{AX} = \mathbf{B}$ where $\mathbf{A} = \mathbf{A}(\mathbf{v})$ and $\mathbf{B} = \mathbf{B}(\mathbf{v})$, we must determine $\mathbf{PX} = \chi\varphi(\xi)$, with $\chi \in \mathcal{M}(n, N_X)$ (see Chapter 4); in the case of a differential equation $x' = ax$ on $(0,T)$, $x(0) = x_0$ with $a = a(\mathbf{v})$, we must determine $Px = \chi(t)\varphi(\xi)$, where $(0,T) \longrightarrow \chi(t) \in \mathcal{M}(1, N_X)$ is an application – i.e. $\chi(t)$ must be determined for infinitely many values of t.

In practice, differential equations are often solved by using *discretizations* that introduce finite-dimensional approximations and a finite number of unknown values, such as finite elements, etc. In this case, the complete determination of $x(t)$ is avoided and we limit to the determination of approximated values $x_1 \approx x(t_1), ..., x_n \approx x(t_n)$ $(0 < t_1 < ... < t_n = T)$: the unknown becomes $\mathbf{X} = (x_i) \in \mathcal{M}(n, 1)$, which is the solution of a system of algebraical equations $\mathbf{F}(\mathbf{X}) = \mathbf{0}$. Such a system may be linear or nonlinear, according to the nature of the original equation and the method of discretization used. In both cases, the methods based on expansions previously presented may be used. Some discretization schemes lead to iterative or progressive methods, which lead naturally to the approach presented in section 5.1.4.

The main difficulty arises, as mentioned above, from the so-called *curse of dimensionality*: the number of steps n may be large (eventually very large). Analogously, N_X may be large, since it grows rapidly with the number of random variables used in the approximation. So, the number of unknowns nN_X may grow excessively and prevent from effective numerical calculation – which is a practical limitation of the methods under consideration.

6.1. The case of linear differential equations

The remarks previously stated show that uncertainty quantification in linear differential equations is closely connected to uncertainty quantification in algebraic linear systems. For instance, let us consider the following ordinary differential equation:

$$\frac{d\mathbf{X}}{dt} = \mathbf{A}\mathbf{X} + \mathbf{B}, \ \mathbf{X}(0) = \mathbf{X}_0, \quad [6.1]$$

where $\mathbf{X}: (0, T) \longrightarrow \mathcal{M}(n, 1)$ is the unknown to be determined, $\mathbf{A} = \mathbf{A}(\mathbf{v}) \in \mathcal{M}(n, n)$, $\mathbf{B} = \mathbf{B}(\mathbf{v}) \in \mathcal{M}(n, 1)$, $\mathbf{X}_0 = \mathbf{X}_0(\mathbf{v}) \in \mathcal{M}(n, 1)$, $\mathbf{v} = (v_1, ..., v_{nr})$ is a random vector that models the uncertainties. In an analogous manner to those presented in section 4.1 (see Chapter 4), we look for an approximation $\mathbf{PX}$ in:

$$S = [\{\varphi_1(\boldsymbol{\xi}), ..., \varphi_{N_X}(\boldsymbol{\xi})\}]^n$$
$$= \left\{\sum_{k=1}^{N_X} \mathbf{D}_k \varphi_k(\boldsymbol{\xi}) : \mathbf{D}_k \in \mathbb{R}^n, 1 \leq k \leq N_X\right\}, \quad [6.2]$$

i.e.:

$$\mathbf{X} \approx \mathbf{PX} = \sum_{k=1}^{p} \chi_k \varphi_k(\boldsymbol{\xi}) \quad \left(\text{i.e. } (\mathbf{PX})_j = \sum_{k=0}^{+\infty} \chi_{jk} \varphi_k(\mathbf{v}) \right), \quad [6.3]$$

where $\boldsymbol{\xi}$ a convenient random variable, $F = \{\varphi_k\}_{k \in \mathbb{N}}$ is a conveniently chosen family of functions, $\varphi(\boldsymbol{\xi}) = (\varphi_1(\boldsymbol{\xi}), ..., \varphi_{N_X}(\boldsymbol{\xi}))^t \in \mathcal{M}(N_X, 1)$ and $\chi = (\chi_{ij}) \in \mathcal{M}(n, N_X)$ is the unknown to be determined. We have

$$\mathbf{X} \approx \mathbf{PX} = \chi \varphi(\boldsymbol{\xi}). \quad [6.4]$$

Let us adopt, for instance, the variational standpoint. Analogously to sections 4.1 and 5.1.3, the differential equation [6.1] is approximated as:

$$\mathbf{PX} \in S, \ E\left(\mathbf{Y}^t . \mathbf{X}(0)\right) = E\left(\mathbf{Y}^t . \mathbf{X}_0\right) \text{ and}$$
$$E\left(\mathbf{Y}^t . \tfrac{d\mathbf{PX}}{dt}\right) = E\left(\mathbf{Y}^t . \mathbf{A} . \mathbf{PX}\right) + E\left(\mathbf{Y}^t . \mathbf{B}\right), \forall \mathbf{Y} \in S. \quad [6.5]$$

Recalling that

$$\mathbf{Y} \in S \iff \mathbf{Y} = \mathbf{D}\varphi(\boldsymbol{\xi}), \mathbf{D} = (D_{ij}) \in \mathcal{M}(n, N_X),$$

on the one hand, we have:

$$E\left(\varphi(\boldsymbol{\xi})^t \mathbf{D}^t \tfrac{d\chi}{dt} \varphi(\boldsymbol{\xi})\right) = E\left(\varphi(\boldsymbol{\xi})^t \mathbf{D}^t \mathbf{A} \varphi(\boldsymbol{\xi})\right)$$
$$+ E\left(\varphi(\boldsymbol{\xi})^t \mathbf{D}^t \mathbf{B}\right), \forall \mathbf{D} \in \mathcal{M}(n, N_X);$$

and, on the other hand,

$$E\left(\varphi(\boldsymbol{\xi})^t \mathbf{D}^t \chi(0) \varphi(\boldsymbol{\xi})\right) = E\left(\varphi(\boldsymbol{\xi})^t \mathbf{D}^t \mathbf{X}_0\right). \quad [6.6]$$

We have:

$$E\left(\varphi(\boldsymbol{\xi})^t \mathbf{D}^t \frac{d\chi}{dt} \varphi(\boldsymbol{\xi})\right) = \sum_{i=1}^{n} \sum_{k,m=1}^{N_X} E\left(\varphi_m(\mathbf{v}) D_{im} \varphi_k(\mathbf{v})\right) \frac{d\chi_{ik}}{dt}, \quad [6.7]$$

$$E\left(\varphi(\mathbf{v})^t \mathbf{D}^t \mathbf{A} \chi \varphi(\mathbf{v})\right) = \sum_{i,j=1}^{n} \sum_{k,m=1}^{N_X} E\left(\varphi_m(\mathbf{v}) D_{im} A_{ij} \varphi_k(\mathbf{v})\right) \chi_{jk}$$

and:

$$E\left(\boldsymbol{\varphi}\left(\mathbf{v}\right)^{t}\mathbf{D}^{t}\mathbf{B}\right)=\sum_{i=1}^{n}\sum_{m=1}^{N_{X}}E\left(\varphi_{m}\left(\mathbf{v}\right)D_{im}B_{i}\right).$$

So, by taking $D_{im} = \delta_{ir}\delta_{ms}$, equation [4.6] leads to, for $1 \leq r \leq n, 1 \leq s \leq N_X$:

$$\sum_{k=1}^{N_X}\left[E\left(\varphi_s\left(\mathbf{v}\right)\varphi_k\left(\mathbf{v}\right)\right)\frac{d\chi_{rk}}{dt}+\sum_{j=1}^{n}E\left(\varphi_s\left(\mathbf{v}\right)A_{rj}\varphi_k\left(\mathbf{v}\right)\right)\chi_{jk}\right]$$
$$=E\left(\varphi_s\left(\mathbf{v}\right)B_r\right). \quad [6.8]$$

Let:

$$\mathcal{M}_{rsjk} = \delta_{jr}E\left(\varphi_s\left(\mathbf{v}\right)\varphi_k\left(\mathbf{v}\right)\right),\ \mathcal{A}_{rsjk} = E\left(\varphi_s\left(\mathbf{v}\right)A_{rj}\varphi_k\left(\mathbf{v}\right)\right),$$
$$\mathcal{C}_{rs}\left(\mathbf{B}\right) = E\left(\varphi_s\left(\mathbf{v}\right)B_r\right). \quad [6.9]$$

We have, for $1 \leq r \leq n, 1 \leq s \leq N_X$,

$$\sum_{j=1}^{n}\sum_{k=1}^{N_X}\left[\mathcal{M}_{rsjk}\frac{d\chi_{jk}}{dt}+\mathcal{A}_{rsjk}\chi_{jk}\right]=\mathcal{C}_{rs}\left(\mathbf{B}\right).$$

Let us consider:

$$ind(j,k) = (k-1)n + j, \quad [6.10]$$

and the matrices $\mathbf{M} = (M_{\alpha\beta})$, $\mathbf{N} = (N_{\alpha\beta}) \in \mathcal{M}(nN_X, nN_X)$, $\mathbf{Y} = (Y_\beta)$ and $\mathbf{Q} = (Q_\beta) \in \mathcal{M}(nN_X, 1)$ such that:

$$M_{\alpha\beta} = \mathcal{M}_{rsjk},\ N_{\alpha\beta} = \mathcal{A}_{rsjk},\ Q_\alpha = \mathcal{C}_{rs}\left(\mathbf{B}\right),\ Y_\beta = \chi_{jk},$$
$$\alpha = ind(r,s),\ \beta = ind(j,k). \quad [6.11]$$

Then, we have:

$$\mathbf{M}\frac{d\mathbf{Y}}{dt} = \mathbf{N}\mathbf{Y} + \mathbf{Q}. \quad [6.12]$$

In addition:

$$E\left(\varphi(\mathbf{v})^t \mathbf{D}^t \mathbf{X}_0\right) = \sum_{i=1}^{n} \sum_{m=1}^{N_X} E\left(\varphi_m(\mathbf{v}) D_{im} X_{0i}\right),$$

$$E\left(\varphi(\boldsymbol{\xi})^t \mathbf{D}^t \chi(0) \varphi(\boldsymbol{\xi})\right) = \sum_{i=1}^{n} \sum_{k,m=1}^{N_X} E\left(\varphi_m(\mathbf{v}) D_{im} \varphi_k(\mathbf{v})\right) \chi_{ik}(0)$$

and equation [6.6] shows that, for $1 \leq r \leq n, 1 \leq s \leq N_X$,

$$\sum_{k=1}^{N_X} [E\left(\varphi_s(\mathbf{v}) \varphi_k(\mathbf{v})\right)] \chi_{rk}(0) = E\left(\varphi_s(\mathbf{v}) X_{0r}\right).$$

Thus, for $1 \leq r \leq n, 1 \leq s \leq N_X$,

$$\sum_{j=1}^{n} \sum_{k=1}^{N_X} \mathcal{M}_{rsjk} \chi_{jk}(0) = \mathcal{C}_{rs}(\mathbf{X}_0).$$

So, by setting $\mathbf{Z}_0 = (Z_{0\beta}) \in \mathcal{M}(nN_X, 1)$ such that:

$$\mathbf{Z}_{0\alpha} = \mathcal{C}_{rs}(\mathbf{X}_0), \alpha = ind(r,s), \quad [6.13]$$

we obtain:

$$\mathbf{MY}(0) = \mathbf{Z}_0. \quad [6.14]$$

Equations [6.12] and [6.14] form a linear system of differential equations that may be solved in order to determine the unknown $\mathbf{Y}$. Once determined $\mathbf{Y}$, we may construct χ by using equation [6.11], and finally, $\mathbf{PX}$.

Matlab implementation is performed as follows: assume that, on the one hand, the value of $\varphi_k(\mathbf{v})$ is furnished by the subprogram phi(k,v), while, on the other hand, the function $\mathbf{f}(t, \mathbf{X}, \mathbf{v})$ is evaluated by a subprogram f(t,X,v). Similarly to the preceding situations, the evaluation of the means involved in the equations is requested and they correspond to scalar products, which may be evaluated by integration or using a sample. Here, we assume that a sample of ns variates from $\mathbf{v}$ is available – using the evaluation of the scalar products by numerical integration is a simple modification of the code below.

Listing 6.1. *Linear differential equation*

```
function chi = expcoef(A,B,phi,n,N_X,vs,X0,time_span)
%
% determines the coefficients of the expansion.
%   by using a sample
%
% IN:
% A : n x n matrix of the linear system - type anonymous
    function
% B : n x 1 matrix of the linear system - type anonymous
    function
% n : number of unknowns - type integer
% N_X : order of the expansion - type integer
% vs : table of values of v - type array of double
% vs(:,i) is a variate from v
% f : second member - type anonymous function
% X0: n x 1 initial vector - type anonymous function
% time_span : ntimes x 1 vector of time moments where the
    solution
%       has to be evaluated - type array of double
%
% OUT:
% chi : ntimes x n x N_X matrix of the coefficients - type
    array of double
%
aux = fixed_variational_matrix(phi, vs);
M = tab4(aux,n,N_X);
[N,Q] = variational_matrices(A,B,phi,n,N_X,vs);
F = @(t,U) second_member(M,N,Q,t,U);
PX0 = zeros(N_X, n);
w = map(X0,vs,n);
for r = 1: n
    Y = w(r,:);
    for s = 1:N_X
        f1 = @(U) phi(s,U);
        Z = map(f1, vs, 1);
        aux = scalprod(Y,Z);
        PX0(r,s) = aux;
    end;
end;
Y0 = zeros(n*N_X,1);
for r = 1: n
    for s = 1: N_X
        alfa = index_map(r,s,n,N_X);
```

```
            Y0(alfa) = PX0(r,s);
        end;
    end;
%
U0 = M \ Y0;
[TT,UU] = ode45(F,time_span,U0);
%
chi = zeros(length(time_span),n,N_X);
for i = 1: length(time_span)
    aux = UU(i,:)';
    auxx = zeros(n,N_X);
    for r = 1: n
        for s = 1: N_X
            alfa = index_map(r,s,n,N_X);
            auxx(r,s) = aux(alfa);
        end;
    end;
    chi(i,:,:) = auxx;
end;
return;
end

function w = second_member(M,N,Q,t,U)
aux = N*U + Q;
w = M \ aux;
return;
end
```

EXAMPLE 6.1.– Let $n = 2$ and the differential equation:

$$\frac{d\mathbf{X}}{dt} = \begin{pmatrix} 0 & -v_1 \\ v_1 & 0 \end{pmatrix} \mathbf{X} \, , \, \mathbf{X}(0) = \begin{pmatrix} v_2 \\ 0 \end{pmatrix}, \text{ on } (0, T).$$

The exact solution is

$$\mathbf{X}(t) = \begin{pmatrix} v_2 \cos(v_1 t) \\ v_2 \sin(v_1 t) \end{pmatrix}.$$

Let us assume that $v = (v_1, v_2)$ is a pair of independent random variables such that v_1 is uniformly distributed on (a_1, b_1) and v_2 is uniformly distributed on (a_2, b_2). We consider the basis $(0 \leq i \leq n_1, 0 \leq j \leq n_2)$:

$$\varphi_k(v) = \left(\frac{v_1 - a_1}{b_1 - a_1}\right)^i \left(\frac{v_2 - a_2}{b_2 - a_2}\right)^j \, , \, k = (j-1)n_1 + i.$$

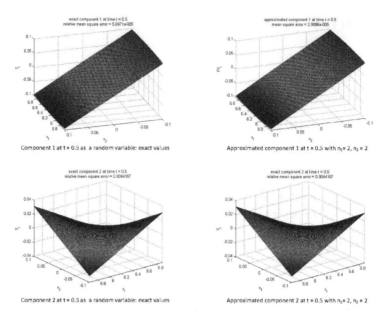

Figure 6.1. *Results for $t = 0.5$ in Example 6.1 ($n_1 = n_2 = 2$). For a color version of the figure, see www.iste.co.uk/souzadecursi/quantification.zip*

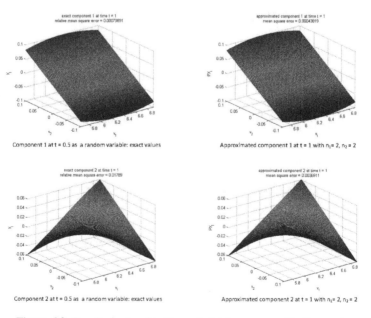

Figure 6.2. *Results for $t = 1$ in Example 6.1 ($n_1 = n_2 = 2$). For a color version of the figure, see www.iste.co.uk/souzadecursi/quantification.zip*

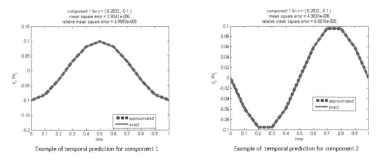

Figure 6.3. *Results for Example 6.1 with $nts = 10$ ($n_1 = n_2 = 2$). For a color version of the figure, see www.iste.co.uk/souzadecursi/quantification.zip*

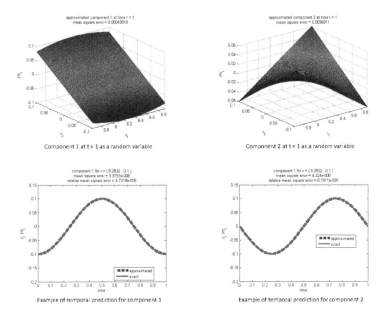

Figure 6.4. *Results for Example 6.1 with $nts = 40$ ($n_1 = n_2 = 2$). For a color version of the figure, see www.iste.co.uk/souzadecursi/quantification.zip*

The results obtained for $T = 1$, $a_1 = 1.8\pi$, $b_1 = 2.2\pi$, $a_2 = -0.1$, $b_1 = 0.1$ and $n_1 = n_2 = 2$ are shown in Figures 6.1–6.4. The differential equation formed by equations [6.12] and [6.14] has been solved by using the internal Matlab function *ode45* and the solution has been evaluated at nts time instants corresponding to $t = \Delta t, 2\Delta t, ..., nts\Delta t = 1$, $\Delta t = T/nts$. The means have been evaluated by numerical integration using the internal function *quad* of Matlab. For $nts = 10$, the global mean quadratic error is of $1E - 4$

and the associated relative error is lower than 0.3%. Figures 6.1–6.2 compare the random variables $\mathbf{X}(t)$, $\mathbf{PX}(t)$ for fixed values of the time t and all the values of $\mathbf{v}$, while Figure 6.3 compares the prediction $\mathbf{PX}(t)$ to the exact value $\mathbf{X}(t)$ at a given point $\mathbf{v}$ and all the values of t. The results are analogous for $nts = 40$. Figure 6.4 shows an example of result for $nts = 40$.

6.2. The case of nonlinear differential equations

The method presented in the previous section may be adapted to nonlinear differential equations. Let us consider:

$$\frac{d\mathbf{X}}{dt} = \mathbf{f}(t?\mathbf{X}), \ \mathbf{X}(0) = \mathbf{X}_0 \,. \quad [6.15]$$

In this case, we have:

$$\mathbf{PX} \in S, \ E\left(\mathbf{Y}^t.\mathbf{X}(0)\right) = E\left(\mathbf{Y}^t.\mathbf{X}_0\right) \text{ and } \\ E\left(\mathbf{Y}^t.\tfrac{d\mathbf{PX}}{dt}\right) = E\left(\mathbf{Y}^t.\mathbf{f}(\mathbf{t},\mathbf{PX})\right), \forall \ \mathbf{Y} \in S. \quad [6.16]$$

Thus, equation [6.1] can be written as:

$$E\left(\varphi\left(\xi\right)^t \mathbf{D}^t \frac{d\chi}{dt}\varphi\left(\xi\right)\right) = E\left(\varphi\left(\xi\right)^t \mathbf{D}^t \mathbf{f}\left(t, \chi\varphi\left(\xi\right)\right)\right),$$
$$\forall \ \mathbf{D} \in \mathcal{M}(n, N_X) \,; \quad [6.17]$$

and, since:

$$E\left(\varphi\left(\mathbf{v}\right)^t \mathbf{D}^t \mathbf{f}\left(t, \chi\varphi\left(\xi\right)\right)\right) = \sum_{i=1}^{n} \sum_{m=1}^{N_X} E\left(\varphi_m\left(\mathbf{v}\right) D_{im} f_i\left(t, \chi\varphi\left(\xi\right)\right)\right),$$

we have:

$$\sum_{i=1}^{n} \sum_{k,m=1}^{N_X} E\left(\varphi_m\left(\mathbf{v}\right) D_{im} \varphi_k\left(\mathbf{v}\right)\right) \frac{d\chi_{ik}}{dt} = \\ \sum_{i=1}^{n} \sum_{m=1}^{N_X} E\left(\varphi_m\left(\mathbf{v}\right) D_{im} f_i\left(t, \chi\varphi\left(\xi\right)\right)\right).$$

So, by taking $D_{im} = \delta_{ir}\delta_{ms}$, for $1 \leq r \leq n, 1 \leq s \leq N_X$, we have:

$$\sum_{k=1}^{N_X} \left[E\left(\varphi_s(\mathbf{v})\,\varphi_k(\mathbf{v})\right) \frac{d\chi_{rk}}{dt} \right] = E\left(\varphi_s(\mathbf{v})\,f_r\left(t, \chi\varphi(\boldsymbol{\xi})\right)\right). \quad [6.18]$$

Let us consider **Y** defined in equation [6.11] and:

$$F_\alpha(t, \mathbf{Y}) = E\left(\varphi_s(\mathbf{v})\,f_r\left(t, \chi\varphi(\boldsymbol{\xi})\right)\right)$$

Then, we have:

$$\mathbf{M}\frac{d\mathbf{Y}}{dt} = \mathbf{F}(t, \mathbf{Y}). \quad [6.19]$$

with the initial condition [6.14]. This approach is implemented in Matlab as follows.

Listing 6.2. *Nonlinear differential equation*

```
function chi = expcoef(f,phi,n,N_X,vs,X0,time_span)
%
%   determines the coefficients of the expansion.
%   by using a sample
%
% IN:
% f : second member f(t,x,v) - type anonymous function
% B : n x 1 matrix of the linear system - type anonymous
     function
% n : number of unknowns - type integer
% N_X : order of the expansion - type integer
% vs : table of values of v - type array of double
% vs(:,i) is a variate from v
% f : second member - type anonymous function
% X0: n x 1 initial vector - type anonymous function
% time_span : ntimes x 1 vector of time moments where the
     solution
%       has to be evaluated - type array of double
%
% OUT:
% chi : ntimes x n x N_X matrix of the coefficients - type
     array of double
%
aux = fixed_variational_matrix(phi, vs);
M = tab4(aux,n,N_X);
F = @(t,U) second_member(f,t,U,vs,phi,M,n,N_X);
```

```matlab
PX0 = zeros(N_X, n);
w = map(X0,vs,n);
for r = 1: n
    Y = w(r,:);
    for s = 1:N_X
        f1 = @(U) phi(s,U);
        Z = map(f1, vs, 1);
        aux = scalprod(Y,Z);
        PX0(r,s) = aux;
    end;
end;
Z = zeros(n*N_X,1);
for r = 1: n
    for s = 1: N_X
        alfa = index_map(r,s,n,N_X);
        Z(alfa) = PX0(r,s);
    end;
end;
%
Y0 = M \ Z;
[T,Y] = ode45(F,time_span,Y0);
%
chi = zeros(length(time_span),n,N_X);
for i = 1: length(time_span)
    aux = Y(i,:)';
    auxx = zeros(n,N_X);
    for r = 1: n
        for s = 1: N_X
            alfa = index_map(r,s,n,N_X);
            auxx(r,s) = aux(alfa);
        end;
    end;
    chi(i,:,:) = auxx;
end;
return;
end

function w = second_member(f,t,U,vs,phi,M,n,N_X)
chi = zeros(n, N_X);
for r = 1: n
    for s = 1: N_X
        alfa = index_map(r,s,n,N_X);
        chi(r,s) = U(alfa);
    end;
end;
ff = @(x,v) f(t,x,v);
w = function_proj(ff,chi,vs,phi);
```

```
aux = mean(w, 1);
w = M \ aux';
return;
end
```

EXAMPLE 6.2.– Let us consider $n = 1$ and the *logistic equation:*

$$\frac{dx}{dt} = \alpha x (1-x), \ x(0) = x_0.$$

Its exact solution is:

$$x(t) = \frac{x_0 \exp(\alpha t)}{1 - x_0 + x_0 \exp(\alpha t)}.$$

Let us assume that $v = (\alpha, x_0)$ is a pair of independent random variables such that α is uniformly distributed on $(0, 1)$, while x_0 is uniformly distributed on $(0.5, 1.5)$. The differential equation [6.19] with initial condition [6.14] has been numerically solved using the function *ode45* of Matlab and the solution has been evaluated at the times $t_i = i\Delta t$, $i = 1, ..., nts$, $\Delta t = T/nts$. The means have been determined by numerical integration using the Matlab function *quad*. Figure 6.5 shows an example of result for $nts = 10$.

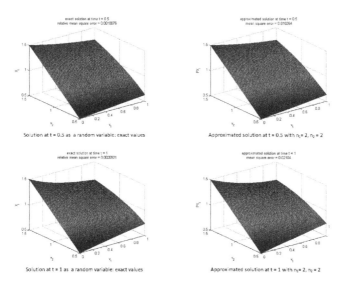

Figure 6.5. *Results obtained in Example 6.2* $(n_1 = n_2 = 2)$

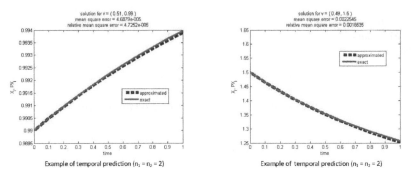

Figure 6.6. *Results obtained in Example 6.2* ($n_1 = n_2 = 2$)

6.3. The case of partial differential equations

6.3.1. *Linear equations*

Partial differential equations (PDEs) may be studied by using the same methods, in particular when they are written in a variational form. For instance, let us consider a functional space V:

a *bilinear* application:

$$V \times V \ni (x,y) \longrightarrow a(x,y) \in \mathcal{M}(1,1),$$

a *linear functional:*

$$V \ni y \longrightarrow b(y) \in \mathcal{M}(1,1),$$

and the variational equation:

$$x \in V \text{ and } a(x,y) = b(y), \forall\, y \in V.$$

Usually, the discretization of this last equation involves a finite family $\{\psi_i\}_{1 \leq i \leq n}$ (such as, for instance, finite elements) and the subspace $V_n = \left[\{\psi_i\}_{1 \leq i \leq n}\right]$. The variational equation is discretized as:

$$x \in V_n \text{ and } a(x,y) = b(y), \forall\, y \in V_n.$$

Let us consider:

$$A_{ij} = a(\psi_j, \psi_i)\ ;\ B_i = b(\psi_i)\ ;\ \mathbf{X} = (x_1, ..., x_n)^t\ ;\ \mathbf{Y} = (y_1, ..., y_n)^t.$$

In this case, the approximated linear variational equation becomes:

$$\mathbf{X} \in \mathbb{R}^n \text{ and } \mathbf{Y}^t \mathbf{A} \mathbf{X} = \mathbf{Y}^t \mathbf{B}, \forall \mathbf{Y} \in \mathbb{R}^n,$$

which corresponds to the linear system:

$$\mathbf{A}\mathbf{X} = \mathbf{B}.$$

Applying the method presented in Chapter 4 (section 4.1), we have:

$$\sum_{j=1}^{n} \sum_{k=1}^{N_X} \mathcal{A}_{rsjk} C_{jk} = \mathcal{B}_{rs}, 1 \leq r \leq n, 1 \leq s \leq N_X,$$

where:

$$\mathcal{A}_{rsjk} = E\left(\varphi_s(\mathbf{v}) A_{rj} \varphi_k(\mathbf{v})\right), \mathcal{B}_{rs} = E\left(\varphi_s(\mathbf{v}) B_r\right).$$

Thus, by using the same variables (see section 4.1) $M = (M_{\alpha\beta}) \in \mathcal{M}(nN_X, nN_X)$, $U = (U_\beta) \in \mathcal{M}(nN_X, 1)$, $N = (N_\alpha) \in \mathcal{M}(nN_X, 1)$ given by:

$$M_{\alpha\beta} = \mathcal{M}_{rsjk}, N_\alpha = \mathcal{N}_{rs}, U_\beta = C_{jk}, \alpha = ind(r, s), \beta = ind(j, k),$$

we have again:

$$\mathbf{M}\mathbf{U} = \mathbf{N}.$$

6.3.2. *Nonlinear equations*

Let us assume that the equation $(x, y) \longrightarrow a(x, y)$ *is not* bilinear, but that only $y \longrightarrow a(x, y)$ is linear. In this case, the approximated variational equation can be written as:

$$\mathbf{X} \in \mathbb{R}^n \text{ and } \mathbf{Y}^t \mathbf{F}(\mathbf{X}) = \mathbf{0}, \forall \mathbf{Y} \in \mathbb{R}^n,$$

where:

$$F_i(\mathbf{X}) = a\left(\sum_{k=1}^{n} x_k \psi_k, \psi_i\right) - b_i$$

i.e.:

$$\mathbf{X} \in \mathbb{R}^n \text{ and } \mathbf{F}(\mathbf{X}) = \mathbf{0}.$$

Let us apply the method introduced in Chapter 5 (section 5.1.3): we have:

$$\mathcal{F}_{rs}(\chi) = 0, 1 \leq r \leq n, 1 \leq s \leq N_X,$$

where:

$$\mathcal{F}_{rs}(\chi) = E\left(\varphi_s(\mathbf{v}) F_r(\chi\varphi(\mathbf{v}))\right).$$

Thus, by introducing the same vectors $G = (G_\alpha) \in \mathcal{M}(nN_X, 1)$, $U = (U_\beta) \in \mathcal{M}(nN_X, 1)$ such that:

$$G_\alpha = \mathcal{F}_{rs}, U_\beta = \chi_{jk}, \alpha = ind(r, s), \beta = ind(j, k),$$

we have:

$$\mathbf{G}(\mathbf{U}) = \mathbf{0}.$$

6.3.3. Evolution equations

Let us consider a variational equation having a time dependence:

$$x \in V \text{ and } \lambda\left(\frac{dx}{dt}, y\right) + a(x, y) = b(t, y), \forall\, y \in V \text{ and } t \in (0, T),$$

where $\mathbb{R}$ is *bilinear*. Let:

$$\Lambda_{ij} = \lambda\left(\psi_j, \psi_i\right).$$

Since:

$$\lambda\left(\frac{dx}{dt}, y\right) = \sum_{i,j=1}^{n} y_i \lambda\left(\psi_j, \psi_i\right) \frac{dx_j}{dt} = \mathbf{Y}^t \Lambda \frac{d\mathbf{X}}{dt},$$

analogously to the preceding situations, we have:

$$\mathbf{Y}^t \Lambda \frac{d\mathbf{X}}{dt} + \mathbf{Y}^t \mathcal{F}(\mathbf{X}) = \mathbf{0} \ , \forall \mathbf{Y} \in \mathbb{R}^n,$$

where:

$$\mathcal{F}_i(t, \mathbf{X}) = a\left(\sum_{k=1}^n x_k \psi_k(\boldsymbol{\xi}), \psi_i(\boldsymbol{\xi})\right) - b_i.$$

This variational equality leads to differential equations defining the approximation. In the following, we consider the linear and the nonlinear situations.

6.3.3.1. *The linear situation*

For linear equations, the approximated variational equation can be written as

$$\Lambda \frac{d\mathbf{X}}{dt} + \mathbf{A}\mathbf{X} = \mathbf{B}.$$

Let us introduce

$$\mathcal{M}_{rsjk} = \delta_{jr} E\left(\varphi_s(\mathbf{v}) \Lambda_{rj} \varphi_k(\mathbf{v})\right),$$
$$\mathcal{A}_{rsjk} = E\left(\varphi_s(\mathbf{v}) A_{rj} \varphi_k(\mathbf{v})\right), \mathcal{C}_{rs}(\mathbf{B}) = E\left(\varphi_s(\mathbf{v}) B_r\right). \quad [6.20]$$

We have, for $1 \leq r \leq n, 1 \leq s \leq N_X$,

$$\sum_{j=1}^n \sum_{k=1}^{N_X} \left[\mathcal{M}_{rsjk} \frac{d\chi_{jk}}{dt} + \mathcal{A}_{rsjk} \chi_{jk}\right] = \mathcal{C}_{rs}(\mathbf{B}).$$

Thus, using the same variables $\mathbf{M} = (M_{\alpha\beta})$, $\mathbf{Q} = (Q_\beta) \in \mathcal{M}(nN_X, 1)$, $\mathbf{N} = (N_{\alpha\beta}) \in \mathcal{M}(nN_X, nN_X)$, $\mathbf{U} = (U_\beta)$, such that

$$M_{\alpha\beta} = \mathcal{M}_{rsjk}, N_{\alpha\beta} = \mathcal{A}_{rsjk}, Q_\alpha = \mathcal{C}_{rs}(\mathbf{B}), U_\beta = \chi_{jk},$$
$$\alpha = ind(r, s), \beta = ind(j, k), \quad [6.21]$$

we have:

$$\mathbf{M}\frac{d\mathbf{U}}{dt} = \mathbf{N}\mathbf{U} + \mathbf{Q}.$$

6.3.3.2. *The nonlinear situation*

In the nonlinear case, the approximated variational equation can be written as:

$$\Lambda \frac{d\mathbf{X}}{dt} = \mathbf{f}(t, \mathbf{X}),$$

where:

$$f_i(t, \mathbf{X}) = b_i - a\left(\sum_{k=1}^{n} x_k \psi_k(\boldsymbol{\xi}), \psi_i(\boldsymbol{\xi})\right).$$

Thus, by considering:

$$F_\alpha(t, \mathbf{U}) = E\left(\varphi_s(\mathbf{v}) f_r(t, \chi\varphi(\boldsymbol{\xi}))\right), \alpha = ind(r, s),$$

we have:

$$\mathbf{M}\frac{d\mathbf{U}}{dt} = \mathbf{F}(\mathbf{t}, \mathbf{U}).$$

6.4. Reduction of Hamiltonian systems

In this section, we consider differential equations *Hamiltonian systems*, i.e. differential equations corresponding to Hamilton's equations of a system. In this case, stochastic methods may be used in order to reduce the number of unknowns and we may generate equations destined to furnish approximate values of a few variables – eventually a single one – without determining the others.

6.4.1. *Hamiltonian systems*

Hamiltonians are usually introduced by using transformations of Lagrangians. Let us consider a mechanical system described by the coordinates $\mathbf{q} = (q_1, ..., q_n)$. Let us denote by t the time variable and by $\dot{\mathbf{q}} = d\mathbf{q}/dt$ the time derivatives. Assume that the system has a kinetic energy $K = K(t, \mathbf{q}, \dot{\mathbf{q}})$ and total potential energy $V = V(t, \mathbf{q})$. The Lagrangian is:

$$L(t, \mathbf{q}, \dot{\mathbf{q}}) = K(t, \mathbf{q}, \dot{\mathbf{q}}) - V(t, \mathbf{q}). \qquad [6.22]$$

The action $\mathcal{A}$ on a time interval $(0, \tau)$ is given by:

$$\mathcal{A}(\mathbf{q}) = \int_0^\tau L(t, \mathbf{q}, \dot{\mathbf{q}}) \, dt,$$

and the equations of motion are written in variational form:

$$D\mathcal{A}(\mathbf{q})(\delta \mathbf{q}) = 0, \ \forall \, \delta \mathbf{q} \text{ such that } \delta \mathbf{q}(0) = \delta q(\tau) = 0,$$

where $D\mathcal{A}$ denotes the *Gateaux derivative* of $\mathcal{A}$. This variational equation corresponds to the Euler–Lagrange's equations:

$$\frac{d}{dt}\left(\frac{\partial L}{\partial \dot{q}_i}\right) = \frac{\partial L}{\partial q_i}, \, 1 \leq i \leq n. \qquad [6.23]$$

Let us introduce the transformation:

$$(\mathbf{q}, \dot{\mathbf{q}}) \xrightarrow{\mathbf{T}} (\mathbf{Q}, \mathbf{P}) \qquad [6.24]$$

which is given by

$$\mathbf{Q}(\mathbf{q}, \dot{\mathbf{q}}) = \mathbf{q} \, ; \, \mathbf{P}(\mathbf{q}, \dot{\mathbf{q}}) = \mathbf{p} \, ; \, p_i = \frac{\partial L}{\partial \dot{q}_i}, 1 \leq i \leq n. \qquad [6.25]$$

$\mathbf{p}$ is the *generalized moment*. Equations [6.23] can be written as:

$$p_i = \frac{\partial L}{\partial \dot{q}_i}, \ \dot{p}_i = \frac{\partial L}{\partial q_i}, 1 \leq i \leq n. \qquad [6.26]$$

Let us denote $\mathbf{T}^{-1} = (\mathbf{Q}^{-1}, \mathbf{P}^{-1})$ the inverse transformation associated with $\mathbf{T}$ (equation [6.24]). The Hamiltonian H of the system is obtained by a *Legendre's transformation* of the Lagrangian L :

$$H(t, \mathbf{q}, \mathbf{p}) = \mathbf{p}^t \mathbf{P}^{-1}(\mathbf{q}, \mathbf{p}) - L\left(t, \mathbf{q}, \mathbf{P}^{-1}(\mathbf{q}, \mathbf{p})\right). \qquad [6.27]$$

This transformation may be interpreted as a *Fenchel's conjugate* of L with respect to $\dot{\mathbf{q}}$:

$$H(t, \mathbf{q}, \mathbf{s}) = \sup \left\{ \mathbf{s}^t \mathbf{x} - L(t, \mathbf{q}, \mathbf{x}) : \mathbf{x} \in \mathbb{R}^n \right\}.$$

By deriving the expression $\mathbf{s}^t\mathbf{x} - L(t, \mathbf{q}, \mathbf{x})$ with respect to x_i, the reader may easily verify that the maximum is attained for $s_i = \partial L / \partial \dot{q}_i = p_i$. We have, on the one hand:

$$H(t, \mathbf{q}, \mathbf{p}) = \mathbf{p}^t \dot{\mathbf{q}} - L(t, \mathbf{q}, \dot{\mathbf{q}}) \ ; \ \dot{\mathbf{q}} = \mathbf{P}^{-1}(\mathbf{q}, \mathbf{p}).$$

Thus:

$$\frac{\partial H}{\partial p_i} = \dot{q}_i + \sum_{j=1}^{n} \underbrace{\left(p_j - \frac{\partial L}{\partial \dot{q}_j}\right)}_{=0} \frac{\partial \dot{q}_j}{\partial p_i} = \dot{q}_i. \qquad [6.28]$$

On the other hand:

$$L(t, \mathbf{q}, \dot{\mathbf{q}}) = \mathbf{p}^t \dot{\mathbf{q}} - H(t, \mathbf{q}, \mathbf{p}) \ ; \ \mathbf{p} = \mathbf{P}(\mathbf{q}, \dot{\mathbf{q}}) .$$

Thus, equation [6.28] shows that:

$$\frac{\partial L}{\partial q_i} = -\frac{\partial H}{\partial q_i} + \sum_{j=1}^{n} \underbrace{\left(\dot{q}_j - \frac{\partial H}{\partial p_j}\right)}_{=0} \frac{\partial p_j}{\partial q_i} = -\frac{\partial H}{\partial q_i}$$

and equation [6.26] yields that

$$\dot{p}_i = -\frac{\partial H}{\partial q_i}.$$

So, the motion's equations may be written as:

$$\dot{q}_i = \frac{\partial H}{\partial p_i} \ , \ \dot{p}_i = -\frac{\partial H}{\partial q_i}, 1 \leq i \leq n. \qquad [6.29]$$

These are the equations of motion in Hamiltonian form, or Hamiltonian motion's equations.

6.4.2. *Reduction of autonomous Hamiltonian systems*

Hamiltonians have many interesting properties such as:

$$\frac{dH}{dt} = \frac{\partial H}{\partial t} + \sum_{j=1}^{n}\left(\frac{\partial H}{\partial q_i}\dot{q}_i + \frac{\partial H}{\partial p_i}\dot{p}_i\right) = \frac{\partial H}{\partial t} + \sum_{j=1}^{n}(-\dot{p}_i\dot{q}_i + \dot{q}_i\dot{p}_i) = \frac{\partial H}{\partial t}.$$

As a result, a Hamiltonian that does not contain explicit dependence on time is constant on the motion, i.e.:

$$\frac{\partial H}{\partial t} = 0 \Longrightarrow \frac{dH}{dt} = 0. \qquad [6.30]$$

A system is said to be autonomous if its equations of motion do not contain explicit dependence on time. For instance, a system such that $K = K(\mathbf{q}, \dot{\mathbf{q}})$ and $V = V(\mathbf{q})$ has a Lagrangien $L(\mathbf{q}, \dot{\mathbf{q}}) = K(\mathbf{q}, \dot{\mathbf{q}}) - V(\mathbf{q})$, and consequently, a Hamiltonian $H(\mathbf{q},\mathbf{p}) = \mathbf{p}^t \mathbf{P}^{-1}(\mathbf{q},\mathbf{p}) - L(\mathbf{q}, \mathbf{P}^{-1}(\mathbf{q},\mathbf{p}))$, which is independent of time. For such a system, equation [6.30] shows that H is time-invariant and

$$f(\mathbf{p},\mathbf{q}) = \exp(-\lambda H(\mathbf{q},\mathbf{p}))$$

defines a strictly positive time-invariant function:

$$f(\mathbf{p},\mathbf{q}) > 0 \text{ and } \frac{df}{dt} = 0.$$

These properties make $f(\mathbf{p},\mathbf{q})$ suitable in order to define an invariant probability:

$$\varphi(\mathbf{q},\mathbf{p}) = \frac{1}{A} f(\mathbf{q},\mathbf{p}),$$

on the phase space (A is a normalizing factor). In addition:

$$\frac{\partial f}{\partial q_i} = -\lambda f(\mathbf{p},\mathbf{q}) \frac{\partial H}{\partial q_i} = \lambda f(\mathbf{p},\mathbf{q}) \dot{p}_i;$$

$$\frac{\partial f}{\partial p_i} = -\lambda f(\mathbf{p},\mathbf{q}) \frac{\partial H}{\partial p_i} = -\lambda f(\mathbf{p},\mathbf{q}) \dot{q}_i.$$

Let us consider $\Omega = \{1, ..., n\}$ and two disjoint subsets $I = \{i_1, ..., i_m\}$, $J = \Omega - I = \{j_1, ..., j_k\}$ ($k = n - m$). Assume that we are interested only in the variables $\{(q_i, p_i) : i \in I\}$ and we want to eliminate the variables $\{(q_j, p_j) : j \in J\}$. We denote:

$$\mathbf{q}_I = (q_{i_1}, q_{i_2}, ..., q_{i_m}), \mathbf{p}_I = (p_{i_1}, p_{i_2}, ..., p_{i_m})$$

and, analogously,

$$\mathbf{q}_J = (q_{j_1}, q_{j_2}, ..., q_{j_k}), \mathbf{p}_J = (p_{j_1}, q_{j_2}, ..., p_{j_k}).$$

A simple idea in order to obtain equations concerning the only variables of interest $(\mathbf{q}_I, \mathbf{p}_I)$ has been introduced by Chorin and consists of considering the approximated equations:

$$\frac{dq_i}{dt} = E\left(\frac{\partial H}{\partial p_i} | (\mathbf{q}_I, \mathbf{p}_I)\right), \quad \frac{dp_i}{dt} = E\left(-\frac{\partial H}{\partial q_i} | (\mathbf{q}_I, \mathbf{p}_I)\right), i \in I, \quad [6.31]$$

where the conditional expectations are evaluated by using the density $\varphi(\mathbf{q}, \mathbf{p})$. By introducing:

$$f_I(\mathbf{q}_I, \mathbf{p}_I) = \int_{\mathbb{R}^k} f(\mathbf{p}, \mathbf{q}) \, d\mathbf{q}_J d\mathbf{p}_J = \int_{\mathbb{R}^k} f(\mathbf{p}, \mathbf{q}) \, dq_{j_1} dp_{j_1} ... dq_{j_k} dp_{j_k},$$

we obtain:

$$E\left(\frac{\partial H}{\partial q_i} | (\mathbf{q}_I, \mathbf{p}_I)\right) = \frac{1}{f_I(\mathbf{q}_I, \mathbf{p}_I)} \int_{S_k} \frac{\partial H}{\partial q_i} f(\mathbf{p}, \mathbf{q}) \, d\mathbf{q}_J d\mathbf{p}_J$$

and:

$$E\left(\frac{\partial H}{\partial p_i} | (\mathbf{q}_I, \mathbf{p}_I)\right) = \frac{1}{f_I(\mathbf{q}_I, \mathbf{p}_I)} \int_{S_k} \frac{\partial H}{\partial p_i} f(\mathbf{p}, \mathbf{q}) \, d\mathbf{q}_J d\mathbf{p}_J.$$

Let:

$$H_I = -\frac{1}{\lambda} \log(f_I(\mathbf{q}_I, \mathbf{p}_I))$$

We have, for $i \in I$:

$$\frac{\partial H_I}{\partial q_i} = \frac{1}{f_I(\mathbf{q}_I, \mathbf{p}_I)} \int_{S_k} \frac{\partial H}{\partial q_i} f(\mathbf{p}, \mathbf{q}) d\mathbf{q}_J d\mathbf{p}_J = E\left(\frac{\partial H}{\partial q_i} | (\mathbf{q}_I, \mathbf{p}_I)\right)$$

and

$$\frac{\partial H_I}{\partial p_i} = \frac{1}{f_I(\mathbf{q}_I, \mathbf{p}_I)} \int_{S_k} \frac{\partial H}{\partial p_i} f(\mathbf{p}, \mathbf{q}) d\mathbf{q}_J d\mathbf{p}_J = E\left(\frac{\partial H}{\partial p_i} | (\mathbf{q}_I, \mathbf{p}_I)\right).$$

Thus:

$$\frac{dq_i}{dt} = \frac{\partial H_I}{\partial p_i}, \quad \frac{dp_i}{dt} = -\frac{\partial H_I}{\partial q_i}, i \in I. \qquad [6.32]$$

This equation shows that the approximated system [6.31] is Hamiltonian and that its Hamiltonian is H_I. Moreover, the equations contain only the variables of interest $(\mathbf{q}_I, \mathbf{p}_I)$.

6.5. Local solution of deterministic differential equations by stochastic simulation

6.5.1. *Ordinary differential equations*

Let us consider the ordinary differential equation:

$$\alpha(x) u''(x) + \beta(x) u'(x) + \gamma(x) u(x) + f(x) = 0 \text{ on } (0, \ell);$$
$$u(0) = u_0; u(\ell) = u_\ell.$$

By setting $\Omega = (0, \ell)$, $\partial \Omega = \{0, \ell\}$:

$$u_{\partial\Omega}(x) = u_0, \text{ if } x = 0; \; u_{\partial\Omega}(x) = u_\ell, \text{ if } x = \ell,$$

we have:

$$\alpha(x) u''(x) + \beta(x) u'(x) + \gamma(x) u(x) + f(x) = 0 \text{ on } \Omega;$$
$$u = u_{\partial\Omega} \text{ on } \partial\Omega.$$

Let us assume that $\alpha \geq 0$: we consider:

$$a(x) = \sqrt{2\alpha(x)};\ b(x) = \beta(x)$$

and the stochastic diffusion:

$$dX_t = a(X_t)\,dW_t + b(X_t)\,dt;\ X(0) = x \in \Omega. \qquad [6.33]$$

Let us introduce the stochastic process:

$$Y(t) = u(X(t))\exp\left(\int_0^t \gamma(X(s))\,ds\right).$$

We have $Y = F\left(t, X(t), \int_0^t \gamma(X(s))\,ds\right)$ where:

$$F(t, x, z) = u(x)\exp(z),$$

so that:

$$\frac{\partial F}{\partial z} = u(x)\exp(z);\ \frac{\partial F}{\partial t} = 0;\ \frac{\partial F}{\partial x} = u'(x)\exp(z);\ \frac{\partial^2 F}{\partial x^2} = u''(x)\exp(z).$$

Thus:

$$dY_t = (A_t dW_t + B_t dt)\exp\left(\int_0^t \gamma(X(s))\,ds\right),$$

where:

$$A_t = a(X_t)u'(X_t);$$
$$B_t = \tfrac{1}{2}[a(X_t)]^2 u''(X_t) + b(X_t)u'(X_t) + \gamma(X_t)u(X_t).$$

Let:

$$\tau = \inf\{t \geq 0 : X(t) \notin \Omega\}$$

for $t < \tau$, we have $X_t \in \Omega$, so that:

$$B_t = \underbrace{\frac{1}{2}\left[a\left(X_t\right)\right]^2 u''\left(X_t\right)}_{\alpha(X_t)} + \underbrace{b\left(X_t\right)u'\left(X_t\right)}_{\beta(X_t)} + \gamma\left(X_t\right)u\left(X_t\right) = -f\left(X_t\right)$$

and consequently:

$$dY_t = \left(A_t dW_t - f\left(X_t\right)dt\right)\exp\left(\int_0^t \gamma\left(X(s)\right)ds\right).$$

Thus:

$$Y(\tau) - Y(0) = \int_0^\tau A_t \exp\left(\int_0^t \gamma\left(X(s)\right)ds\right) dW_t -$$
$$\int_0^\tau f(X_t) \exp\left(\int_0^t \gamma\left(X(s)\right)ds\right) dt.$$

Since:

$$E\left(\int_0^\tau A_t \exp\left(\int_0^t \gamma\left(X(s)\right)ds\right) dW_t\right) = 0 \text{ and } Y_0 = u(x),$$

we have:

$$u(x) = E(Z(\tau)); \; Z(\tau) = Y(\tau) + \int_0^\tau f(X_t) \exp\left(\int_0^t \gamma\left(X(s)\right)ds\right) dt.$$

For $t < \tau$, we have $X_t \in \Omega$ and there exists $\varepsilon > 0$ such that for $\tau < t < \tau + \varepsilon$, we have $X_t \notin \Omega$, so that $X_\tau \in \partial\Omega$ and:

$$u(x) = E(Z(\tau)),$$

where:

$$Z(\tau) = u_{\partial\Omega}(X_\tau) \exp\left(\int_0^\tau \gamma(X(t))\,dt\right)$$

$$+ \int_0^\tau f(X(t)) \exp\left(\int_0^t \gamma(X(s))\,ds\right) dt.$$

This result may be exploited for the numerical determination of $u(x)$. We discretize [6.33] by using a standard method. For instance:

$$X_{n+1} - X_n = a(X_n)\Delta W_n + b(X_n)\Delta t;\ X_0 = x \in \Omega.\ \text{(Euler)}$$

Such a discretization furnishes a sequence of values $\{X_n\}_{n \geq 0}$, which may be used in order to determine:

$$n = \min\{i > 0 : X_i \notin \Omega\},$$

which gives an approximated realization of $Z(\tau)$:

$$Z(\tau) \approx u_{\partial\Omega}(X_n) \exp\left(\Delta t \sum_{j=0}^{n-1} \gamma(X_j)\right) +$$

$$\Delta t \sum_{i=0}^{n-1} \left[f(X_i) \exp\left(\Delta t \sum_{j=0}^{i-1} \gamma(X_j)\right)\right].$$

By repeating this procedure nr times, we generate a sample of $Z(\tau)$, formed by the values $Z_1, ..., Z_{nr}$ and we may estimate:

$$u(x) \approx \frac{1}{nr} \sum_{i=0}^{nr} Z_i.$$

Matlab implementation is performed as follows: assume that a(x),b(x),c(x),f(x) are furnished by the subprograms a(x), b(x), c(x), f(x), respectively. In addition, assume that a subprogram interior(x) returns either the value false (or 0) if $x \notin (0,\ell)$ or the value true (or 1) if $x \in (0,\ell)$. Finally, assume that the boundary values are furnished by a subprogram ub(x) such that $ub(0) = u_0$ and $ub(\ell) = u_\ell$. The following is an example of Matlab code.

Listing 6.3. *Deterministic boundary value problem*

```
function v = ode_stoc_part(a,b,c,f,interior,u_b,xc,ns,delta_t)
%
% determines the solution of the ODE
% (a(x)^2/2)*u''(x) + b(x)*u'(x) + c(x)*u(x) + f(x) = 0 on (xl
    ,xu)
%       u(xl) = ul  , u(xu) = uu
% at the points of vector xc
% by performing ns simulations with step delta_t
%
% IN:
% a,b,c,f : coefficients - type anonymous function
% interior : true, if x is interior; false otherwise - type
    anonymous function
% u_b : boundary value of the solution - type anonymous
    function
% xc : points where the solution has to be evaluated - type
    array of double
% ns : number of simulations - type integer
% deltat_t : time step - type double
%
% OUT:
% v : estimation of the value - type array of double
%
v = zeros(size(xc));
aa = @(x,t) a(x);
bb = @(x,t) b(x);
for n = 1: length(xc)
    x = xc(n);
    s = 0;
    for i = 1:ns
        aux = Z_variate(aa,bb,c,f,delta_t,interior,u_b,x);
        s = s + aux;
    end;
    v(n) = s/ns;
end;
return;
end

function v = Z_variate(a,b,gama,f,dt,interior,u_boundary,xa)
%
% furnishes one variate from Z
%
x = xa;
t = 0;
inreg = interior(x,t);
s_gama = 0;
```

```
s_f = 0;
ndimw = length(x);
while inreg
    s_f = s_f + f(x)*exp(dt*s_gama);
    s_gama = s_gama + gama(x);
    [xn, tn] = new_point_ito(a,b,x,t,ndimw,dt);
    x = xn;
    t = tn;
    inreg = interior(x,t);
end;
v = u_boundary(x)*exp(dt*s_gama) + dt*s_f;
return;
end

function [xn, tn] = new_point_ito(a,b,x,t,ndimw,dt)
%
% furnishes the new point corresponding to one step of
% the numerical simulation of Ito's diffusion
%   dX_t = a(X_t, t) dW_t + b(X_t, t) dt
%
% IN:
% a: the first coefficient — type anonymous function
% b: the second coefficient — type anonymous function
% x: the actual point — type double
% t: the actual time — type double
% ndimw: dimension of the Wiener process
% dt: the time step — type double
%
% OUT:
% xn : the new value of X — type array of double
% tn : the new time — type double
%
dw = randn(ndimw,1)*sqrt(dt);
dx = a(x,t)*dw + b(x,t)*dt;
xn = x + dx;
tn = t + dt;
return;
end
```

EXAMPLE 6.3.– Let us consider the differential equation:

$$u'' - \pi^2 u = 0, \text{ on } (0,1); \ u(0) = 1, u(1) = e.$$

The exact solution is $u(x) = \exp(x)$. Using the method above with $ns = 2{,}500$, we obtain the results shown in Table 6.1. The relative error in mean quadratic norm is about 3% in all the simulations.

x	0.1	0.25	0.5	0.75	0.9
$\Delta t = 1e-3$	1.1488	1.3712	1.735	2.171	2.4633
$\Delta t = 2.5e-4$	1.1391	1.3317	1.7237	2.1635	2.4952
$\Delta t = 1e-4$	1.1369	1.3543	1.705	2.1859	2.4893
exact	1.1052	1.2840	1.6487	2.1170	2.4596

Table 6.1. *Results for Example 6.3*

6.5.2. *Elliptic partial differential equations*

Let us consider the PDE (we use Einstein's convention about the sum of repeated indexes $1 \leq i, j \leq n$):

$$\alpha_{ij}(x) \frac{\partial^2 u}{\partial x_i \partial x_j} + \beta_i(x) \frac{\partial u}{\partial x_i} + \gamma(x) u(x) + f(x) = 0 \text{ on } \Omega;$$
$$u = u_{\partial\Omega} \text{ on } \partial\Omega.$$

Let us assume that:

$$\alpha = \frac{1}{2} a a^t \iff \alpha_{ij} = \frac{1}{2} a_{ik} a_{jk}. \qquad [6.34]$$

We consider the multidimensional stochastic diffusion:

$$dX_t = a(X_t) dW_t + b(X_t) dt \quad (b = \beta) \; ; \; X(0) = x \in \Omega. \qquad [6.35]$$

and the stochastic process:

$$Y(t) = u(X(t)) \exp\left(\int_0^t \gamma(X(s)) ds \right).$$

We have $Y = F\left(t, X(t), \int_0^t \gamma(X(s)) ds \right)$ where $F: \mathbb{R} \times \mathbb{R}^n \times \mathbb{R} \longrightarrow \mathbb{R}$ verifies:

$$F(t, x, z) = u(x) \exp(z).$$

In this case:

$$\frac{\partial F}{\partial z} = u(x)\exp(z) \,;\, \frac{\partial F}{\partial t} = 0 \,;\, \frac{\partial F}{\partial x_i} = \frac{\partial u}{\partial x_i}(x)\exp(z) \,;$$
$$\frac{\partial^2 F}{\partial x_i \partial x_j} = \frac{\partial^2 u}{\partial x_i \partial x_j}(x)\exp(z) \,,$$

so that:

$$dY_t = ((A_t)_i\, d(W_t)_i + B_t dt)\exp\left(\int_0^t \gamma(X(s))\, ds\right),$$

where:

$A_t = a(X_t)\nabla u(X_t);$
$B_t = \frac{1}{2}a_{ik}(X_t)a_{jk}(X_t)\frac{\partial^2 u}{\partial x_i \partial x_j}(X_t) + b_i(X_t)\frac{\partial u}{\partial x_i}(X_t) + \gamma(X_t)u(X_t)$.

In an analogous manner, we consider:

$$\tau = \inf\{t \geq 0 : X(t) \notin \Omega\}.$$

For $t < \tau$, we have $X_t \in \Omega$, so that, for $t < \tau$:

$$B_t = -f(X_t)$$

and we also have:

$$u(x) = E(Z(\tau)),$$

where:

$$Z(\tau) = u_{\partial\Omega}(X_\tau)\exp\left(\int_0^\tau \gamma(X(t))\, dt\right)$$
$$+ \int_0^\tau f(X(t))\exp\left(\int_0^t \gamma(X(s))\, ds\right) dt.$$

Thus, $u(\mathbf{x})$ may be approximated by the method presented in the previous section. Matlab implementation is performed by modifying the subprograms evaluating $a(\mathbf{x})$, $b(\mathbf{x})$ a(x), b(x), respectively): here, $\mathbf{x}$ is a multidimensional vector instead of a real number. Analogously to the preceding situation, let us assume that, on the one hand, a subprogram interior(x) returns either the value false (or 0) if $x \notin \Omega$ or the value true (or 1) if $x \in \Omega$, and, on the other hand, the boundary values are furnished by a subprogram ub(x). Then, an example of Matlab code is:

Listing 6.4. *Deterministic elliptic PDE*

```
function v = ell_stoc_part(a,b,c,f,interior,u_b,xc,ns,delta_t)
%
%  determines the solution of the PDE
%  (a_i(x)a_j(x)/2)*D_ij u) + b_i(x)*D_i u(x) + c(x)*u(x) + f(x
     ) = 0 on
%  \Omega
%       u(x) = u_b(x) on \partial\Omega
%
%  at the points of vector xc
%  by performing ns simulations with step delta_t
%
% IN:
%  a,b,c,f : coefficients - type anonymous function
%  interior : true, if x is interior; false otherwise - type
     anonymous function
%  u_b : boundary value of the solution - type anonymous
     function
%  xc : points where the solution has to be evaluated - type
     array of double
%    xc(:,i) is a point of evaluation
%  ns : number of simulations - type integer
%  deltat_t : time step - type double
%
% OUT:
%  v : estimation of the value - type array of double
%
v = zeros(1,size(xc,2));
aa = @(x,t) a(x);
bb = @(x,t) b(x);
for n = 1: length(xc)
    x = xc(:,n);
    s = 0;
    for i = 1:ns
        aux = Z_variate(aa,bb,c,f,delta_t,interior,u_b,x);
        s = s + aux;
    end;
```

```
        v(n) = s/ns;
end;
return;
end
```

EXAMPLE 6.4.– Let us consider $\Omega = \{\mathbf{x} = (x_1, x_2, x_3) \in \mathbb{R}^3 : x_i > 0$ for $i = 1, 2, 3$ and $x_1^2 + x_2^2 + x_3^2 < 1\}$ and the PDE:

$$\Delta u = 0, \text{ on } \Omega; \ u(x) = x_1^2 - x_2^2/2 - x_3^2/2 \text{ on } \partial\Omega.$$

The exact solution is $u(x) = x_1^2 - x_2^2/2 - x_3^2/2$. We evaluate the solution at the points $(2r, r, r)/\sqrt{(6)}$ for different values of r. Using the method above with $\Delta t = 1e - 3$, we obtain the results shown in Table 6.2. The relative error is about 0.4% in all the simulations.

r	0.1	0.25	0.5	0.75	0.9
$ns = 2,500$	0.0051	0.0323	0.1211	0.2915	0.39745
$ns = 10,000$	0.0052	0.0312	0.1239	0.2806	0.3966
$ns = 25,000$	0.0047	0.0311	0.1246	0.2813	0.4068
exact	0.0050	0.0313	0.1250	0.2813	0.4050

Table 6.2. *Results for Example 6.4*

Results obtained using $ns = 2,500$ and different values of Δt are shown in Table 6.3.

r	0.1	0.25	0.5	0.75	0.9
$\Delta t = 1e - 3$	0.0051	0.0323	0.1211	0.2915	0.39745
$\Delta t = 1e - 4$	0.0054	0.0299	0.1272	0.2911	0.4037
exact	0.0050	0.0313	0.1250	0.2813	0.4050

Table 6.3. *Results for Example 6.4*

6.5.3. *Parabolic partial differential equations*

Let us consider the PDE (we use Einstein's convention about the sum of repeated indexes $1 \leq i, j \leq n$):

$$\frac{\partial u}{\partial t}(x,t) = \alpha_{ij}(x,t) \frac{\partial^2 u}{\partial x_i \partial x_j} + \beta_i(x,t) \frac{\partial u}{\partial x_i}$$
$$+ \gamma(x,t) u(x,t) + f(x,t) \text{ on } \Omega \times (0,T);$$
$$u = u_{\partial\Omega} \text{ on } \partial\Omega; \ u(x,0) = u_0(x) \text{ on } \Omega.$$

We assume that [6.34] is satisfied and we consider the multidimensional stochastic diffusion:

$$dX_t = a(X_t, S_t)\,dW_t + b(X_t, S_t)\,dt \quad (b = \beta) \,;\, dS_t = -dt; \quad [6.36]$$

$$X(0) = x \in \Omega;\, S(0) = t \quad [6.37]$$

and the stochastic process:

$$Y(t) = u(X(t), S(t)) \exp\left(\int_0^t \gamma(X(s), S(s))\,ds\right).$$

By setting $\widetilde{X} = (X, S)$, we have $Y = F\left(t, \widetilde{X}(t), \int_0^t \gamma\left(\widetilde{X}(s)\right)ds\right)$ where $F : \mathbb{R} \times \mathbb{R}^{n+1} \times \mathbb{R} \longrightarrow \mathbb{R}$ verifies

$$F(t, \widetilde{x}, z) = u(\widetilde{x})\exp(z).$$

In this case, for

$$\tau = \inf\left\{t \geq 0 : \widetilde{X}(t) \notin \Omega \times (0, T)\right\}$$

we have:

$$u(x) = E(Z(\tau)),$$

where:

$$Z(\tau) = u_{\partial\Omega}(X_\tau, S_\tau) \exp\left(\int_0^\tau \gamma(X(t), S(t))\,dt\right)$$

$$+ \int_0^\tau f(X(t), S(t)) \exp\left(\int_0^t \gamma(X(s), S(s))\,ds\right)dt.$$

Thus, we may approximate $u(x,t)$ by the method previously introduced. We note that:

$$\tau = \inf\{t \geq 0 : X(t) \notin \Omega \text{ ou } S(t) \notin (0,T)\}.$$

For Matlab implementation, let us assume that subprograms evaluating $a(\mathbf{x},t), b(\mathbf{x},t), \gamma(\mathbf{x},t)$ are available (a(x,t), b(x,t), gama(x,t), respectively). The subprogram interior(x,t) returns either the value false (or 0) if $x \notin \Omega$ or $t \notin (0,T)$. It returns the value true (or 1) if $x \in \Omega$ and $t \in (0,T)$. Moreover, the boundary values are furnished by a subprogram ub(x) and the initial values by a subprogram u0(x). Then, an example of Matlab code is:

Listing 6.5. *Deterministic parabolic PDE*

```
function v = for_stoc_part(a,b,c,f,interior ,u_b,u_0,xtc,ns,
    delta_t)
%
%   determines the solution of the PDE
%   D_t u = (a_i(x)a_j(x)/2)*D_ij u) + b_i(x)*D_i u(x) + c(x)*u(
    x) + f(x)    on
%   \Omega
%        u(x) = u_b(x) on \partial\Omega
%
%   at the points of vector xc
%   by performing ns simulations with step delta_t
%
% IN:
% a,b,c,f : coefficients - type anonymous function
% interior : true, if x is interior; false otherwise - type
    anonymous function
% u_b : boundary value of the solution - type anonymous
    function
% u_0 : initial value of the solution - type anonymous function
% xtc : points where the solution has to be evaluated - type
    array of double
%     xtc(1:ndim,n) = spatial coordinates of the point of the
    point number n
%     xtc(ndim+1,n) = temporal coordinate of the point number n
% ns : number of simulations - type integer
% deltat_t : time step - type double
%
% OUT:
% v : estimation of the value - type array of double
%
```

```
v = zeros(1,size(xtc,2));
ndim = size(xtc,1) - 1;
for n = 1: length(xtc)
    x = xtc(1:ndim,n);
    t = xtc(ndim + 1, n);
    s = 0;
    for i = 1:ns
        aux = Z_variate(a,b,c,f,delta_t,interior,u_b,u_0,x,t);
        s = s + aux;
    end;
    v(n) = s/ns;
    disp(['v = ',num2str(v(n)),', ',num2str(xtc(1,n)^2 - 0.5*(
        xtc(2,n)^2 + xtc(3,n)^2) + t)]);
end;
return;
end

function v = Z_variate(a,b,gama,f,dt,interior, u_boundary,u_0,
    xa,ta)
%
% furnishes one variate from Z
%
x = xa;
t = ta;
inreg = interior(x,t);
s_gama = 0;
s_f = 0;
ndimw = length(x);
while inreg
    s_f = s_f + f(x)*exp(dt*s_gama);
    s_gama = s_gama + gama(x);
    [xn, tn] = new_point_ito(a,b,x,t,ndimw,dt);
    inreg = interior(xn,tn);
    x = xn;
    t = tn;
end;
if t <= 0
    v = u_0(x)*exp(dt*s_gama) + dt*s_f;
else
    v = u_boundary(x,t)*exp(dt*s_gama) + dt*s_f;
end;
return;
end

function [xn, tn] = new_point_ito(a,b,x,t,ndimw,dt)
%
% furnishes the new point corresponding to one step of
```

```
% the numerical simulation of Ito's diffusion
%   dX_t = a(X_t, t) dW_t + b(X_t, t) dt
%   DT_t = -dt
%
% IN:
% a: the first coefficient - type anonymous function
% b: the second coefficient - type anonymous function
% x: the actual point - type double
% t: the actual time - type double
% ndimw: dimension of the Wiener process
% dt: the time step - type double
%
% OUT:
% xn : the new value of X - type array of double
% tn : the new time - type double
%
dw = randn(ndimw,1)*sqrt(dt);
dx = a(x,t)*dw + b(x,t)*dt;
xn = x + dx;
tn = t - dt;
return;
end
```

EXAMPLE 6.5.– Let us consider $\Omega = \{\mathbf{x} = (x_1, x_2, x_3) \in \mathbb{R}^3 : x_i > 0$ for $i = 1, 2, 3$ and $x_1^2 + x_2^2 + x_3^2 < 1\}$ and the PDE:

$\frac{\partial u}{\partial t}(x,t) = \Delta u + 1$, on Ω; $u(x,t) = x_1^2 - x_2^2/2 - x_3^2/2 + t$, on $\partial\Omega$; $u(x,0) = x_1^2 - x_2^2/2 - x_3^2/2$, on Ω.

The exact solution is $u(x,t) = x_1^2 - x_2^2/2 - x_3^2/2 + t$. We evaluate the solution at the points $(x = (2r, r, r), t(r))/\sqrt{(6)}$ for different values of r. Using the method above with $\Delta t = 1e-3$, we obtain the results shown in Table 6.4. The relative error is about 0.4% in all the simulations.

r	0.1	0.25	0.5	0.75	0.9
$t(r)$	0.1	0.2	0.3	0.4	0.5
$ns = 2,500$	0.1050	0.2328	0.4213	0.6824	0.9069
$ns = 10,000$	0.1045	0.2331	0.4254	0.6820	0.9042
$ns = 25,000$	0.1051	0.2310	0.4254	0.6823	0.9060
exact	0.1050	0.2313	0.4250	0.6813	0.9050

Table 6.4. *Results for Example 6.5*

Results obtained using $ns = 2,500$ and different values of Δt are shown in Table 6.5.

r	0.1	0.25	0.5	0.75	0.9
$t(r)$	0.1	0.2	0.3	0.4	0.5
$\Delta t = 1e-3$	0.1050	0.2328	0.4213	0.6824	0.9069
$\Delta t = 1e-4$	0.1046	0.2332	0.4285	0.6838	0.9058
exact	0.1050	0.2313	0.4250	0.6813	0.9050

Table 6.5. *Results for Example 6.4*

6.5.4. *Finite difference schemes*

Finite difference approximation schemes may be interpreted as probabilistic schemes: let us consider a sequence of random variables $\{X_n\}_{n \geq 0}$ such that the law of X_{n+1} conditional to X_n is independent of n and given by:

$$f(x \mid X_n = y) = \pi^h(x, y),$$

where $h > 0$ is a parameter.

Let $\{X_t^h\}_{t \geq 0}$ be the stochastic process obtained by linear interpolation of $\{X_n\}_{n \geq 0}$ with a step h:

$$X^h(nh) = X_n;\ t \in (nh, (n+1)h) \Longrightarrow X^h(t) = X_n + \left(\frac{t-nh}{h}\right)(X_{n+1} - X_n).$$

Let us consider:

$$a_{ij}^h(x) = \frac{1}{h} \int_{\|y-x\| \leq 1} (y_i - x_i)(y_j - x_j) \pi^h(x, y) \ell(dy),$$

$$b_i^h(x) = \frac{1}{h} \int_{\|y-x\| \leq 1} (y_i - x_i) \pi^h(x, y) \ell(dy),$$

$$A^h(u)(x) = \frac{1}{h} \int_{\mathbb{R}^n} (u(y) - u(x)) \pi^h(x, y) \ell(dy).$$

We have (see [DAU 89])

THEOREM 6.1.– If

i) $a^h(x) \longrightarrow a(x)$ and $b^h(x) \longrightarrow b(x)$ uniformly on all compact subset of $\mathbb{R}^n$ for $h \longrightarrow 0+$;

ii) $\{a^h(x)\}_{h>0}$ and $\{b^h(x)\}_{h>0}$ are bounded, independently of x;

iii) $\forall\, \varepsilon > 0 : \sup\left\{ \dfrac{1}{h} \displaystyle\int_{\mathbb{R}^n - B_\varepsilon(x)} \pi^h(x,y)\,\ell(dy) : x \in \mathbb{R}^n \right\} \longrightarrow 0$ to $h \longrightarrow 0+$;

and then, for $h \longrightarrow 0+$, X_t^h converges in distribution to X_t such that:

$$dX_t = a(X_t)\,dW_t + b(X_t)\,dt;\quad X(0) = x$$

and

$$A^h(u)(x) \longrightarrow A(u)(x) = \left.\frac{d}{dt}E(u(X_t)|X_0 = x)\right|_{t=0}.\ \blacksquare$$

Let us consider the equation:

$$\Delta u = -f \text{ on } \Omega \subset \mathbb{R}^2;\ u = u_{\partial\Omega} \text{ on } \partial\Omega.$$

When using the finite difference discretization scheme defined by:

$$\frac{u_{i+1,j} - 2u_{i,j} + u_{i-1,j}}{h_1^2} + \frac{u_{i,j+1} - 2u_{i,j} + u_{i,j-1}}{h_2^2} = -f_{i,j},$$

we have:

$$\frac{1}{h_1^2}u_{i+1,j} + \frac{1}{h_1^2}u_{i-1,j} + \frac{1}{h_2^2}u_{i,j+1} + \frac{1}{h_2^2}u_{i,j-1} - 2\left(\frac{1}{h_1^2} + \frac{1}{h_2^2}\right)u_{i,j} = -f_{i,j},$$

i.e.:

$$\Pi_{i+1,j}u_{i+1,j} + \Pi_{i-1,j}u_{i-1,j} + \Pi_{i,j+1}u_{i,j+1} + \Pi_{i,j-1}u_{i,j-1} - u_{i,j} = -hf_{i,j},$$

where:

$$h = \frac{h_1^2 h_2^2}{2(h_1^2 + h_2^2)}; \Pi_{i+1,j} = \Pi_{i-1,j} = \frac{h}{h_1^2}; \Pi_{i,j+1} = \Pi_{i,j-1} = \frac{h}{h_2^2}.$$

We observe that:

$$\Pi_{i+1,j} + \Pi_{i-1,j} + \Pi_{i,j+1} + \Pi_{i,j-1} = 1.$$

Let $\{e_1, e_2\}$ be the canonical basis of $\mathbb{R}^2$ and:

$$\pi^h(x, y) = \Pi_{i-1,j}\delta(x - e_1 - y) + \Pi_{i+1,j}\delta(x + e_1 - y)$$
$$+ \Pi_{i,j-1}\delta(x - e_2 - y) + \Pi_{i,j+1}\delta(x + e_2 - y).$$

We have:

$$A^h(u)(x) = \Pi_{i+1,j}u(x + e_1) + \Pi_{i-1,j}u(x - e_1)$$
$$+ \Pi_{i,j+1}u(x + e_2) + \Pi_{i,j-1}u(x - e_2);$$

$a_{ij}^h(x) = 0$, if $i \neq j$; $a_{11}^h(x) = \Pi_{i+1,j} + \Pi_{i-1,j}$;
$a_{22}^h(x) = \Pi_{i,j+1} + \Pi_{i,j-1}$; $b^h = 0$.

Applying the theorem establishes, we obtain:

$$\left.\frac{d}{dt}E(u(X_t)|X_0 = x)\right|_{t=0} = \lim_{h \to 0+} A^h(u),$$

so that we may approximate the solution by using X_t^h. We have:

$$E(u(X_{i+1})|X_i) - u(X_i) \approx hA^h(u)(X_i) = hf(X_i).$$

By making the sum of these equalities up to the index $n - 1$ and taking the mean of the result:

$$u(x) \approx E\left(u_{\partial\Omega}(X_n) + h\sum_{i=0}^{n-1} f(X_i)\right)$$

6.6. Statistics of dynamical systems

The analysis of dynamical systems affected by uncertainty is a subject in development. In mechanics, for instance, the variability of material parameters, geometry, initial conditions or boundary conditions introduces uncertainty in the dynamical systems. For instance, a pendulum may be affected by various uncertainties concerning its length, mass, rigidity, damping initial position, velocity, etc. All these variabilities affect its dynamics and some of them may have a significant impact on its natural frequencies and its stability.

Specific difficulties arise when we are interested in periodic motions, since the usual parameters – such as period, eigenvalues of the linearized system, etc. – become random variables. For instance, some eigenvalues may be have a sign that depends on the values taken by these parameters, with consequences on the stability. In addition, typical curves such as limit cycles, periodic orbits and Poincaré sections become random objects.

As in the other situations considered in this text, we are interested in the determination of the probability distributions and statistics of these elements and some associated parameters. This kind of analysis involves particular difficulties: curves such as limit cycles, orbits, etc., belong to infinite-dimensional functional spaces and we need to construct probabilities in these spaces. In addition, these probabilities must be operational in the sense that they have to be used in order to provide numerical evaluation of means, variances and probabilities of events. We may find in the literature a traditional procedure furnished by cylindrical measures (see, for example, [SCH 69, BAD 70, BAD 74]), but this approach is not operational enough in order to furnish numerical methods for the evaluation of the quantities of interest. It may be convenient to adopt the point of view of section 1.9, which leads to the construction of statistics into a natural manner. It is assumed in the following that the reader has the capacity to construct these operational infinite-dimensional probabilities: the presented results use the approach of section 1.9.

Let us consider a general dynamical system on $\mathbb{R}^n$, which is defined by the equations:

$$\begin{cases} \frac{d\mathbf{x}}{dt} = \mathbf{f}(\mathbf{x}, t) \ \forall t \in [O\ ;\ T],\ \mathbf{x} \in \mathbb{R}^n \\ \mathbf{x}(0) = \mathbf{x}_0 \qquad \mathbf{x}_0 \in \mathbb{R}^n \end{cases} \qquad [6.38]$$

A simple example is given by the basic harmonic oscillator formed by a spring and a mass (oscillator 1):

$$\begin{cases} \dot{x}_1(t) = x_2(t) & \forall t \in [O\,;\,T] \\ \dot{x}_2(t) = -\frac{k}{m} x_1(t) &, \\ \mathbf{x}(0) = \mathbf{X}_0 & \text{random} \end{cases} \quad [6.39]$$

where $\mathbf{x} = (x_1\,;\,x_2)^t$, k is the spring's stiffness, m is the mass and $\dot{x}_1 = dx_1/dt$. Figures 6.7 and 6.8 (see [CRO 10]) show the phase portrait and the trajectories for different values of $\mathbf{x}_0$ with $k = 1$ and $m = 1$. It is interesting to note that the limit cycle varies with $\mathbf{x}_0$.

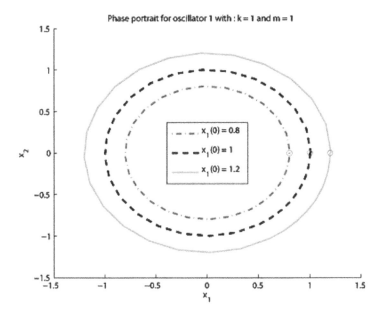

Figure 6.7. *Phase portrait of oscillator 1 for different initial conditions*

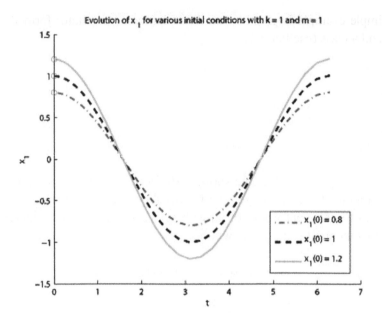

Figure 6.8. *Evolution of $x_1(t)$ from oscillator 1 for different initial conditions*

A more complex situation is furnished by the Van der Pol's oscillator [DER 28], which has been extensively studied in the literature. One of its interesting properties is the existence of a limit cycle that is independent of the initial conditions. Van der Pol's equations are given by (oscillator 2):

$$\begin{cases} \dot{x}_1 = x_2 & \forall t \in [0\,;\,T] \\ \dot{x}_2 = \varepsilon(1 - x_1^2)x_2 - \Omega_0^2 x_1 \\ \mathbf{x}(0) = \mathbf{x}_0 \end{cases} \quad , \qquad [6.40]$$

where ε is a parameter measuring the nonlinearity of the system and Ω_0 is a positive random variable. Ω_0 influences the properties of the system, as it may be observed in the figures below: the limit cycles are shown in Figure 6.9 and the trajectories for x_1 are exhibited in Figure 6.10 – we use $\varepsilon = 1$. Table 6.6 shows the values of the period (see [CRO 10]).

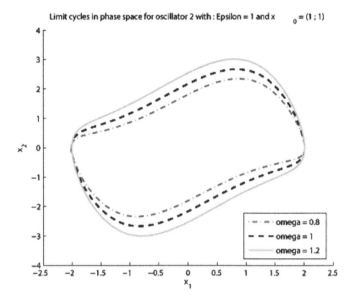

Figure 6.9. *Periodic orbits for oscillator 2*

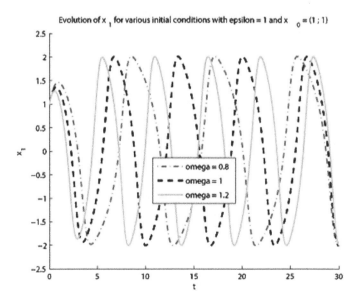

Figure 6.10. *Evolution of $x_1(t)$ for oscillator 2*

Ω_0	period	$\max_t x_1(t)$	$\min_t x_1(t)$	$\max_t x_2(t)$	$\min_t x_2(t)$
0.8	8.58	2.01	-2.01	2.35	-2.35
1	6.66	2.00	-2.01	2.67	-2.67
1.2	5.46	2.01	-2.00	3.01	-3.00

Table 6.6. *Characteristics of the periodic orbits for different values of Ω_0*

6.6.1. *Evaluation of statistics of the limit cycle*

A limit cycle is a periodic orbit (so, a closed curve) so that a limit cycle is described by a continuous function $y: I \to \mathbb{R}^n$, where I is an interval of $\mathbb{R}$. Since we are not interested in temporal aspects, we may, without loss of generality, consider $I = [0; 1]$. As exposed in section 1.9, we consider curves belonging to a separable Hilbert space V, possessing a Hilbert basis $\Phi = \{\phi_n\}_{n \geq 1}$ and we use a representation of the limit cycles in this basis:

$$\mathbf{Y}(\omega)(\cdot) = \sum_{i \geq 1} \mathbf{A}_i(\omega) \phi_i(\cdot) \quad [6.41]$$

Thus, the limit cycle is a random variable taking its values on V, which may be identified to the sequence $\{\mathbf{A}_i(\omega)\}_{i \geq 1}$ of the coefficients of its development in the Hilbert basis under consideration. Usually, the mean of a random variable X having a law μ and taking values on Ω is given by:

$$\mathbb{E}[X] = \int_\Omega x(\omega) d\mu(\omega). \quad [6.42]$$

Here, μ is defined on V – which is an infinite-dimensional space – and the classical definitions given by Riemann or Lebesgue do not apply. The adapted tool to handle this situation is furnished by Bochner's functional spaces (see, for instance, [BOC 33]), which corresponds to the approach presented in section 1.9: initially, the integral is defined for simple functions. Then, it is extended to sequences of simple functions and their limits. The theory established in this field furnishes conditions that make possible the equality (see, for example, [DIE 77]):

$$\mathbb{E}[Y(\omega)] = \lim_{n \to \infty} \mathbb{E}\left[\sum_{i=1}^n A_i \phi_i\right]$$
$$= \lim_{n \to \infty} \sum_{i=1}^n \mathbb{E}[A_i] \phi_i.$$

In practice, we use an approximation of $Y(\omega)$ by its projection on a finite-dimensional space V_N having dimension N:

$$p_N Y = \sum_{i=1}^{N} A_i \phi_i, \qquad [6.43]$$

where $p_N(\cdot)$ is the orthogonal projection from V onto V_N. In this case:

$$p_N\left(\mathbb{E}\left[Y(\omega)\right]\right) = \sum_{i=1}^{N} \mathbb{E}\left[A_i\right] \phi_i \qquad [6.44]$$

Figures 6.11–6.12 show the results obtained by collocation, using a basis of linear finite elements and a sample of 20 periodic orbits. For oscillator 1, $x_0 = (1; U)$ where U is uniformly distributed on $[0.5\,;\,1.5]$. For Van der Pol's oscillator, we consider $\Omega_0 = U$ (see [CRO 10]).

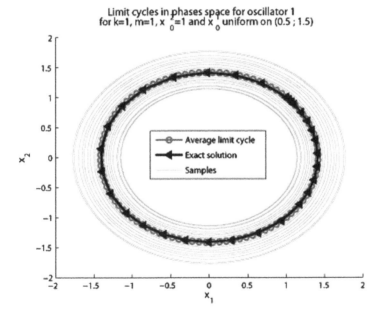

Figure 6.11. *Mean periodic orbit and sample of periodic orbits in the phase space of oscillator 1*

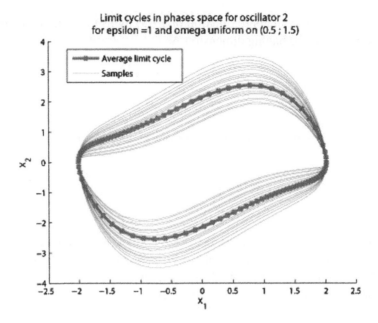

Figure 6.12. *Mean periodic orbit and sample of periodic orbits in the phase space of oscillator 2*

In the case of oscillator 1, the exact periodic orbit is:

$$\mathbf{Y}_\omega(t) = \begin{pmatrix} \cos(t) + U\sin(t) \\ -\sin(t) + U\cos(t)) \end{pmatrix} \qquad [6.45]$$

and it is possible to determine the exact mean. We observe that the mean furnished by the method agrees with the exact result.

If we are interested in the distributions of the coefficients $\{\mathbf{A}_i(\omega)\}_{i\geq 1}$, we may use the methods previously introduced. Let us define:

$$\mathbf{PY} = \sum_{n\geq 1}\sum_{j\geq 1} a_{n,j}\psi_j(\{\boldsymbol{\xi}_k(\omega)\}_{k\geq 1})\phi_n(.) \qquad [6.46]$$

and look for the deterministic coefficients $a_{n,j} \in \mathbb{R}^n$ – which may be determined by one of the methods previously presented, such as collocation. Figure 6.13 shows the results obtained for Van der Pol's oscillator by using a sample of 40 limit cycles and a Hilbert basis formed by Hermite polynomials, with a second-order approximation and an expansion variable $\boldsymbol{\xi} = \xi_1$ uniformly distributed one $[0; \ 1]$. The figure shows the 95% confidence interval for the periodic orbit (dotted line, results obtained by collocation, see [CRO 10]).

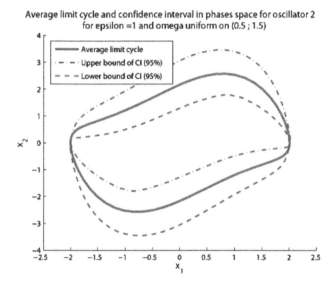

Figure 6.13. *Mean periodic orbit and its 95% confidence interval for oscillator 2*

7
Optimization Under Uncertainty

Optimization looks for the best values of some parameters with regard to a given cost, under given restrictions. If some of these elements involve parameters collected in a vector **v**, the solution **x** becomes a function **x** (**y**) of **v**. Thus, when stochastic or variable parameters are part of the optimization problem, the solutions become stochastic or variable too.

For instance, let us consider the simple situation where we are looking for the real number x that minimizes the function $F(x) = (x - v_1)^2$ under the restriction $x \geq v_2$: the obvious solution is

$$x = \begin{cases} v_1, \text{ if } v_1 \geq v_2 \\ v_2, \text{ otherwise} \end{cases}, \quad F(x) = \begin{cases} 0, \text{ if } v_1 \geq v_2 \\ (v_2 - v_1)^2, \text{ otherwise} \end{cases}.$$

If v_1 and v_2 are random variables, the optimal point x and the optimal value $F(x)$ are also random variables. For instance, if they vary both on $(0, 1)$, x varies on $(-1, 1)$ and $F(x)$ varies on $(0, 1)$.

An example of catastrophic variations is furnished by the minimization of $F(x_1, x_2) = ((v_1 - 1)^2 + 1)x_1^2 - 4(v_1 - 1)x_1 x_2 + 2x_1 - 2x_2 + ((v_1 - 1)^2 + 1)x_2^2$. In this case, the obvious solution is for $v_1 \neq 0$, $x_1 = -1/v_1^2$, $x_2 = 1/v_1^2$. For $v_1 = 0$, there are infinitely many solutions. If the parameter v_1 varies on $[-\varepsilon, \varepsilon]$, the values of x_1 may vary from $-\infty$ to $-1/\varepsilon^2$ and those of x_2 from $1/\varepsilon^2$ to ∞.

These sample examples show that the variability of the optima has to be taken into account in order to ensure security and reliability. In this chapter, we are interested in the determination of the distribution of the optimal point. An alternative approach that looks for a solution satisfying restrictions with a given probability is the realibility-based optimization, presented in Chapter 8.

7.1. Representation of the solutions in unconstrained optimization

A deterministic unconstrained optimization problem (UOP) is:

$$\mathbf{x} = \arg\min \{F(\mathbf{y}) : \mathbf{y} \in \mathbb{R}^n\}$$

where $F : \mathcal{M}(n,1) \longrightarrow \mathbb{R}$ is a real-valued function and $\mathbf{x} = (x_i) \in \mathcal{M}(n,1)$ is the unknown. However, in many practical situations, F depends also on parameters affected by some uncertainty. In such a situation, the UOP above formulated becomes a UOP under uncertainty. For instance, we may consider that the uncertainty is represented by an uncertain vector $\mathbf{v} = (v_1, ..., v_{nr})$ and that $F = F(\mathbf{x}, \mathbf{v})$. Thus, the UOP becomes

$$\mathbf{X} = \arg\min \{F(\mathbf{Y}, \mathbf{v}) : \mathbf{Y} \in \mathbb{R}^n\} \qquad [7.1]$$

The situation of equation [7.1] may be studied by different techniques, such as stochastic programming, reliability-based design optimization and fuzzy programming. Stochastic and reliability approaches model the uncertainty by using random variables, while the fuzzy approach considers uncertain parameters as fuzzy numbers. Usually, stochastic and reliability approaches are used when the uncertainty may be properly addressed by known random processes or variables, while the fuzzy approach is used in situations where the uncertainty cannot be easily quantified. Taking into account the presence of $\mathbf{v}$, $\mathbf{X}$ becomes uncertain.

Here, we are interested in the situation where $\mathbf{X}$ is considered a random variable and our goal is the numerical determination of its probability distribution; this approach is different from the approaches cited above: stochastic programming generally involves the minimization of statistics characteristics of F – for instance, its mean or variance; reliability-based design optimization introduces probabilistic constraints and looks for the minimization of F subject to these probabilistic constraints; fuzzy optimization produces fuzzy vectors and not probability distributions. We observe that the knowledge of the probability distribution of $\mathbf{X}$ allows the determination of statistics of the solution $\mathbf{X}$ and of the optimal value of F.

Analogously to the previous situations, we consider an approximation **PX** of **X** in a convenient subspace S of random variables (see section 3). For instance, let $F = \{\varphi_k\}_{k \in \mathbb{N}}$ be a convenient family of functions, $N_X \in \mathbb{N}^*$, $\boldsymbol{\xi}$ is a convenient random variable, $\boldsymbol{\varphi}(\boldsymbol{\xi}) = (\varphi_1(\boldsymbol{\xi}), ..., \varphi_p(\boldsymbol{\xi}))^t \in \mathcal{M}(N_X, 1)$ and

$$\mathbf{PX} = \boldsymbol{\chi}\boldsymbol{\varphi}(\boldsymbol{\xi}) = \sum_{k=1}^{N_X} \boldsymbol{\chi}_k \varphi_k(\boldsymbol{\xi}) \left(\text{i.e. } (\mathbf{PX})_j = \sum_{k=0}^{N_X} \chi_{jk} \varphi_k(\boldsymbol{\xi}) \right), \quad [7.2]$$

which corresponds to

$$S = [\{\varphi_1(\boldsymbol{\xi}), ..., \varphi_{N_X}(\boldsymbol{\xi})\}]^n$$

$$= \left\{ \sum_{k=1}^{N_X} \mathbf{D}_k \varphi_k(\boldsymbol{\xi}) : \mathbf{D}_k \in \mathbb{R}^n, 1 \leq k \leq N_X \right\}. \quad [7.3]$$

As previously, the unknown to be determined is $\boldsymbol{\chi} = (\chi_{ij}) \in \mathcal{M}(n, N_X)$. For instance, equation [7.1] may be approximated as

$$\mathbf{PX} = \arg\min \{F(\mathbf{Y}, \mathbf{v}) : \mathbf{Y} \in S\} \quad [7.4]$$

and $\boldsymbol{\chi}$ may be determined by solving equation [7.4] by one of the approaches discussed in the following.

7.1.1. *Collocation*

Analogously to the preceding situations, we may consider a sample $\mathcal{X} = (\mathbf{X}_1, ..., \mathbf{X}_{ns})$ of ns variates from $\mathcal{X}$ is available, and solve the system of equations:

$$\mathbf{PX}(\boldsymbol{\xi}_i) = \mathbf{X}_i, \quad i = 1, ..., ns. \quad [7.5]$$

This approach is implemented in Matlab by using the programs given in section 3.3.

EXAMPLE 7.1.– Let us consider the situation where $\mathbf{x} = (x_1, x_2)$ and

$$F(\mathbf{x}, v) = (x_1 - v_1 v_2)^2 + (x_2 - v_1 - v_2)^2.$$

In this case:

$$X = \begin{pmatrix} v_1 v_2 \\ v_1 + v_2 \end{pmatrix}.$$

We consider (v_1, v_2) as a couple of independent variables, with v_i uniformly distributed over (a_i, b_i). The approximations use

$$\varphi_k(\mathbf{v}) = \left(\frac{v_1 - a_1}{b_1 - a_1}\right)^r \left(\frac{v_2 - a_2}{b_2 - a_2}\right)^s, \quad k = sn_1 + r, \qquad [7.6]$$

where $n_1 > 0$, $n_2 > 0$, $0 \leq r \leq n_1$, $0 \leq s \leq n_2$). The calculations use $a_1 = 0$, $b_1 = 1$, $a_2 = -1$, $b_2 = 1$, $n_1 = 3$, $n_2 = 3$. The results are shown in Figures 7.1 and 7.2.

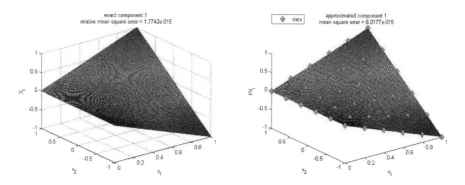

Figure 7.1. *Results obtained for X_1 in Example 7.1 (ns = 6 equally spaced v_i). For a color version of the figure, see www.iste.co.uk/souzadecursi/quantification.zip*

EXAMPLE 7.2.– Let us consider the situation where $\mathbf{x} = (x_1, x_2)$ and

$$F(\mathbf{x}, v) = ((2v_1 + 3)x_1 + (v_1 + 3)x_2 - v_1)^2$$
$$+ ((v_1 + 1)x_1 + (v_1 + 2)x_2 - 1)^2.$$

In this case,

$$X = \frac{1}{v_1^2 + 3v_1 + 3} \begin{pmatrix} v_1^2 + v_1 - 3 \\ -v_1^2 + v_1 + 3 \end{pmatrix}.$$

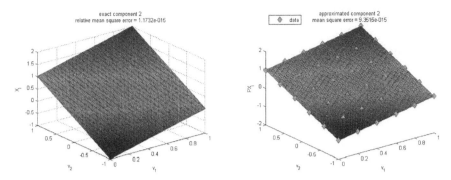

Figure 7.2. *Results obtained for X_2 in Example 7.1 ($ns = 11$ equally spaced v_i). For a color version of the figure, see www.iste.co.uk/souzadecursi/quantification.zip*

We consider v_1 as uniformly distributed on $(-1, 1)$, $ns = 11$ equally spaced values of v_i, with $v_1 = -1$, $v_{ns} = 1$. The procedure is applied with $N_X = 5$ and a polynomial basis

$$\varphi_k(v) = \left(\frac{v+1}{2}\right)^k.$$

The results are shown in Figure 7.3.

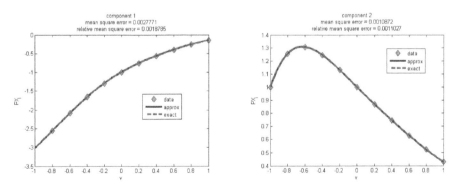

Figure 7.3. *Results obtained in Example 7.2 (numerical quadrature). For a color version of the figure, see www.iste.co.uk/souzadecursi/quantification.zip*

7.1.2. *Moment matching*

As explained in Chapter 3, a sample may be used in order to determine the coefficients such that the approximated moments $\mathbf{M}^{\mathbf{a}}(\ell) = \left(M_1^a, ..., M_q^a\right)$ match the empirical moments $\mathbf{M}^{\mathbf{e}} = \left(M_1^e, ..., M_q^e\right)$ (*moment matching method*). This may be accomplished either by solving the nonlinear equations $\mathbf{M}^{\mathbf{a}}(\ell) = \mathbf{M}^{\mathbf{e}}$ or by minimizing a pseudo-distance $d\left(\mathbf{M}^{\mathbf{a}}(\ell), \mathbf{M}^{\mathbf{e}}\right)$.

This approach is implemented in Matlab by using the programs given in section 3.2.

EXAMPLE 7.3.– Let us consider the situation described in Example 7.1 and a sample formed by np^2 points $\left((v_1)_i, (v_2)_j\right)$, with, for each r, np values $(v_r)_s$ equally spaced over (a_r, b_r). We consider the empirical moments $E\left(x_1^i x_2^j\right)$, with $i, j \leq m$. Some results are shown in Figure 7.4. We observe that the approach is effective to calculate when $\boldsymbol{\xi} \neq \mathbf{v}$: when the values of $\mathbf{v}$ are not known, but only the values of x_1 and $x2$, we may introduce $(\xi_r)_s$ equally spaced over (a_r, b_r) and consider $x_1 = x_1(\boldsymbol{\xi})$, $x_2 = x_2(\boldsymbol{\xi})$. For instance, we consider a random sample np^2 points $\left((v_1)_i, (v_2)_j\right)$, increasingly ordered for each component. The equally spaced values of $\boldsymbol{\xi}$ are used in order to fit the empirical moments.

EXAMPLE 7.4.– Let us consider the situation described in Example 7.2 and a sample formed by $ns = 11$ equally spaced values of v_i. The results furnished by the moment fitting method are shown in Figure 7.5. It is interesting to notice that, for $N_X = 7$, the variables themselves are correctly approximated (Figure 7.6); for the other values of N_X, only the distribution is fitted.

7.1.3. *Stochastic optimization*

The unknown coefficients may also be determined by minimizing one of the statistics of F. For instance, we may minimize its mean:

$$\chi = \arg\min \left\{ E\left(F\left(\mathbf{Y}\varphi(\boldsymbol{\xi}), \mathbf{v}\right)\right) : \mathbf{Y} \in \mathcal{M}(n, N_X) \right\}. \quad [7.7]$$

This problem may be solved by *stochastic optimization methods*, such as stochastic quasi-gradient methods (see, for instance, [KLE 01]).

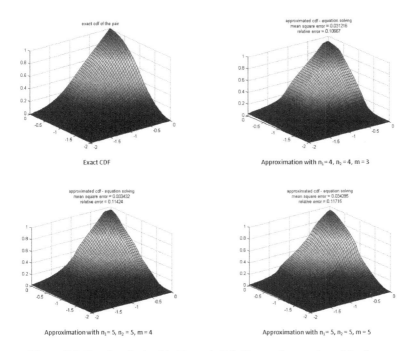

Figure 7.4. *Results obtained in Example 7.3. For a color version of the figure, see www.iste.co.uk/souzadecursi/quantification.zip*

EXAMPLE 7.5.– Let us consider the situation described in Example 7.3. We consider the resolution of equation [7.7] with the expectations approximated by the empirical means using the sample points. Some results are shown in Figure 7.7. We observe that the method is effective to calculate even for $\xi \neq \mathbf{v}$.

EXAMPLE 7.6.– Let us consider the situation described in Example 7.4 and the resolution of equation [7.7] with the expectations approximated by the empirical means obtained by using the sample points. The results are shown in Figure 7.8. The results for $\xi \neq \mathbf{v}$ are shown in Figure 7.9.

7.1.4. *Adaptation of iterative methods*

The approach used in section 5.1.4 may also be used here. For instance, let us consider the iterative solution of equation [7.1] by using iteration function Ψ:

$$\mathbf{X}^{(p+1)} = \Psi\left(\mathbf{X}^{(p)}\right). \qquad [7.8]$$

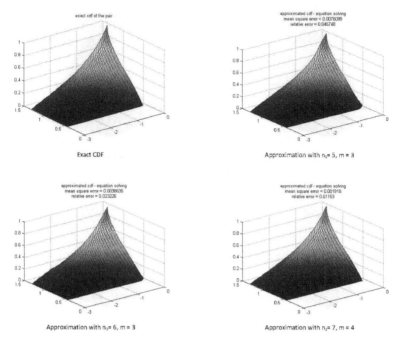

Figure 7.5. *Results obtained in Example 7.4 (ξ uniformly distributed, random $\mathbf{v}$). For a color version of the figure, see www.iste.co.uk/souzadecursi/quantification.zip*

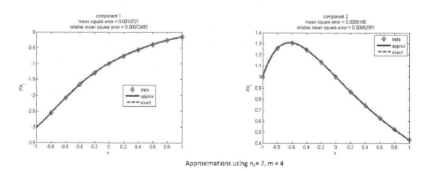

Figure 7.6. *Results obtained in Example 7.4 (ξ uniformly distributed, random $\mathbf{v}$). For a color version of the figure, see www.iste.co.uk/souzadecursi/quantification.zip*

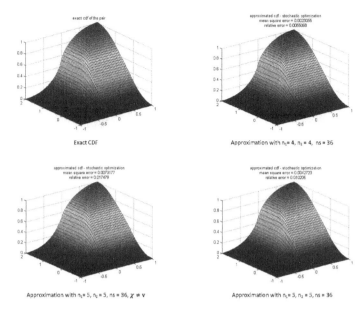

Figure 7.7. *Results obtained in Example 7.5 (uniformly distributed data). For a color version of the figure, see www.iste.co.uk/souzadecursi/quantification.zip*

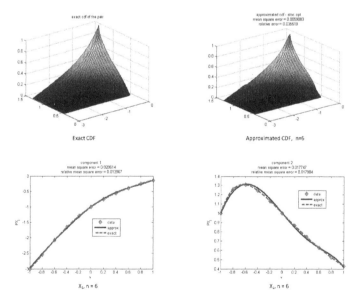

Figure 7.8. *Results obtained in Example 7.6 (uniformly distributed data). For a color version of the figure, see www.iste.co.uk/souzadecursi/quantification.zip*

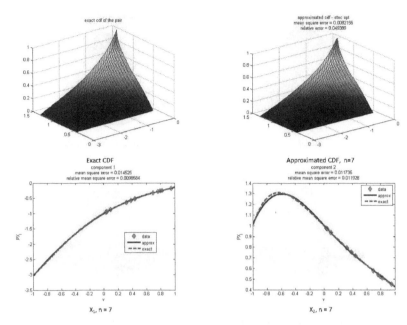

Figure 7.9. *Results obtained in Example 7.5 (ξ uniformly distributed, random $\mathbf{v}$). For a color version of the figure, see www.iste.co.uk/souzadecursi/quantification.zip*

Analogously to section 5.1.4, we may generate a sequence $\{\chi^{(p)}\}_{p \geq 0}$ starting from an initial guess $\chi^{(0)}$ by solving a sequence of linear systems

$$\mathbf{AC} = \mathbf{B}. \qquad [7.9]$$

where

$$A_{\alpha\beta} = \mathcal{A}_{rsjk}, B_\alpha = \mathcal{B}_{rs}, C_\beta = \chi^{(p+1)}_{jk},$$
$$\alpha = ind(r,s), \beta = ind(j,k), \qquad [7.10]$$
$$\mathcal{A}_{rsjk} = \delta_{jr} E\left(\varphi_s\left(\boldsymbol{\xi}\right)\varphi_k\left(\boldsymbol{\xi}\right)\right),$$

$$\mathcal{B}_{rs} = E\left(\varphi_s\left(\boldsymbol{\xi}\right) \Psi_r\left(\sum_{k=1}^{N_X} \chi_k^{(p)} \varphi_k\left(\boldsymbol{\xi}\right)\right)\right). \qquad [7.11]$$

This approach is implemented in Matlab by a simple modification of the code given in section 5.1.4: we modify program iteration_sample in order to furnish χ instead of $\Delta\chi$. An example of code is given below:

Listing 7.1. *UQ by adaptation in unconstrained optimization*

```
function chi = expcoef(f_iter,chi_ini,vs,phi,nitmax,errmax)
%
%    determines the coefficients of the expansion.
%    by using a sample
%
% IN:
% f_iter  : Iteration function    - type anonymous function
% chi_ini : initial n x N_X matrix of the coefficients  - type
          array of double
% vs : table of values of v - type array of double
% vs(:,i) is a variate from X
% phi : basis function phi(k,v)  - type anonymous function
% nitmax = max iteration number  - type integer
% errmax = max precision  - type double
%
% OUT:
% chi : n x N_X matrix of the coefficients  - type array of
        double
%
M = fixed_variational_matrix(phi, vs);
A = tab4(M,n,N_X);
chi = chi_ini;
nit = 0;
not_stop = true;
while not_stop
    nit = nit + 1;
    chi_new = iteration_sample(f_iter,chi,vs,phi,A);
    delta_chi = chi_new - chi;
    chi = chi_new;
    err = norm(delta_chi);
    not_stop = nit < nitmax && err > errmax;
end;
return;
end
%
function w = function_proj(f,chi,vs,phi)
%
% maps f( PX(v) , v ) on the sample vs from v
% f is assumed to have the same dimension as the
% number of lines of chi
%
% IN:
```

```
% f : the function to be evaluated − type anonymous function
% chi : n x N_X matrix of the coefficients − type array of
    double
% vs : table of values of v − type array of double
% vs(:,i) is a variate from v
% phi : basis function phi(k,v) − type anonymous function
%
% OUT:
% w : n x ns table of values of f − type array of double
%
ns = size(vs,1);
n = size(chi,1);
PXs = projection(chi,vs,phi);
w = zeros(n,ns);
for i = 1: ns
    w(:,i) = f(PXs(:,i),vs(:,i));
end;
return;
end
%
function A = tab4(M,n,N_X)
%
%   generates the table A from the table of
%   scalar products of the basis functions
%     M_ij = E(phi_i(v)phi_j(v))
%
% IN:
% M: N_X x N_X table of scalar products − type array of double
% n: number of unknowns (length of X) − type integer
% N_X : order of the expansion − type integer
%
% OUT:
% A = nN_X x nN_X table − type array of double
% contains A(alpha, beta)
%
aaaa = zeros(n,N_X,n,N_X);
for r = 1: n
    for s = 1: N_X
        for j = 1: n
            for k=s:N_X
                if r == j
                    aux = M(s,k);
                    aaaa(r,s,j,k) = aux;
                    aaaa(j,k,r,s) = aux;
                end;
            end;
        end;
```

```
        end;
end;
nn = n*N_X;
A = zeros(nn, nn);
for r = 1: n
    for s = 1: N_X
        alffa = index_map(r,s);
        for j = 1: n
            for k = 1: N_X
                betta = index_map(j,k,n,N_X);
                A(alffa, betta) = aaaa(r,s,j,k);
            end;
        end;
    end;
end;
return;
end

%
function B = tab2(N,n,N_X)
%
%   generates the table B from table N
%     N_rs = E(phi_s(v)f_r(PX(v)))
%
% IN:
% N: N_X x n table of scalar products - type array of double
% n: number of unknowns (length of X) - type integer
% N_X : order of the expansion - type integer
%
% OUT:
% B = nN_X x 1 table - type array of double
% contains B(alpha)
%
nn = n*N_X;
B = zeros(nn, nn);
for r = 1: n
    for s = 1: N_X
        alffa = index_map(r,s,n,N_X);
        B(alffa) = N(r,s);
    end;
end;
return;
end
%
function M = fixed_variational_matrix(phi, vs)
%
% generates the matrix M such that
```

```
%   M_ij = E(phi_i(v)phi_j(v))
%
% IN:
% phi : basis function phi(k,v) - type anonymous function
% vs  : table of values of v - type array of double
% vs(:,i) is a variate from X
%
% OUT:
% M: N_X x N_X table of scalar products - type array of double
%
M = zeros(N_X, N_X);
for i = 1: N_X
    f1 = @(U) phi(i,U);
    Y = map(f1, vs,1);
    A(i,i) = scalprod(Y,Y);
    for j = i+1:N_X
        f1 = @(U) phi(j,U);
        Z = map(f1, vs, 1);
        aux = scalprod(Y,Z);
        M(i,j) = aux;
        M(j,i) = aux;
    end;
end;
return;
end
%
function N = iteration_variational_matrix(phi, f, chi, vs,n,N_X
    )
%
% generates the matrix N such that
%   N_rs = E(phi_s(v)f_r(X(v)))
% assumes that f furnishes a vector of length n
%  and the number of lines of Xs is also n
%
% IN:
% phi : basis function phi(k,v) - type anonymous function
% f : Iteration function  - type anonymous function
% chi : n x N_X matrix of the coefficients - type array of
      double
% vs : table of values of v - type array of double
% vs(:,i) is a variate from X
% n: number of unknowns (length of X) - type integer
% N_X : order of the expansion - type integer
%
% OUT:
% N: N_X x n table of scalar products - type array of double
%
```

```
N = zeros(N_X, n);
w = function_proj(f,chi,vs,phi);
for r = 1: n
    Y = w(r,:);
    for s = 1:N_X
        f1 = @(U) phi(s,U);
        Z = map(f1, vs, 1);
        aux = scalprod(Y,Z);
        N(r,s) = aux;
    end;
end;
return;
end
%
function chi = iteration_sample(f_iter,chi_old,vs,phi,A)
%
% evaluates the new coefficients
%
% IN:
% f_iter : Iteration function    - type anonymous function
% chi_old : n x N_X matrix of the coefficients - type array of
%     double
% vs : table of values of v - type array of double
% vs(:,i) is a variate from X
% phi : basis function phi(k,v) - type anonymous function
% A = nN_X x nN_X table - type array of double
% contains A(alpha, beta)
%
% OUT:
% delta_chi : n x N_X matrix of the coefficients - type array
%     of double
% chi = chi_old + delta_chi
%
n = size(chi_old,1);
N_X = size(chi_old,2);
N = iteration_variational_matrix(phi, f_iter, chi_old, vs,n,N_X
   );
B = tab2(N,n,N_X);
C = A \B;
chi = zeros(size(chi_old));
for r = 1: n
    for s = 1: N_X
        alffa = index_map(r,s,n,N_X);
        chi(r,s) = C(alffa);
    end;
end;
return;
```

```
end
%
function v = index_map(i,j,n,N_X)
v = i + (j-1)*n;
return;
end
```

EXAMPLE 7.7.– Let us consider the situation described in Example 7.3. We consider gradient descent using a fixed step $\mu = 0.1$, with a starting point determined by the intrinsic Nelder–Meade method furnished by Matlab. As in the previous situations, the empirical means are approximately evaluated by using sample points forming a uniform grid on the region, as described in Example 7.3. The results obtained for $np = 11$ are shown in Figure 7.10.

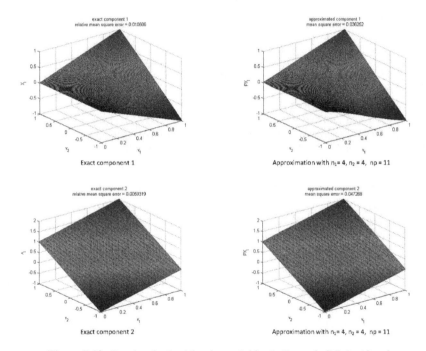

Figure 7.10. *Results obtained for the variables in Example 7.7 (uniformly distributed data). For a color version of the figure, see www.iste.co.uk/souzadecursi/quantification.zip*

EXAMPLE 7.8.– We consider again the situation described in Example 7.4. As in the previous example, we consider gradient descent using a fixed step $\mu = 0.1$, with a starting point determined by the intrinsic Nelder–Meade method furnished by Matlab, the empirical means evaluated by using the sample of $ns = 11$ equally spaced points, as described in Example 7.4. The results are shown in Figure 7.12.

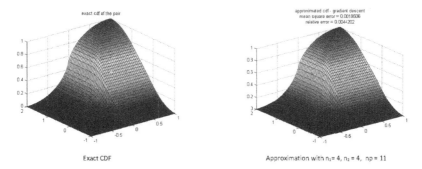

Figure 7.11. *Results obtained for the CDF in Example 7.7 (uniformly distributed data). For a color version of the figure, see www.iste.co.uk/souzadecursi/quantification.zip*

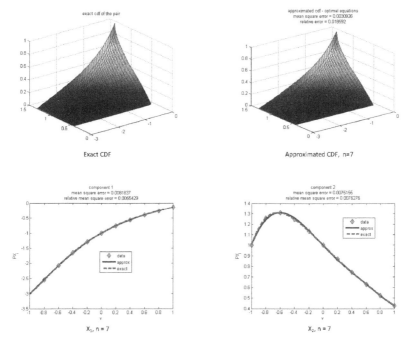

Figure 7.12. *Results obtained in Example 7.8 (uniformly distributed data). For a color version of the figure, see www.iste.co.uk/souzadecursi/quantification.zip*

7.1.5. *Optimal criteria*

In some situations, namely *for a convex objective function F*, optimal criteria may be established under the form of systems of algebraical equations and the methods exposed in section 5.1 may be used. For instance, we may consider the equations

$$\nabla F(x, v) = 0, \qquad [7.12]$$

which are satisfied by the optimal point. The methods presented in section 5.1 may be applied in order to determine an approximation **PX**, as described above. Namely, the Matlab implementation is identical.

EXAMPLE 7.9.– We consider the situation described in Example 7.3. The resolution of equation [7.12] is performed by the variational approach, with the expectations approximated by the empirical means using the sample points. The results obtained for the variables themselves are shown in Figure 7.13. A comparison of the cumulative functions is furnished in Figure 7.14.

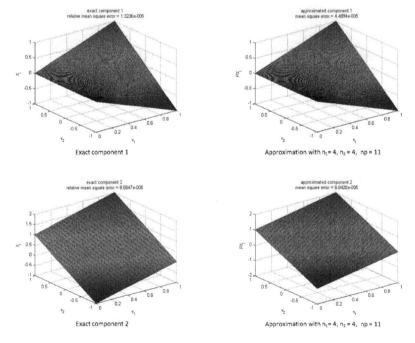

Figure 7.13. *Results obtained for the variables in Example 7.9 (uniformly distributed data). For a color version of the figure, see www.iste.co.uk/souzadecursi/quantification.zip*

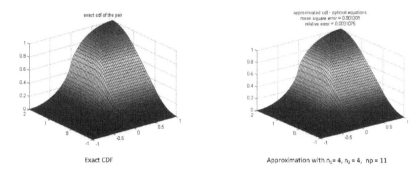

Figure 7.14. *Results obtained for the CDF in Example 7.10 (uniformly distributed data). For a color version of the figure, see www.iste.co.uk/souzadecursi/quantification.zip*

EXAMPLE 7.10.– We consider now the situation described in Example 7.4. The resolution of equation [7.12] is performed again by the variational approach, with the expectations approximated by the empirical means using the $ns = 11$ sample points. The results are shown in Figure 7.15.

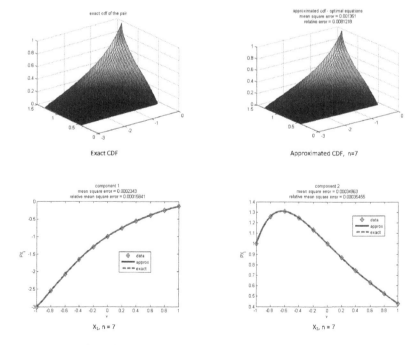

Figure 7.15. *Results obtained in Example 7.10 (uniformly distributed data). For a color version of the figure, see www.iste.co.uk/souzadecursi/quantification.zip*

7.2. Stochastic methods in deterministic continuous optimization

This section presents stochastic numerical algorithms for the determination of the points of global minima of deterministic regular functions, i.e. for the global optimization of continuous functions.

This area is a subject of extensive research, and the readers may find in the literature a wide variety of approaches which prevents any tentative of exhaustive presentation and condemns all texts to partial presentation – the present one is not an exception and it is possible that the readers will not find below their preferred algorithm.

Our focus is on the construction of numerical methods: but more than the simple description of algorithms, this chapter aims to give the readers the basic principles for the construction of stochastic algorithms – which are based on the fundamental theorems presented in the following. The examples presented must be considered as illustrations of the methodology: for a specific situation, the readers must consider the probable existence of adapted algorithms, specifically conceived for the situation under analysis and exploiting particularities of the problem in order to increase the efficiency – these algorithms may be used as seed algorithms and combined to those presented by using the procedures described in the following.

The theorems presented may be interpreted as follows: generating random points with a strictly positive probability for any region of the working space leads to a strictly positive probability of obtaining a random point in a given neighborhood of a point of global minimum. In addition, the probability of generating a random point in such a neighborhood on a large number of draws increases (namely, it converges to one for an infinite number of trials).

From these observations, a first, and simple, idea results for the minimization of a regular function $f : \mathbb{R}^k \longrightarrow \mathbb{R}$. We may, for instance, randomly generate nr points $\{x_i : 1 \leq i \leq nr\}$ and estimate the value m of the minimum by using the values $\{f(x_i) : 1 \leq i \leq nr\}$. In practice, it may be more convenient to consider the generation of trial points in a "controlled" neighborhood of a point: we generate the point x_{i+1} by means of a displacement Δx_i of the actual point x_i. In this case, we obtain an algorithm which is as follows:

1) *Initialization*: it is given a starting point x_0 and a maximum iteration number nm. Set iteration number to zero: $n \longleftarrow 0$;

2) *Iterations*: the actual point is x_n and we determine x_{n+1} in two substeps

– *drawing*: randomly generate an increment Δx_n

– *dynamics*: determine x_{n+1} from x_n and $t_n = x_n + \Delta x_n$. For instance, use an *elitist dynamics* given by

$$x_{n+1} = \arg\min\ \{f(x_n), f(t_n)\}.$$

3) *Stopping test*: if $n < nm$, then increment $n : n \longleftarrow n+1$ and return to step 2. Otherwise, stop iterations and estimate $m \approx f(x_n)$ and $x^* \approx x_n$.

The epithet *elitist* is justified by the systematic rejection of the elements that do not improve the value of f. We have

$$x_{n+1} = \begin{cases} x_n, \text{ if } f(x_n) < f(t_n) \\ t_n, \text{ if } f(x_n) \geq f(t_n) \end{cases}.$$

Usually, this dynamics is written under the synthetic form

$$x_{n+1} = D_n x_n + (1 - D_n) t_n,$$

where

$$P(D_n = 0 | f(x_n) < f(t_n)) = 0;\ P(D_n = 1 | f(x_n) < f(t_n)) = 1$$

$$P(D_n = 0 | f(x_n) \geq f(t_n)) = 1;\ P(D_n = 1 | f(x_n) \geq f(t_n)) = 0$$

By assuming that a subprogram alea(n,nt); furnishes an $n \times nt$ table of random values, Matlab implementation reads as:

Listing 7.2. *Stochastic descent*

```
function [xsol, fsol] = stoc_desc(x0,ntmax,ncmax,alea)
%
% performs nc cycles involving ntmax trial points
% of stochastic descent starting from x0
% assumes that x0 is n x 1
%
```

```
%IN:
% x0: n x 1 vector - type array of double
% ntmax: maximum number of trial points by cycle - type integer
% ncmax: maximum number of cycles - type integer
% alea: generator of random points - type anonymous function
% alea(n,nt) furnishes a n x nt table of random values
%
dynamics = @elitist_dynamics;
[xsol, fsol] = stoc_iterations(x0,ntmax,ncmax, alea, dynamics);
return;
end
function [xsol, fsol] = stoc_iterations(x0,ntmax,ncmax, alea, dynamics)
%
% performs nc cycles involving ntmax trial points
% starting from x0
% assumes that x0 is n x 1
%
%IN:
% x0: n x 1 vector - type array of double
% ntmax: maximum number of trial points by cycle - type integer
% ncmax: maximum number of cycles - type integer
% alea: generator of random points - type anonymous function
% alea(n,nt) furnishes a n x nt table of random values
% dynamics: the dynamics to be used - type anonymous function
%
xac = x0;
fac = f(x0);
for nc = 1: ncmax
    xc = xac;
    fc = fac;
    for nit = 1: ntmax
        xt = trial_point(xc,1, alea);
        ft = f(xt);
        [xc, fc] = dynamics(xc,fc,xt,ft,nit);
    end;
    xac = xc;
    fac = fc;
end;
xsol = fac;
fsol = fac;
return;
end

function xt = trial_point(xac,nt, alea)
%
% generates nt trial points by perturbation of
```

```
% the actual point xac
% assumes that xac is n x 1
%
%IN:
% xac: n x 1 vector – type array of double
% nt: number of trial points – type integer
% alea: generator of randon points – type anonymous function
% alea(n,nt) furnishes a n x nt table of random values
%
n = length(xac);
xt = zeros(n,nt);
dx = alea(n,nt);
for i = 1: nt
    xt(:,i) = xac + dx(:,i);
end;
return;
end

function [xd, fd] = elitist_dynamics(xac,fac,xt,ft,nit)
%
% applies the dynamics by selection of the best point
%
if ft < fac
    xd = xt;
    fd = ft;
else
    xd = xac;
    fd = fac;
end;
return;
end
```

This simple algorithm is usually referred to as *stochastic descent method*. It is concerned with a convergence theorem with quite general assumptions, which include non-convex objective functions f:

$$\inf \left\{ f(x) : x \in \mathbb{R}^k \right\} = m \in \mathbb{R} ; \qquad [7.13]$$

$$\forall \lambda > m : S_\lambda = \left\{ x \in \mathbb{R}^k : f(x) < \lambda \right\}$$
is bounded, with non-empty interior; $\qquad [7.14]$

$$\forall \lambda > m : S_\lambda^c = \left\{ x \in \mathbb{R}^k : f(x) \geq \lambda \right\} \text{ has a non-empty interior;} \qquad [7.15]$$

Assumptions [7.14] and [7.15] are satisfied, for instance, when f is continuous and coercive or when f is continuous, and we are interested in the minimum on a bounded subset. Assumption [7.15] guarantees that $P(x_n \notin S_\lambda) > 0$ and may be weakened by considering the values $\lambda \in (m, m + \varepsilon)$ for which this assumption is satisfied. However, the convergence may be slow, even extremely slow, in large dimensions. In practice, the efficiency of a method purely based on generation of trial points is poor, even extremely poor, in large dimensions. Thus, modifications of this basic method have been introduced and have led to other versions of the fundamental theorem. For instance, the use of the *dynamics of Metropolis* instead of the elitist one leads to the *algorithm of simulated annealing*, which is based on version 2 of the fundamental theorem. In the *dynamics of Metropolis*,

$$P(D_n = 0 f(x_n) < f(t_n)) = c_n;$$
$$P(D_n = 1 f(x_n) < f(t_n)) = 1 - c_n;$$

$$P(D_n = 0\, f(x_n) \geq f(t_n)) = 1;\ P(D_n = 1\, f(x_n) \geq f(t_n)) = 0;$$

$$c_n = \exp\left(-\left(\frac{f(t_n) - f(x_n) + \alpha}{\theta_n} + \beta\right)\right),$$

where $\alpha \geq 0, \beta \geq 0$ and $\{\theta_n\}_{n \in \mathbb{N}}$ is a sequence of strictly positive real numbers such that $\theta_n \longrightarrow 0$ "slowly enough" (in general, as $1/\sqrt{\log(n)}$). In this case,

$$x_{n+1} = \begin{cases} \begin{Bmatrix} x_n, \text{ with probability } 1 - c_n \\ t_n, \text{ with probability } c_n \end{Bmatrix}, \text{ if } f(x_n) < f(t_n) \\ \\ t_n, \text{ if } f(x_n) \geq f(t_n) \end{cases}.$$

Matlab implementation consists of a small modification of the preceding code:

Listing 7.3. *Stochastic descent with Metropolis' dynamics*

```
function [xsol, fsol] = stoc_desc(x0,ntmax,ncmax,alea,a,b,c,
    aalfa,bbeta)
%
% performs nc cycles involving ntmax trial points
% of stochastic descent starting from x0
% assumes that x0 is n x 1
%
%IN:
% x0: n x 1 vector - type array of double
% ntmax: maximum number of trial points by cycle - type integer
% ncmax: maximum number of cycles - type integer
% alea: generator of random points - type anonymous function
% alea(n,nt) furnishes a n x nt table of random values
%
teta = @(nit) teta_fading(a,b,c,nit);
dynamics = @(xc,fc,xt,ft,nit) metropolis_dynamics(xc,fc,xt,ft,
    aalfa,bbeta,nit,teta);
[xsol, fsol] = stoc_iterations(x0,ntmax,ncmax, alea, dynamics);
return;
end

function [xd, fd] = metropolis_dynamics(xac,fac,xt,ft,a,b,nit,
    teta)
%
% applies the dynamics of Metropolis
%
if ft < fac
    xd = xt;
    fd = ft;
else
    teta_n = teta(nit);
    aux = -(b + (ft - fac + a)/teta_n);
    c = exp(aux);
    test = rand();
    if test < c
        xd = xt;
        fd = ft;
    else
        xd = xac;
        fd = fac;
    end;
end;
return;
end
```

We may also consider a hybrid approach by a combination of deterministic descent methods and the stochastic descent methods based on version 3 of the fundamental theorem.

7.2.1. *Version 1: Stochastic descent*

In this section, we present the fundamental result that justifies the stochastic descent method.

THEOREM 7.1.– Let $\{U_n\}_{n \in \mathbb{N}}$ be a sequence of random variables such that

$$\forall n \geq 0 : U_{n+1} \leq U_n \text{ a.s.}$$

$$\forall n \geq 0 : U_n \geq m \text{ a.s.}$$

$$\forall \lambda > m : \exists \alpha(\lambda) > 0 \text{ such that } P(U_{n+1} \leq \lambda \mid U_n \geq \lambda) \geq \alpha(\lambda). \quad [7.16]$$

Then:

$$U_n \longrightarrow m \text{ a.s.} \blacksquare$$

We remark that condition [7.16] implies that $P(U_n \geq \lambda) > 0$. In the following, we establish that this inequality is satisfied when the probability density of the trial points drawn is strictly positive on $\mathbb{R}^n$.

The proof of the theorem uses the following lemmas:

LEMMA 7.1.– Let $\{E_n\}_{n \in \mathbb{N}} \subset \Omega$ be a family of almost sure events. Then, $E = \bigcap_{n \in \mathbb{N}} E_n$ is almost sure. $\blacksquare$

PROOF.– Let $E_n^c = \Omega - E_n$ be the complement of the event E_n. We have

$$\forall n \geq 0 : P(E_n^c) = 1 - P(E_n) = 1 - 1 = 0.$$

Let $E^c = \Omega - E$ be the complement of the event E. We have

$$E^c = \left(\bigcap_{n \in \mathbb{N}} E_n\right)^c = \bigcup_{n \in \mathbb{N}} E_n^c,$$

so that

$$P\left(\bigcup_{n \in \mathbb{N}} E_n^c\right) \leq \sum_{n=0}^{+\infty} P(E_n^c) = \sum_{n=0}^{+\infty} 0 = 0$$

and $P(E) = 1 - P(E^c) = 1 - 0 = 1$. ∎

LEMMA 7.2.– Let $\{U_n\}_{n \in \mathbb{N}}$ be a sequence of random variables such that

$$\forall n \geq 0 : U_{n+1} \leq U_n \text{ a.s.}$$

$$\forall n \geq 0 : U_n \geq m \text{ a.s.}$$

then there exists a random variable U_∞ such that $U_n \longrightarrow U_\infty$ a.s. and $U_\infty \geq m$ a.s. In addition, $U_n \geq U_\infty$ a.s. for any $n \geq 0$. ∎

PROOF.–

1) Let us consider

$$A_n = \{\omega \in \Omega : U_{n+1}(\omega) \leq U_n(\omega)\},$$

$$B_n = \{\omega \in \Omega : U_n(\omega) \geq m\}.$$

By denoting $A_n^c = \Omega - A_n$, $B_n^c = \Omega - B_n$ (complements of A_n and B_n, respectively), we have $P(A_n^c) = P(B_n^c) = 0$. Let us introduce $E_n = A_n \cap B_n$. Then, $E_n^c = (A_n \cap B_n)^c = A_n^c \cup B_n^c$, so that

$$P(E_n^c) = P(A_n^c \cup B_n^c) \leq P(A_n^c) + P(B_n^c) = 0 + 0 = 0.$$

and E_n is almost sure, since $P(E_n) = 1 - P(E_n^c) = 1 - 0 = 1$. Let us consider

$$E = \{\omega \in \Omega : m \leq U_{n+1}(\omega) \leq U_n(\omega), \forall n \geq 0\}.$$

Then, $E = \bigcap_{n \in \mathbb{N}} E_n$, and it yields from 7.1 that E is almost sure.

2) Let $\omega \in E$. Then, $\{U_n(\omega)\}_{n \in \mathbb{N}} \subset \mathbb{R}$ is decreasing and bounded, so that there exists $U_\infty(\omega)$ such that

$$U_n(\omega) \longrightarrow U_\infty(\omega) \text{ for } n \longrightarrow +\infty.$$

In this way, we define a numerical function $U_\infty : \Omega \longrightarrow \mathbb{R}$. The basic properties of measurable applications (see, for instance, [HAL 64], p. 184) show that U_∞ is a random variable. In addition, $U_\infty(\omega) \geq m$, since $m \leq U_n(\omega), \forall n \geq 0$.

3) Let

$$F = \{\omega \in \Omega : U_n(\omega) \longrightarrow U_\infty(\omega) \text{ for } n \longrightarrow +\infty \}.$$

We have $E \subset F$, so that $P(F) \geq P(E) = 1$ and F is almost sure. Let

$$G = \{\omega \in \Omega : U_\infty(\omega) \geq m \}.$$

We have $E \subset G$, so that $P(G) \geq P(E) = 1$ and G is almost sure.

4) Let us consider

$$H_n = \{\omega \in \Omega : U_n(\omega) \geq U_\infty(\omega) \}$$

Let $\omega \in E$. Since $\{U_n(\omega)\}_{n \in \mathbb{N}} \subset \mathbb{R}$ is decreasing and bounded, we have $U_\infty(\omega) \leq U_n(\omega), \forall n \geq 0$. Thus, $E \subset H_n$, so that $P(H_n) \geq P(E) = 1$. So, H_n is almost sure for any $n \geq 0$. ∎

LEMMA 7.3.– Let U_∞ be a random variable such that

$$\forall \lambda > m : U_\infty \leq \lambda \text{ a.s.,} \qquad [7.17]$$

$$U_\infty \geq m \text{ a.s.} \qquad [7.18]$$

Then, $U_\infty = m$ a.s. ∎

PROOF.–

1) Let us introduce

$$A = \{\omega \in \Omega : U_\infty(\omega) = m \}$$

$$B = \{\omega \in \Omega : U_\infty(\omega) > m\}$$

Assumption [7.18] shows that:

$$P(A \cup B) = 1.$$

Since

$$A \cap B = \emptyset,$$

we have

$$P(A \cup B) = P(A) + P(B).$$

So,

$$P(A) + P(B) = 1.$$

2) Let us consider

$$B_n = \left\{\omega \in \Omega : U_\infty(\omega) \geq m + \frac{1}{n}\right\}.$$

We have

$$\omega \in B \implies \exists\, n > 0 \text{ such that } U_\infty(\omega) \geq m + \frac{1}{n} \implies \omega \in \bigcup_{n \in \mathbb{N}} B_n.$$

Thus, on the one hand,

$$B \subset \bigcup_{n \in \mathbb{N}} B_n.$$

On the other hand,

$$B_n \subset B, \forall\, n \geq 0 \implies \bigcup_{n \in \mathbb{N}} B_n \subset B,$$

so that

$$B = \bigcup_{n \in \mathbb{N}} B_n.$$

3) The first assumption of Lemma (equation [7.17]) yields that B_n is almost impossible, $\forall n \geq 0$.

Thus:

$$P(B) = P\left(\bigcup_{n \in \mathbb{N}} B_n\right) \leq \sum_{n=0}^{+\infty} P(B_n) = 0$$

and B is almost impossible. So,

$$P(A) = 1 - P(B) = 1 - 0 = 1$$

and A is almost sure. □

LEMMA 7.4.– Under the assumptions of Theorem 7.1, we have

$$\forall \lambda > m : P(U_n \geq \lambda) \longrightarrow 0 \quad \text{quando} \quad n \longrightarrow +\infty. \blacksquare$$

PROOF.– Let us consider the random variable Z_n given by

$$Z_n = 0, \text{ if } U_n \geq \lambda, \ Z_n = 1, \text{ if } U_n < \lambda.$$

We have

$$P(Z_{n+1} = i) = P(Z_{n+1} = i, Z_n = 0) + P(Z_{n+1} = i, Z_n = 1),$$

so that

$$P(Z_{n+1} = i) = P(Z_n = 0) P(Z_{n+1} = i \mid Z_n = 0) \\ + P(Z_n = 1) P(Z_{n+1} = i \mid Z_n = 1)$$

and, by taking

$$\pi_n = \begin{pmatrix} P(Z_n = 0) \\ P(Z_n = 1) \end{pmatrix},$$

we obtain

$$\pi_{n+1} = A\pi_n, A_{ij} = P(Z_{n+1} = i | Z_n = j) \ (0 \leq i, j \leq 1).$$

Since the sequence is decreasing a.s.,

$$A_{01} = P(Z_{n+1} = 0 | Z_n = 1) = P(U_{n+1} \geq \lambda | U_n < \lambda) = 0.$$

Thus, from the assumptions of the theorem,

$$A_{00} = P(Z_{n+1} = 0 | Z_n = 0) = 1 - P(U_{n+1} < \lambda | U_n \geq \lambda) \leq 1 - \alpha(\lambda).$$

Thus:

$$\forall n \geq 0 : (\pi_{n+1})_0 = A_{00} (\pi_n)_0 + A_{01} (\pi_n)_1$$
$$= A_{00} (\pi_n)_0 \leq (1 - \alpha(\lambda)) (\pi_n)_0 .$$

This equation of recurrence shows that

$$\forall n \geq 0 : (\pi_{n+1})_0 \leq (1 - \alpha(\lambda))^{n+1} (\pi_0)_0.$$

Consequently, given that $\alpha(\lambda) > 0$,

$$(\pi_{n+1})_0 \longrightarrow 0 \text{ for } n \longrightarrow +\infty . \ \blacksquare$$

LEMMA 7.5.– Under the assumptions of Theorem 7.1, there exists U_∞ such that $U_n \longrightarrow U$. In addition,

$$\forall \lambda > m : P(U_\infty \geq \lambda) = 0. \ \blacksquare$$

PROOF.–

1) Lemma 7.2 shows that there exists U_∞ such that $U_n \longrightarrow U$ a.s. Since $U_n \longrightarrow U_\infty$ a.s., we have also $U_n \longrightarrow U_\infty$ p. and, as a result, $U_n \longrightarrow U$ D.

2) Let F_n be the distribution of U_n and F_∞ be the distribution of U_∞. The convergence in distribution implies that

$$P(U_n < \lambda) = F_n(\lambda) \longrightarrow F_\infty(\lambda) = P(U_\infty < \lambda),$$

so that

$$P(U_n \geq \lambda) = 1 - F_n(\lambda) \longrightarrow 1 - F_\infty(\lambda) = P(U_\infty \geq \lambda).$$

Thus:

$$P(U_\infty \geq \lambda) = \lim_{n \longrightarrow +\infty} P(U_n \geq \lambda) = 0. \square$$

PROOF OF THE THEOREM 7.1.–

1) It yields from Lemma 7.2 that there exists a random variable U_∞ such that $U_n \longrightarrow U_\infty$ a.s. and $U_\infty \geq m$ a.s. As in the proof of these lemmas, we consider

$$E = \{\omega \in \Omega : m \leq U_{n+1}(\omega) \leq U_n(\omega), \forall\, n \geq 0\}.$$

E is almost sure and $U_n(\omega) \longrightarrow U_\infty(\omega)$, for any $\omega \in E$.

2) Let $\lambda > m$: it follows from Lemma 7.5 that $U_\infty < \lambda$ a.s. Consequently, it follows from Lemma 7.3 that $U_\infty = m$ a.s. $\square$

NOTE 7.1.– An alternative proof that $U_\infty = m$ a.s. is the following:

– Let $\lambda > m$ and consider

$$A_n(\lambda) = \{\omega \in E : U_n(\omega) < \lambda\}.$$

We have

$$A_n(\lambda) \subset A_{n+1}(\lambda), \forall\, n \geq 0.$$

– Let

$$A(\lambda) = \{\omega \in E : U_\infty(\omega) < \lambda\}.$$

Since $U_\infty(\omega) \leq U_n(\omega)$ (Lemma 7.2), we have

$$\omega \in A_n(\lambda) \Longrightarrow U_\infty(\omega) \leq U_n(\omega) < \lambda \Longrightarrow \omega \in A(\lambda).$$

Thus:

$$A_n(\lambda) \subset A(\lambda), \forall n \geq 0.$$

So, by taking $p_n(\lambda) = P(A_n(\lambda))$, we have

$$P(A(\lambda)) \geq p_n(\lambda), \forall n \geq 0. \quad [7.19]$$

– Denoting $A_n^c(\lambda) = \Omega - A_n(\lambda)$ and $B_n(\lambda) = A_{n+1}(\lambda) \cap A_n^c(\lambda)$, we have

$$A_n(\lambda) \cap B_n(\lambda) = \varnothing.$$

In addition:

$$A_{n+1}(\lambda) = A_{n+1}(\lambda) \cap \Omega = A_{n+1}(\lambda) \cap \underbrace{(A_n(\lambda) \cup A_n^c(\lambda))}_{=\Omega}$$

$$= \underbrace{(A_{n+1}(\lambda) \cap A_n(\lambda))}_{=A_n(\lambda)} \cup \underbrace{(A_{n+1}(\lambda) \cap A_n^c(\lambda))}_{=B_n(\lambda)},$$

so that

$$A_{n+1}(\lambda) = A_n(\lambda) \cup B_n(\lambda)$$

and

$$P(A_{n+1}(\lambda)) = P(A_n(\lambda)) + P(B_n(\lambda)). \quad [7.20]$$

– We have

$$\forall n \geq 0 : P(A_{n+1}(\lambda) | A_n^c(\lambda)) = P(U_{n+1} \leq \lambda | U_n \geq \lambda) \geq \alpha(\lambda) > 0,$$

so that, for any $n \geq 0$: on the one hand, $p_n(\lambda) > 0$ and, on the other hand,

$$P(B_n(\lambda)) = P(A_n^c(\lambda)) P(A_{n+1}(\lambda) | A_n^c(\lambda)) =$$
$$(1 - p_n(\lambda)) P(A_{n+1}(\lambda) | A_n^c(\lambda)) \geq (1 - p_n(\lambda)) \alpha(\lambda).$$

– Thus, equation [7.20] shows that

$$\forall n \geq 0 : p_{n+1}(\lambda) \geq p_n(\lambda) + (1 - p_n(\lambda)) \alpha(\lambda).$$

By setting $q_n(\lambda) = 1 - p_n(\lambda)$, we have

$$\forall n \geq 0 : q_{n+1}(\lambda) \leq (1 - \alpha(\lambda))q_n(\lambda).$$

This equation of recurrence shows that

$$\forall n \geq 0 : q_n(\lambda) \leq (1 - \alpha(\lambda))^n q_0(\lambda).$$

Consequently, given that $\alpha(\lambda) > 0$,

$$q_n(\lambda) \longrightarrow 0 \text{ e } p_n(\lambda) \longrightarrow 1 \text{ for } n \longrightarrow +\infty.$$

– So, equation [7.19] shows that $P(A(\lambda)) = 1$ and, as a result, $U_\infty < \lambda$ a.s.

– It follows from Lemma 7.3 that $U_\infty = m$ a.s. $\square$

This theorem suggests a method for the minimization of a continuous function f: let us consider the following:

1) $Z : \Omega \longrightarrow \mathbb{R}^k$ a continuous random vector having a probability density $\varphi_Z : \mathbb{R}^k \longrightarrow \mathbb{R}$ such that ($|\bullet|$ is the Euclidean norm)

$$\forall r > 0 : \inf \{\varphi_Z(z) : |z| \leq r\} \geq a(r) > 0$$

2) $x_0 : \Omega \longrightarrow \mathbb{R}^k$ a given random vector.

Let us define a sequence $\{x_n\}_{n \in \mathbb{N}}$, $x_n : \Omega \longrightarrow \mathbb{R}^k$ of random vectors as follows:

$$\forall n \geq 0 : x_{n+1}(\omega) = \arg\min\{f(x_n(\omega)), f(t_n(\omega))\}, \quad t_n = x_n + Z;$$

and a second sequence $\{U_n\}_{n \in \mathbb{N}}$, $U_n : \Omega \longrightarrow \mathbb{R}$ given by

$$U_n(\omega) = f(x_n(\omega)).$$

We observe that the conditional cumulative function of t_n is

$$\varphi_n(t | x_n = x) = \varphi_Z(t - x) > 0$$

The conditional density of probability of t_n is

$$\varphi_n(t\,|\,x_n = x) = \varphi_Z(t-x)$$

and the density of probability of x_n is

$$\phi_n(t) = \int_{\mathbb{R}} \varphi_n(t\,|\,x_n = x)\,dx = \int_{\mathbb{R}} \varphi_Z(t-x)\,dx.$$

Thus:

$$\forall\, r > 0 : \phi_n(t) \geq \int_{|t-x|\leq r} \varphi_Z(t-x)\,dx \geq a(r)\,\ell(B_r) > 0,$$

where ℓ is the Lebesgue's measure and B_r is the ball of radius r ($B_r = \{z \in \mathbb{R}^n : |z| \leq r\}$).

We have

THEOREM 7.2.– $U_n \longrightarrow m$ a.s. ∎

PROOF.–

1) By construction:

$$\forall\, n \geq 0 : U_{n+1} \leq U_n \text{ a.s.}$$

and

$$\forall\, n \geq 0 : U_n \geq m \text{ a.s.}$$

2) So, on the one hand,

$$\int_{S_\lambda} \varphi_n(t\,|\,x_n = x)\,dt \geq \int_{B(\lambda)} \varphi_n(t\,|\,x_n = x)\,dt =$$
$$\int_{B(\lambda)} \varphi_Z(t-x)\,dt \geq \int_{B(\lambda)} a(\eta_\lambda)\,dt$$

and (ℓ is the Lebesgue's measure)

$$\int_{S_\lambda} \varphi_n(t\,|x_n = x)\,dt \geq a(\eta_\lambda)\,\ell(B(\lambda))$$

3) On the other hand,

$$P(x_{n+1} \in S_\lambda, x_n \notin S_\lambda) = \int_{S_\lambda^c} \mu_n(dx) \int_{S_\lambda} \varphi_n(t\,|x_n = x)\,dt$$
$$\geq a(\eta_\lambda)\,\ell(B(\lambda)) \int_{S_\lambda^c} \mu_n(dx),$$

where F_n is the distribution of x_n and μ_n is the measure associated with F_n. Thus:

$$P(x_{n+1} \in S_\lambda, x_n \notin S_\lambda) \geq a(\eta_\lambda)\,\ell(B(\lambda))\,P(x_n \notin S_\lambda).$$

From assumption [7.15], there exist $y_\lambda \in S_\lambda^c$ and $r_\lambda > 0$ such that

$$C(\lambda) = \left\{x \in \mathbb{R}^k : |x - y| \leq r_\lambda\right\} \subset S_\lambda^c.$$

So,

$$P(x_n \notin S_\lambda) = \int_{S_\lambda^c} \phi_n(x)\,dx \geq \int_{C(\lambda)} \phi_n(x)\,dx \geq a(r_\lambda)\,\ell(B_{r_\lambda}) > 0$$

and, consequently,

$$P(x_{n+1} \in S_\lambda\,|x_n \notin S_\lambda) = \frac{P(x_{n+1} \in S_\lambda, x_n \notin S_\lambda)}{P(x_n \notin S_\lambda)} \geq a(\eta_\lambda)\,\ell(B(\lambda)).$$

4) By setting

$$\alpha(\lambda) = a(\eta_\lambda)\,\ell(B(\lambda)),$$

we have

$$P(U_{n+1} \leq \lambda\,|U_n \geq \lambda) \geq \alpha(\lambda) > 0.$$

5) The result follows from Theorem 7.1. ∎

As previously mentioned, the algorithm associated with this theorem is the stochastic descent method:

1) *Initialization*: It is given a starting point x_0 and a maximum iteration number nm. Set iteration number to zero: $n \longleftarrow 0$.

2) *Iterations*: the actual point is x_n and we determine x_{n+1} by using two substeps

– *drawing*: we generate a variate Z_n from Z and we set $t_n = x_n + Z_n$.

– *dynamics*: determine x_{n+1} as follows:

$$x_{n+1} = \begin{Bmatrix} x_n, \text{ if } f(x_n) < f(t_n) \\ t_n, \text{ if } f(x_n) \geq f(t_n) \end{Bmatrix}.$$

3) *Stopping test*: if $n < nm$, then increment $n : n \longleftarrow n+1$ and return to step 2. Otherwise, stop iterations and estimate $m \approx f(x_n)$ and $x^* \approx x_n$.

7.2.2. Version 2: Dynamics of Metropolis

As previously observed, the efficiency of the stochastic descent is poor. As a result, we may find in the literature modifications of the basic stochastic descent tending to improve its efficiency. For instance, a first idea comes from the observation that the elitist dynamics rarely modifies the actual point: the trial t_n is rejected, except when it corresponds to an improvement in the value of the objective function f. Thus, in practice, the elitist dynamics confines to exploration of neighborhoods of the actual point by a large number of trials. These observations suggest that an improvement may consist of using a dynamics leading to a larger exploration of the working space, by accepting a controlled degradation of the objective function. An example of such a modification is furnished by the *dynamics of Metropolis*, which consists of accepting a degradation of the values of f with a probability that rapidly decreases (for instance, exponentially decreases) with the value of the degradation. The convergence of the resulting method of optimization is based on the following theorem:

THEOREM 7.3.– Let $\{U_n\}_{n \in \mathbb{N}}$ be a sequence of random variables such that

$$\forall n \geq 0 : P(U_{n+1} > U_n) \leq \beta_n \text{ com } \sum_{n=0}^{+\infty} \beta_n < \infty$$

$$\forall n \geq 0 : U_n \geq m \text{ a.s.}$$

$\forall \lambda > m : \exists \alpha(\lambda) > 0$ such that $P(U_{n+1} \leq \lambda \,|\, U_n \geq \lambda) \geq \alpha(\lambda).$ [7.21]

Then:

$$U_n \longrightarrow m \text{ a.s.} \blacksquare$$

As previously observed, Condition [7.21] implies that $P(U_n \geq \lambda) > 0$. As in the preceding situation, this inequality is satisfied by random generation of trial points such that the probability density is strictly positive everywhere on $\mathbb{R}^n$.

The proof uses the following lemmas:

LEMMA 7.6.– Let $\{E_n\}_{n \in \mathbb{N}} \subset \Omega$ be a family of events such that $P(E_n^c) \leq \beta_n$, with $\sum_{n=0}^{+\infty} \beta_n < \infty$. If $E \subset \Omega$ is an event such that

$$\forall n \geq 0 : \bigcap_{i=n}^{+\infty} E_i \subset E$$

then E is almost sure. $\blacksquare$

PROOF.– Initially, we observe that the convergence of the series ($\sum_{n=0}^{+\infty} \beta_n < \infty$), imply that the series of the residuals

$$R_n = \sum_{i=n}^{+\infty} \beta_i \longrightarrow 0 \text{ for } n \longrightarrow +\infty.$$

Let $E_n^c = \Omega - E_n$, $E^c = \Omega - E$ be the complementary event of E. We have

$$E^c \subset \left(\bigcap_{i=n}^{+\infty} E_i\right)^c = \bigcup_{i=n}^{+\infty} E_i^c,$$

so that

$$P(E^c) \leq P\left(\bigcup_{i=n}^{+\infty} E_i^c\right) \leq \sum_{i=n}^{+\infty} P(E_i^c) \leq \sum_{i=n}^{+\infty} \beta_i = R_n.$$

Thus:

$$P(E^c) \leq \lim_{n \to +\infty} R_n = 0,$$

and $P(E) = 1 - P(E^c) = 1 - 0 = 1$. ∎

LEMMA 7.7.– Let $\{U_n\}_{n \in \mathbb{N}}$ be a sequence of random variables such that

$$\forall n \geq 0 : P(U_{n+1} > U_n) \leq \beta_n \text{ with } \sum_{n=0}^{+\infty} \beta_n < \infty$$

$$\forall n \geq 0 : U_n \geq m \text{ a.s.}$$

Then there exists a random variable U_∞ such that $U_n \longrightarrow U_\infty$ a.s. and $U_\infty \geq m$ a.s. ∎

PROOF.–

1) Let us consider

$$A_n = \{\omega \in \Omega : U_{n+1}(\omega) \leq U_n(\omega)\},$$

$$B_n = \{\omega \in \Omega : U_n(\omega) \geq m\}.$$

By denoting $A_n^c = \Omega - A_n$, $B_n^c = \Omega - B_n$ (complementary events of A_n and B_n, respectively), we have $P(A_n^c) = \beta_n$ and $P(B_n^c) = 0$. Let us introduce $E_n = A_n \cap B_n$. Then, $E_n^c = (A_n \cap B_n)^c = A_n^c \cup B_n^c$, so that

$$P(E_n^c) = P(A_n^c \cup B_n^c) \leq P(A_n^c) + P(B_n^c) \leq \beta_n + 0 = \beta_n,$$

and E_n is almost sure, since $P(E_n) = 1 - P(E_n^c) = 1 - 0 = 1$. Let us consider

$$E = \{\omega \in \Omega : \exists\, n_0(\omega) \text{ such that } m \leq U_{n+1}(\omega) \leq U_n(\omega), \forall\, n \geq n_0(\omega)\}.$$

We have $E = \bigcup_{n=0}^{+\infty} \left(\bigcap_{i=n}^{+\infty} E_i \right)$, so that it yields from Lemma 7.6 that E is almost sure.

2) Let $\omega \in E$. Then, there exists $n \geq 0$ such that $\omega \in \bigcap_{i=n}^{+\infty} E_i$, so that

$$\forall\, i \geq n : m \leq U_{i+1}(\omega) \leq U_i(\omega).$$

Thus, $\{U_i(\omega)\}_{i \geq n} \subset \mathbb{R}$ is decreasing and bounded, so that there exists $U_\infty(\omega)$ verifying

$$U_i(\omega) \longrightarrow U_\infty(\omega) \text{ for } i \longrightarrow +\infty.$$

3) In this way, we define a numerical function $U_\infty : \Omega \longrightarrow \mathbb{R}$. The basic properties of the measurable applications (see, for instance, [HAL 64], p. 184) show that U_∞ is a random variable. In addition, $U_\infty(\omega) \geq m$, since $m \leq U_i(\omega), \forall\, i \geq n$.

4) Let

$$F = \{\omega \in \Omega : U_n(\omega) \longrightarrow U_\infty(\omega) \text{ for } n \longrightarrow +\infty\}.$$

We have $E \subset F$, so that $P(F) \geq P(E) = 1$ and F is almost sure.

5) Let

$$G = \{\omega \in \Omega : U_\infty(\omega) \geq m\}.$$

We have $E \subset G$, so that $P(G) \geq P(E) = 1$ and G is almost sure. ∎

LEMMA 7.8.– Under the assumptions of Theorem 7.3, we have

$$\forall\, \lambda > m : P(U_n \geq \lambda) \longrightarrow 0 \text{ for } n \longrightarrow +\infty. \blacksquare$$

PROOF.–

1) Let us consider the random variable Z_n given by

$$Z_n = 0, \text{ if } U_n \geq \lambda, \ Z_n = 1, \text{ if } U_n < \lambda.$$

We have

$$P(Z_{n+1} = i) = P(Z_{n+1} = i, Z_n = 0) + P(Z_{n+1} = i, Z_n = 1),$$

so that

$$P(Z_{n+1} = i) = P(Z_n = 0) P(Z_{n+1} = i \,|Z_n = 0)$$

$$+ P(Z_n = 1) P(Z_{n+1} = i \,|Z_n = 1)$$

and, by taking

$$\pi_n = \begin{pmatrix} P(Z_n = 0) \\ P(Z_n = 1) \end{pmatrix},$$

we obtain

$$\pi_{n+1} = A\pi_n, \ A_{ij} = P(Z_{n+1} = i \,|Z_n = j) \ (0 \leq i, j \leq 1).$$

2) Let us consider the events

$$F = \{\omega \in \Omega : U_{n+1}(\omega) \geq \lambda \text{ and } U_n(\omega) < \lambda\},$$

$$G = \{\omega \in \Omega : U_{n+1}(\omega) \geq U_n(\omega)\}.$$

We have $F \subset G$, so that $P(F) \leq P(G) \leq \beta_n$. As a result,

$$P(Z_{n+1} = 0, Z_n = 1) = P(U_{n+1} \geq \lambda, U_n < \lambda) \leq \beta_n$$

and

$$A_{01}(\pi_n)_1 = P(Z_{n+1} = 0, Z_n = 1) \leq \beta_n.$$

3) Moreover,

$$A_{00} = P(Z_{n+1} = 0 \mid Z_n = 0) = 1 - P(U_{n+1} < \lambda \mid U_n \geq \lambda) \leq 1 - \alpha(\lambda).$$

Thus:

$$\forall n \geq 0 : (\pi_{n+1})_0 = A_{00}(\pi_n)_0 + A_{01}(\pi_n)_1 \leq (1 - \alpha(\lambda))(\pi_n)_0 + \beta_n$$

and, therefore,

$$\forall n \geq k : (\pi_{n+1})_0 \leq (1 - \alpha(\lambda))^{n+1-k}(\pi_k)_0 + \sum_{i=k}^{n} \underbrace{(1 - \alpha(\lambda))^{n-i}}_{\leq 1} \beta_i.$$

Thus,

$$\forall n \geq k : (\pi_{n+1})_0 \leq (1 - \alpha(\lambda))^{n+1}(\pi_k)_0 + \sum_{i=k}^{n} \beta_i,$$

which implies that

$$\forall n \geq k : (\pi_{n+1})_0 \leq (1 - \alpha(\lambda))^{n+1}(\pi_k)_0 + \sum_{i=k}^{+\infty} \beta_i$$

4) Hence

$$\forall k \geq 0 : \limsup_n (\pi_{n+1})_0 \leq \sum_{i=k}^{+\infty} \beta_i.$$

Due to the convergence of the series, we have

$$R_k = \sum_{i=k}^{+\infty} \beta_i \longrightarrow 0 \text{ for } k \longrightarrow +\infty.$$

So

$$\limsup_n (\pi_{n+1})_0 \leq 0.$$

Furthermore, $(\pi_{n+1})_0 \geq 0, \forall\, n \geq 0$, so that

$$\limsup_n (\pi_{n+1})_0 = 0.$$

5) Since

$$\liminf_n (\pi_{n+1})_0 \leq \limsup_n (\pi_{n+1})_0,$$

we also have

$$\liminf_n (\pi_{n+1})_0 \leq 0.$$

Since $(\pi_{n+1})_0 \geq 0, \forall\, n \geq 0$, we also have

$$\liminf_n (\pi_{n+1})_0 = 0.$$

6) The equality of both the values of lim inf and lim sup shows that

$$(\pi_{n+1})_0 \longrightarrow 0 \text{ for } n \longrightarrow +\infty. \blacksquare$$

This result implies that

LEMMA 7.9.– Under the assumptions of Theorem 7.3, there exists U_∞ such that $U_n \longrightarrow U$. Moreover,

$$\forall \lambda > m : P(U_\infty \geq \lambda) = 0. \blacksquare$$

PROOF.–

1) Lemma 7.2 shows that there exists U_∞ such that $U_n \longrightarrow U$ a.s. Since $U_n \longrightarrow U_\infty$ a.s., we also have $U_n \longrightarrow U_\infty$ p. and, therefore, $U_n \longrightarrow U$ D.

2) Let F_n be the cumulative distribution of U_n and F_∞ be the cumulative distribution of U_∞. The convergence in distribution shows that

$$P(U_n < \lambda) = F_n(\lambda) \longrightarrow F_\infty(\lambda) = P(U_\infty < \lambda),$$

so that

$$P(U_n \geq \lambda) = 1 - F_n(\lambda) \longrightarrow 1 - F_\infty(\lambda) = P(U_\infty \geq \lambda).$$

Thus:
$$P(U_\infty \geq \lambda) = \lim_{n \longrightarrow +\infty} P(U_n \geq \lambda) = 0. \blacksquare$$

PROOF OF THEOREM 7.3.–

1) It yields from Lemma 7.7 the existence of a random variable U_∞ such that $U_n \longrightarrow U_\infty$ a.s. and $U_\infty \geq m$ a.s. As in the proof of this lemma, we consider

$$E = \{\omega \in \Omega : m \leq U_{n+1}(\omega) \leq U_n(\omega), \forall n \geq 0\}.$$

E is almost sure and $U_n(\omega) \longrightarrow U_\infty(\omega)$, for any $\omega \in E$.

2) Let $\lambda > m$: it follows from Lemma 7.9 that $U_\infty < \lambda$ a.s. Thus, it follows from Lemma 7.3 that $U_\infty = m$ a.s. $\blacksquare$

This theorem suggests a method for the minimization of the objective function f. Let us consider the following:

1) $Z : \Omega \longrightarrow \mathbb{R}^k$ a continuous random vector having a probability density $\varphi_Z : \mathbb{R}^k \longrightarrow \mathbb{R}$ such that ($|\bullet|$ is the Euclidean norm)

$$\forall r > 0 : \inf \{\varphi_Z(z) : |z| \leq r\} \geq a(r) > 0;$$

2) $x_0 : \Omega \longrightarrow \mathbb{R}^k$ a random vector;

3) two strictly positive real numbers $\alpha > 0$ and $\beta > 0$;

4) $\{\theta_n\}_{n \in \mathbb{N}}$ a sequence of strictly positive real numbers $\theta_n > 0, \forall n \geq 0$ such that $\sum_{n=0}^{+\infty} \exp\left(-\dfrac{\alpha}{\theta_n}\right) < \infty.$

We define the sequence of random variables $\{x_n\}_{n \in \mathbb{N}}$, $x_n : \Omega \longrightarrow \mathbb{R}^k$ as follows:

$$\forall n \geq 0 : x_{n+1}(\omega) = D_n(\omega) x_n(\omega) + (1 - D_n(\omega)) t_n(\omega), \quad t_n = x_n + Z;$$

$$P(D_n = 0 | f(x_n) < f(t_n)) = 1 - \beta_n; \quad P(D_n = 1 | f(x_n) < f(t_n)) = \beta_n;$$

$$P(D_n = 0\ f(x_n) \geq f(t_n)) = 1;\ P(D_n = 1\ f(x_n) \geq f(t_n)) = 0;$$

$$\beta_n = \exp\left(-\left(\frac{\alpha}{\theta_n} + \beta\right)\right).$$

The sequence $\{U_n\}_{n \in \mathbb{N}}$, $U_n : \Omega \longrightarrow \mathbb{R}$ is given by

$$U_n(\omega) = f(x_n(\omega)).$$

As in the preceding section, the conditional probability density of t_n is

$$\varphi_n(t\,|x_n = x) = \varphi_Z(t - x)$$

and the density of probability of x_n is

$$\phi_n(t) = \int_\mathbb{R} \varphi_n(t\,|x_n = x)\,dx = \int_\mathbb{R} \varphi_Z(t - x)\,dx.$$

This last equation shows that

$$\forall\, r > 0 : \phi_n(t) \geq a(r)\,\ell(B_r) > 0,$$

where ℓ is the Lebesgue's measure and B_r is the ball of radius r ($B_r = \{z \in \mathbb{R}^n : |z| \leq r\}$).

We have

THEOREM 7.4.– $U_n \longrightarrow m$ a.s. ∎

PROOF.–

1) First, we observe that

$$\beta_n = \exp\left(-\left(\frac{\alpha}{\theta_n} + \beta\right)\right) \leq \exp\left(-\frac{\alpha}{\theta_n}\right),$$

so that

$$\sum_{n=0}^{+\infty} \beta_n < \infty.$$

2) By construction:

$$\forall n \geq 0 : P(U_{n+1} > U_n) \leq \beta_n$$

and

$$\forall n \geq 0 : U_n \geq m \text{ a.s.}$$

3) In an analogous manner to those used in the proof of Theorem 7.2, we have

$$P(U_{n+1} \leq \lambda \mid U_n \geq \lambda) \geq \alpha(\lambda) > 0.$$

4) The result follows from Theorem 7.3. ∎

In practice, the following alternative is often used:

$$P(D_n = 0 \; f(x_n) < f(t_n)) = c_n; \quad P(D_n = 1 \; f(x_n) < f(t_n)) = 1 - c_n;$$

$$P(D_n = 0 \; f(x_n) \geq f(t_n)) = 1; \quad P(D_n = 1 \; f(x_n) \geq f(t_n)) = 0;$$

$$c_n = \exp\left(-\left(\frac{f(t_n) - f(x_n) + \alpha}{\theta_n} + \beta\right)\right).$$

We observe that

$$c_n \leq \exp\left(-\left(\frac{\overbrace{f(t_n) - f(x_n)}^{\geq 0} + \alpha}{\theta_n} + \beta\right)\right)$$

$$\leq \exp\left(-\left(\frac{\alpha}{\theta_n} + \beta\right)\right) \leq \exp\left(-\frac{\alpha}{\theta_n}\right),$$

so that Theorem 7.4 applies yet. The algorithm associated with this choice is the *stochastic descent method with dynamics of Metropolis*, which corresponds to the *simulated annealing algorithm*:

1) Initialization: they are given a starting point x_0 and a number of trials nm. Set iteration number to zero: $n \longleftarrow 0$;

2) *Iterations*: the actual point is x_n and we determine x_{n+1} by using two substeps:

— *drawing*: randomly generate a variate Z_n from Z and set $t_n = x_n + Z_n$.

— *dynamics*: determine x_{n+1} : if $f(x_n) \geq f(t_n)$ then $x_{n+1} = t_n$. Otherwise: generate a variate a uniformly distributed on $(0,1)$; if $a < c_n$ then $x_{n+1} = t_n$; else $x_{n+1} = x_n$.

3) *Stopping test*: if $n < nm$, then increment $n : n \longleftarrow n+1$ and return to step 2. Otherwise, stop iterations and estimate $m \approx f(x_n)$ and $x^* \approx x_n$.

NOTE 7.2.– The readers may find alternative convergence results in the literature, namely involving proofs based on stochastic diffusions (see, for instance, [AZE 88] and [GEM 86]).

7.2.3. *Version 3: Hybrid methods*

Despite the modifications introduced, the method presented in the previous section remains entirely based on the random generation of trial points. In order to obtain a significant improvement in efficiency, namely for a large number of variables, a simple idea consists of combining the random generation of trial points with a deterministic method. For instance, let us consider descent iterations which read as follows:

$$x_{n+1} = Q_n(x_n). \qquad [7.22]$$

For instance, gradient descent with a fixed step reads as

$$Q_n(x) = x - \mu \nabla f(x).$$

In terms of algorithms, the iterations read as follows:

1) *Initialization*: It is given a starting point x_0 and a maximum iteration number nm. Set iteration number to zero: $n \longleftarrow 0$;

2) *Iterations*: the actual point is x_n and we determine x_{n+1} by using two substeps:

— *descent*: we generate a new point $t_0 = Q_n(x_n)$ by using the descent method.

– *dynamics*: determine x_{n+1} from x_n and t_0. For instance, use an *blind dynamics* by setting $x_{n+1} = t_0$.

3) *Stopping test*: if $n < nm$, then increment $n : n \longleftarrow n + 1$ and return to step 2. Otherwise, stop iterations and estimate $m \approx f(x_n)$ and $x^* \approx x_n$.

As observed, we may combine the descent iterations with the random generation of trial points. For instance, we may introduce an intermediary step in the basic iterations:

1) *Initialization*: It is given a starting point x_0 and a maximum iteration number nm. Set iteration number to zero: $n \longleftarrow 0$.

2) *Iterations*: the actual point is x_n and we determine x_{n+1} by using three substeps:

– *descent*: we generate a new point $t_0 = Q_n(x_n)$ by using the descent method;

– *drawing*: for $i = 1, ..., nr$: randomly generate an increment $(\Delta x_n)_i$ and set $t_i = t_0 + (\Delta x_n)_i$;

– *dynamics*: determine x_{n+1} from x_n and $t_i, i = 0, ..., nr$. For instance, use an *elitist dynamics* as follows:

$$x_{n+1} = \arg\min \ \{f(x_n), f(t_i) : 0 \leq i \leq nr\}.$$

3) *Stopping test*: if $n < nm$, then increment $n : n \longleftarrow n + 1$ and return to step 2. Otherwise, stop iterations and estimate $m \approx f(x_n)$ and $x^* \approx x_n$.

This kind of combination between a descent method and a stochastic method may be interpreted as a *stochastic perturbation* of a deterministic descent method. The basic general algorithm is as follows:

1) *Initialization*: It is given a starting point x_0, a maximum of trials nr and a maximum iteration number nm. Set iteration number to zero: $n \longleftarrow 0$;

2) *Iterations*: the actual point is x_n and we determine x_{n+1} by using three substeps:

– *deterministic*: generate a new point t_0 by applying the deterministic method to x_n;

– *perturbation:*

- *generation*: for $i = 1, ..., nr$: randomly generate a perturbation $(\Delta x_n)_i$ and set $t_i = t_0 + (\Delta x_n)_i$;
- *selection*: determine $\tilde{x}_n = \arg\min \{ f(t_i) : 0 \leq i \leq nr \}$.

– *dynamics*: determine x_{n+1} by using x_n and $\tilde{x}_n$. For instance, use the elitist dynamics.

3) *Stopping test*: if $n < nm$, then increment $n : n \longleftarrow n + 1$ and return to step 2. Otherwise, stop iterations and estimate $m \approx f(x_n)$ and $x^* \approx x_n$.

Below is an example of Matlab implementation:

Listing 7.4. *Stochastic Perturbation of a descent method*

```
function [xsol, fsol] = stoc_pert_iterations(x0,Q, nitmax,
    npert,alea, dynamics_pert,dynamics_it )
%
% performs nitmax iterations using iteration function Q
% involving npertmax perturbations at each iteration
% starting from x0
% assumes that x0 is n x 1
%
%IN:
% x0: n x 1 vector - type array of double
% Q : iteration function - type anonymous function
% nitmax: maximum iteration number - type integer
% npert: mnumber of perturbations - type integer
% alea: generator of random points - type anonymous function
% alea(n,nt) furnishes a n x nt table of random values
% dynamics_pert: the dynamics to be used to select the
    perturnbation - type anonymous function
% dynamics_it: the dynamics to be used to select the new point
    - type anonymous function
%
xac = x0;
fac = f(x0);
for nit = 1: nitmax
    xc = Q(xac);
    fc = f(xc);
    for np = 1: npert
        xt = trial_point(xc,1, alea);
        ft = f(xt);
        [xc, fc] = dynamics_pert(xc,fc, xt,ft, nit);
    end;
```

```
    [xac, fac] = dynamics_it(xac, fac, xc, fc, nit);
end;
xsol = xac;
fsol = fac;
return;
end
```

EXAMPLE 7.11.– Let us consider $\overline{\mathbf{x}} = (1, 2, \ldots, n)$ and the Griewank's function

$$F(\mathbf{x}) = 1 + \frac{1}{200}\|\mathbf{x} - \overline{\mathbf{x}}\|^2 - \prod_{i=1}^{n} \cos\left(\frac{x_i - \overline{x}_i}{\sqrt{i}}\right).$$

We apply the method with $nr = 10$, $n = 5$. The starting point is $\mathbf{0}$ and we use a gradient descent with step $\mu = 0.1$. Stochastic perturbations are generated by variates from $N(0, \sigma)$, with $\sigma = 0.1$. The evolution of the objective function is shown in Figure 7.16.

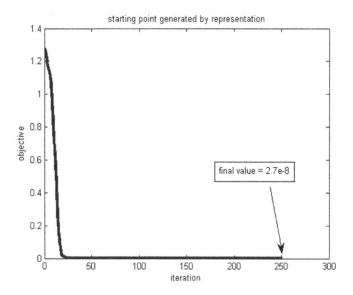

Figure 7.16. *Results for Griewank's function of dimension 5 with stochastic perturbation*

This algorithm corresponds to the iterations

$$x_{n+1} = Q_n(x_n) + P_n, \qquad [7.23]$$

where P_n is a random variable. These iterations converge to a point of global minimum under quite general assumptions, such that

$$\forall M > 0 : \exists\, a(M) \text{ such that } |x| \leq M \implies |Q_n(x)| \leq a(M), \forall\, n \geq 0.$$

[7.24]

In addition to the preceding assumptions on f (equations [7.13] and [7.14]), the random variable P_n has to be conveniently chosen:

$$\forall\, \lambda > m : P(U_{n+1} \leq \lambda \,|U_n \geq \lambda) \geq \alpha(n, \lambda) > 0 \,;\, \sum_{n=0}^{+\infty} \alpha(n, \lambda) = +\infty.$$

[7.25]

These choices are based on the following result:

THEOREM 7.5.– Let $\{U_n\}_{n \in \mathbb{N}}$ be a sequence of random variables such that

$$\forall\, n \geq 0 : U_{n+1} \leq U_n \quad \text{a.s.}$$

$$\forall\, n \geq 0 : U_n \geq m \quad \text{a.s.}$$

$\forall\, \lambda > m : \exists\, \alpha(n, \lambda) > 0$ such that $P(U_{n+1} \leq \lambda \,|U_n \geq \lambda) \geq \alpha(n, \lambda)$ [7.26]

and

$$\forall\, \lambda > m : \sum_{n=0}^{+\infty} \alpha(n, \lambda) = +\infty.$$

Then:

$$U_n \longrightarrow m \quad \text{a.s.} \quad \blacksquare$$

We observe again that Condition [7.26] implies that $P(U_n \geq \lambda) > 0$. This inequality is satisfied by trial points having a strictly positive probability density everywhere on $\mathbb{R}^n$.

The proof of the theorem uses the following lemma:

LEMMA 7.10.– Under the assumptions of Lemma 7.5, we have

$$\forall \lambda > m : P(U_n \geq \lambda) \longrightarrow 0 \quad \text{for} \quad n \longrightarrow +\infty. \quad \blacksquare$$

PROOF.–

1) Let us consider the random variable Z_n given by

$$Z_n = 0, \text{ if } U_n \geq \lambda, \ Z_n = 1, \text{ if } U_n < \lambda.$$

We have

$$P(Z_{n+1} = i) = P(Z_{n+1} = i, Z_n = 0) + P(Z_{n+1} = i, Z_n = 1),$$

so that

$$P(Z_{n+1} = i) = P(Z_n = 0) P(Z_{n+1} = i \,|\, Z_n = 0) \\ + P(Z_n = 1) P(Z_{n+1} = i \,|\, Z_n = 1)$$

and, by setting

$$\pi_n = \begin{pmatrix} P(Z_n = 0) \\ P(Z_n = 1) \end{pmatrix},$$

we obtain

$$\pi_{n+1} = A\pi_n, \ A_{ij} = P(Z_{n+1} = i \,|\, Z_n = j) \ (0 \leq i, j \leq 1).$$

2) Since the sequence is decreasing a.s., we have

$$P(Z_{n+1} = 0 \,|\, Z_n = 1) = 0, \ P(Z_{n+1} = 1 \,|\, Z_n = 1) = 1.$$

Thus:

$$A_{11} (\pi_n)_1 = (\pi_n)_1.$$

3) Moreover,

$$A_{10} = P(Z_{n+1} = 1 \,|\, Z_n = 0) = P(U_{n+1} < \lambda \,|\, U_n \geq \lambda) \geq \alpha(\lambda, n),$$

so that

$$\forall n \geq 0 : (\pi_{n+1})_1 = A_{10} (\pi_n)_0 + A_{11} (\pi_n)_1 \geq \alpha(\lambda, n) (\pi_n)_0 + (\pi_n)_1$$

and therefore:

$$\forall n \geq 0 : (\pi_{n+1})_1 \geq \alpha(\lambda, n)(1 - (\pi_n)_1) + (\pi_n)_1.$$

4) Since $\alpha(\lambda, n)(1 - (\pi_n)_1) \geq 0$, this inequality shows that

$$\forall n \geq 0 : (\pi_{n+1})_1 \geq (\pi_n)_1.$$

So, $\{(\pi_n)_1\}_{n \in \mathbb{N}}$ is increasing and upper-bounded by 1. Thus, there exists $p \leq 1$ such that

$$(\pi_n)_1 \longrightarrow p \text{ for } n \longrightarrow +\infty.$$

Moreover,

$$\forall n \geq 0 : p \geq (\pi_n)_1.$$

5) We have

$$\alpha(\lambda, n)(1 - (\pi_n)_1) \geq \alpha(\lambda, n)(1 - p),$$

so that

$$\forall n \geq 0 : (\pi_{n+1})_1 \geq \alpha(\lambda, n)(1 - p) + (\pi_n)_1.$$

Thus:

$$\forall n \geq 0 : (\pi_{n+1})_1 \geq (\pi_0)_0 + (1 - p) \sum_{i=0}^{n} \alpha(\lambda, i),$$

and

$$\forall n \geq 0 : 1 \geq (\pi_0)_0 + (1 - p) \sum_{i=0}^{n} \alpha(\lambda, i),$$

6) Assume that $p < 1$. In this case, we have

$$\sum_{i=0}^{n} \alpha(\lambda, i) \leq \frac{1 - (\pi_0)_0}{1 - p},$$

so that

$$+\infty = \sum_{i=0}^{+\infty} \alpha(i, \lambda) \leq \frac{1-(\pi_0)_0}{1-p},$$

which is a contradiction. Consequently, $p = 1$ and we have $(\pi_{n+1})_1 \longrightarrow 1$ for $n \longrightarrow +\infty$, so that

$$(\pi_{n+1})_0 \longrightarrow 0 \text{ for } n \longrightarrow +\infty. \blacksquare$$

LEMMA 7.11.– Under the assumptions of Theorem 7.5, there exists U_∞ such that $U_n \longrightarrow U$ D. Moreover,

$$\forall \lambda > m : P(U_\infty \geq \lambda) = 0. \blacksquare$$

PROOF.–

1) Lemma 7.2 establishes the existence of U_∞ such that $U_n \longrightarrow U$ a.s. Since $U_n \longrightarrow U_\infty$ a.s., we also have $U_n \longrightarrow U_\infty$ p. and, therefore, $U_n \longrightarrow U$ D.

2) Let F_n be the cumulative distribution of U_n and F_∞ be the cumulative distribution of U_∞. The convergence in distribution implies that

$$P(U_n < \lambda) = F_n(\lambda) \longrightarrow F_\infty(\lambda) = P(U_\infty < \lambda),$$

so that

$$P(U_n \geq \lambda) = 1 - F_n(\lambda) \longrightarrow 1 - F_\infty(\lambda) = P(U_\infty \geq \lambda).$$

Thus:

$$P(U_\infty \geq \lambda) = \lim_{n \to +\infty} P(U_n \geq \lambda) = 0. \blacksquare$$

PROOF OF THEOREM 7.5.–

1) It follows from Lemma 7.2 that there exists a variable U_∞ such that $U_n \longrightarrow U_\infty$ a.s. and $U_\infty \geq m$ a.s. Analogously to the proof of this lemma,

we consider

$$E = \{\omega \in \Omega : m \leq U_{n+1}(\omega) \leq U_n(\omega), \forall n \geq 0\}.$$

E is almost sure and $U_n(\omega) \longrightarrow U_\infty(\omega)$, for any $\omega \in E$.

2) Let $\lambda > m$: it follows from lemma 7.11 that $U_\infty < \lambda$ a.s. Consequently, it yields from lemma 7.3 that $U_\infty = m$ a.s. ∎

This theorem suggests an optimization method. Let us consider the following:

1) $Z : \Omega \longrightarrow \mathbb{R}^k$ a continuous random vector having a probability density $\varphi_Z : \mathbb{R}^k \longrightarrow \mathbb{R}$ such that ($|\bullet|$ is the Euclidean norm)

$$\forall r > 0 : \inf \{\varphi_Z(z) : |z| \leq r\} \geq a(r) > 0;$$

2) a random vector $x_0 : \Omega \longrightarrow \mathbb{R}^k$;

3) $\{\lambda_n\}_{n \in \mathbb{N}}$ a sequence of strictly positive real numbers such that $\lambda_n > 0$, $\forall n \geq 0$ and $\sum_{n=0}^{+\infty} \frac{1}{\lambda_n^k} a\left(\frac{\xi}{\lambda_n}\right) = +\infty, \forall \xi > 0.$

The last condition is satisfied, for instance, when $0 < \lambda_n \leq M$, $\forall n \geq 0$ and $\sum_{n=0}^{+\infty} a\left(\frac{\xi}{\lambda_n}\right) = +\infty, \forall \xi > 0$. In this case:

$$\sum_{n=0}^{+\infty} \frac{1}{\lambda_n^k} a\left(\frac{\xi}{\lambda_n}\right) \geq \frac{1}{M^k} \sum_{n=0}^{+\infty} a\left(\frac{\xi}{\lambda_n}\right) = +\infty.$$

An alternative way to satisfy this condition consists of taking $0 < \lambda_n \leq M$, $\forall n \geq 0$ and $\xi \longmapsto a(\xi)$ decreasing and strictly positive on $(0, +\infty)$. In this case:

$$a\left(\frac{\xi}{\lambda_n}\right) \geq a\left(\frac{\xi}{M}\right) > 0 \Longrightarrow \sum_{n=0}^{+\infty} a\left(\frac{\xi}{\lambda_n}\right) = +\infty.$$

Let us define a sequence of random variables $\{x_n\}_{n \in \mathbb{N}}$, $x_n : \Omega \longrightarrow \mathbb{R}^k$ as follows:

$$\forall n \geq 0 : x_{n+1}(\omega) = \arg\min \{f(x_n(\omega)), f(t_n(\omega))\},$$
$$t_n = Q_n(x_n) + \lambda_n Z;$$

and $\{U_n\}_{n \in \mathbb{N}}$, $U_n : \Omega \longrightarrow \mathbb{R}$ given by

$$U_n(\omega) = f(x_n(\omega)).$$

In this case, the conditional probability density of t_n is

$$\varphi_n(t \mid x_n = x) = \frac{1}{\lambda_n^k} \varphi_Z \left(\frac{t - Q_n(x)}{\lambda_n} \right).$$

and the density of t_n is

$$\phi_n(t) = \frac{1}{\lambda_n^k} \int_{\mathbb{R}} \varphi_Z \left(\frac{t - Q_n(x)}{\lambda_n} \right) dx$$

$$\geq \frac{1}{\lambda_n^k} \int_{\left| \frac{t - Q_n(x)}{\lambda_n} \right| \leq r} \varphi_Z \left(\frac{t - Q_n(x)}{\lambda_n} \right) dx,$$

so that

$$\forall r > 0 : \phi_n(t) \geq \frac{1}{\lambda_n^k} \int_{\left| \frac{t - Q_n(x)}{\lambda_n} \right| \leq r} \varphi_Z \left(\frac{t - Q_n(x)}{\lambda_n} \right) dx$$

$$\geq \frac{1}{\lambda_n^k} a(r) \ell(B_r) > 0,$$

where ℓ is the Lebesgue's measure and B_r is a ball of radius r ($B_r = \{z \in \mathbb{R}^n : |z| \leq r\}$).

We have

THEOREM 7.6.– $U_n \longrightarrow m$ a.s. ∎

PROOF.–

1) By construction:

$$\forall n \geq 0 : U_{n+1} \leq U_n \text{ a.s.}$$

$$\forall n \geq 0 : U_n \geq m \text{ a.s;}$$

2) Let $\lambda > m$. Let us consider $S_\lambda = \{x \in \mathbb{R}^k : f(x) < \lambda\}$ and $S_\lambda^c = \{x \in \mathbb{R}^k : f(x) \geq \lambda\}$. From assumption [7.14], there exist $x_\lambda \in S_\lambda$ and $\eta_\lambda > 0$ such that

$$B(\lambda) = \{x \in \mathbb{R}^k : |x - x_\lambda| \leq \eta_\lambda\} \subset S_\lambda$$

3) Let $\gamma > f(x_0)$: S_γ is bounded so that there is $M_0 > 0$ such that $|x| \leq M_0$ for all $x \in S_\gamma$. Thus, $|Q_n(x)| \leq b(M_0)$ for all $x \in S_\gamma$. So

$$\int_{S_\lambda} \varphi_n(t \mid x_n = x) \, dt \geq \int_{B(\lambda)} \varphi_n(t \mid x_n = x) \, dt$$

$$= \frac{1}{\lambda_n^k} \int_{B(\lambda)} \varphi_Z \left(\frac{t - Q_n(x)}{\lambda_n} \right) dt$$

verifies (ℓ is the Lebesgue's measure)

$$\int_{S_\lambda} \varphi_n(t \mid x_n = x) \, dt \geq \frac{1}{\lambda_n^k} \int_{B(\lambda)} b \left(\frac{\eta_\lambda + b(M_0)}{\lambda_n} \right) dt$$

$$\geq \frac{1}{\lambda_n^k} a \left(\frac{\eta_\lambda + b(M_0)}{\lambda_n} \right) \ell(B(\lambda)).$$

4) Let us denote by F_n the cumulative distribution of x_n and by μ_n the measure associated with F_n. We have

$$P(x_{n+1} \in S_\lambda, x_n \notin S_\lambda) = \int_{S_\lambda^c} \mu_n(dx) \int_{S_\lambda} \varphi_n(t \mid x_n = x) \, dt$$

$$\geq \int_{C(\lambda)} \mu_n(dx) \int_{S_\lambda} \varphi_n(t \mid x_n = x) \, dt$$

so that

$$P(x_{n+1} \in S_\lambda, x_n \notin S_\lambda) \geq \frac{1}{\lambda_n^k} a\left(\frac{\eta_\lambda + b(M_0)}{\lambda_n}\right) \ell(B(\lambda)) \int_{S_\lambda^c} \mu_n(dx),$$

that is

$$P(x_{n+1} \in S_\lambda, x_n \notin S_\lambda) \geq \frac{1}{\lambda_n^k} a\left(\frac{\eta_\lambda + b(M_0)}{\lambda_n}\right) \ell(B(\lambda)) \, P(x_n \notin S_\lambda).$$

By assumption [7.15], there are $y_\lambda \in S_\lambda^c$ and $r_\lambda > 0$ such that

$$C(\lambda) = \left\{x \in \mathbb{R}^k : |x - y| \leq r_\lambda\right\} \subset S_\lambda^c.$$

Thus:

$$P(x_n \notin S_\lambda) = \int_{S_\lambda^c} \phi_n(x)\,dx \geq \int_{C(\lambda)} \phi_n(x)\,dx \geq a(r_\lambda)\ell(B_{r_\lambda}) > 0$$

and, consequently,

$$P(x_{n+1} \in S_\lambda \,|\, x_n \notin S_\lambda) = \frac{P(x_{n+1} \in S_\lambda, x_n \notin S_\lambda)}{P(x_n \notin S_\lambda)}$$
$$\geq \frac{1}{\lambda_n^k} a\left(\frac{\eta_\lambda + b(M_0)}{\lambda_n}\right) \ell(B(\lambda)).$$

5) By setting

$$\alpha(n, \lambda) = \frac{1}{\lambda_n^k} a\left(\frac{\eta_\lambda + b(M_0)}{\lambda_n}\right) \ell(B(\lambda)),$$

we have

$$P(U_{n+1} \leq \lambda \,|\, U_n \geq \lambda) \geq \alpha(n, \lambda) > 0$$

and

$$\sum_{n=0}^{+\infty} \alpha(n, \lambda) = +\infty$$

6) The result follows from Theorem 7.5. ∎

The algorithm associated with this theorem is the *stochastic perturbation method*:

1) *Initialization*: let be given a starting point x_0, a number of trial points nr and a maximum iteration number nm. Set iteration number to zero: $n \longleftarrow 0$;

2) *Iterations*: the actual point is x_n and we determine x_{n+1} by using three substeps:

 – *descent*: generate $t_0 = Q_n(x_n)$;

 – *drawing*: for $i = 1, ..., nr$: randomly generate a variate Z_i from Z and set $t_i = t_0 + \lambda_n Z_i$;

 – *dynamics*: determine $x_{n+1} = \arg\min\{f(x_n), f(t_i) : i = 0, ..., nr\}$

3) *Stopping test*: if $n < nm$, then increment $n : n \longleftarrow n+1$ and return to step 2. Otherwise, stop iterations and estimate $m \approx f(x_n)$ and $x^* \approx x_n$.

NOTE 7.3.– In this situation, the dynamics of Metropolis may be used in step 2.3.

NOTE 7.4.– The readers may find other developments in the literature, such as methods for non-differentiable objective functions and methods for constrained optimization (see, for instance, [POG 94, AUT 97, DE 04b, DE 04a, MOU 06] and [ESS 09]).

7.3. Population-based methods

The methods presented in the last section may be applied to a set, i.e. a *population*, of initial points. For instance, let us consider an initial population formed by np elements:

$$\Pi_0 = \{x_0^1, ..., x_0^{np}\}.$$

A simple idea consists of applying the methods to each element x_0^i in such a manner that a sequence of populations $\{\Pi_n\}_{n \in \mathbb{N}}$ is generated:

$$\Pi_n = \{x_n^1, ..., x_n^{np}\}.$$

In such an approach, each x_0^i is used as a starting point, and the resulting algorithm is often referred to as *multistart method*. Assuming that the table xpop contains npop starting points (each column xpop(:,i) corresponds to a starting point), it may be implemented in Matlab as follows:

Listing 7.5. *Multistart method*

```
function [xsol, fsol] = multistart_pert(f,xpop,Q, nitmax, npert
    , alea, dynamics_pert, dynamics_it )
xpopsol = zeros(size(xpop));
npop = size(xpop,2);
fpopsol = zeros(npop,1);
for np = 1: npop
    x0 = xpop(:,np);
    [xs, fs] = stoc_pert_iterations(f,x0,Q, nitmax, npert,alea,
        dynamics_pert, dynamics_it );
    xpopsol(:,np) = xs;
    fpopsol(np) = fs;
end;
[m, ind] = min(fpopsol);
xsol = xpopsol(:,ind);
fsol = m;
return;
end
```

More sophisticated approaches may involve combinations of the population members. For instance, we may consider the following algorithm:

1) *Initialization*: They are given the initial population Π_0 formed by np elements, a number of trial points nr and a maximum iteration number nm. Set iteration number to zero: $n \longleftarrow 0$;

2) *Iterations*: the actual population is Π_n and we evaluate Π_{n+1} in three substeps:

— *deterministic*: generate $M_n^0 = \{t_0^1, ..., t_0^{np}\}$, where $t_0^i = Q_n(x_n^i)$;

— *drawing*: for $i = 1, ..., np$ and $j = 1, ..., nr$: randomly generate an increment $(\Delta x_n)_j^i$ and set $t_j^i = t_0^i + (\Delta x_n)_j^i$. This substep generates $M_n^1 = \left\{t_j^i : i = 1, ..., np \text{ and } j = 0, ..., nr\right\}$;

— *dynamics*: determine the new population Π_{n+1} as a part of $\Pi_n \cup M_n$, $M_n = M_n^0 \cup M_n^1$. For instance, the *elitist dynamics* consists of the selection of the np best elements of $\Pi_n \cup M_n$.

3) *Stopping test*: if $n < nm$, then increment $n : n \longleftarrow n+1$ and return to step 2. Otherwise, stop iterations and estimate m by the best value of f on Π_n and x^* as one of the elements corresponding to this value.

Various modifications of this basic algorithm may be found in the literature. For instance, it is possible to define M_n by choosing the best element $\left\{t_j^i : j = 0, ..., nr\right\}$ for each fixed i: the reunion of the results for $i = 1, ..., np$ forms M_n. We may also introduce supplementary substeps. For instance, we may introduce a substep where the available elements are combined in order to generate new elements. An example is furnished by the generation of random affine combinations of the elements of Π_n: we may generate the supplementary set

$$C_n = \{\alpha^{ij} x_n^i + \beta^{ij} x_n^j + \gamma^{ij} : i, j, \alpha^{ij}, \beta^{ij}, \gamma^{ij} \text{ random}\}$$

The elements of C_n may be used in a substep placed at the point chosen by the user. For instance, we may modify the dynamics in order to select the np best elements of $\Pi_n \cup M_n \cup C_n$.

An example of implementation is given below:

Listing 7.6. *Population-based method involving affine combinations*

```
function [xpop, fpop, err] = pop_based_iterations(f,xpop_ini,Q,
       nitmax, npert,alea, dynamics_pert,dynamics_it,cmin, cmax,
       n_r)
xpop = xpop_ini;
npop = size(xpop,2);
fpop = zeros(npop,1);
for i = 1: npop
    fpop(i) = f(xpop(:,i));
end;
for nit = 1: nitmax
    [xpop_r , fpop_r] = affine_comb(xpop,cmin, cmax, n_r,f);
    npop_r = length(fpop_r);
    nn = npop_r +npop;
    xpop_m = zeros(size(xpop,1), nn);
    xpop_m(:,1:npop_r) = xpop_r;
    xpop_m(:,npop_r+1:npop_r+npop) = xpop;
    [xpop_m, fpop_m] = mute_pop(f,xpop_m,Q, nitmax, npert,alea,
           dynamics_pert,dynamics_it);
    [xpop , fpop, err] = select_new_pop(xpop_r,fpop_r,xpop_m,
           fpop_m,xpop,fpop);
```

```
end;
return;
end

function [xpop_m, fpop_m] = mute_pop(f,xpop,Q, nitmax, npert,
    alea, dynamics_pert,dynamics_it)
%
% generates the perturbations of the population xpop
%
npop = size(xpop,2);
xpop_m = zeros(size(xpop));
fpop_m = zeros(npop,1);
for np = 1: npop
    x0 = xpop(:,np);
    [xs, fs] = stoc_pert_iterations(f,x0,Q, nitmax, npert,alea,
        dynamics_pert,dynamics_it );
    xpop_m(:,np) = xs;
    fpop_m(np) = fs;
end;
return;
end

function [xpop_r , fpop_r] = affine_comb(xpop,cmin, cmax, n_r,f
    )
%
% generates n_r affine combinations of each element in xpop
% coefficients are randomly chosen in (cmin, cmax)
%
%IN:
% xpop: n x npop table of the actual population — type array of
    double
% cmin, cmax : 3 x 1 table of real numbers — type array of
    double
% n_r: number of combinations by element — type integer
%
% OUT:
% xpop_r: the combinations — type array of double
% fpop_r: values of the objective — type array of double
%
npop = size(xpop,2);
nn = n_r*npop;
xpop_r = zeros(size(xpop,1), nn);
for i = 1: npop
    x = xpop(:,i);
    ind = randperm(npop,n_r);
    for j = 1: n_r
        a = cmin + rand(size(cmax)).*(cmax - cmin);
```

```
                xpop_r(:,(i-1)*n_r + j) = a(1)*x + a(2)*xpop(:, ind(j))
                    + a(3);
        end;
    end;
    fpop_r = zeros(nn,1);
    for i = 1: nn
        fpop_r(i) = f(xpop_r(:,i));
    end;
    return;
    end


    function [xpop_n , fpop_n, err] = select_new_pop(xpop_r,fpop_r,
        xpop_m, fpop_m,xpop_old,fpop_old)
    %
    % selects the best npop elements from the available ones
    %
    npop_r = size(xpop_r,2);
    npop_m = size(xpop_m,2);
    npop = size(xpop_old,2);
    nn = npop_r + npop_m + npop;
    xpop_a = zeros(size(xpop_old,1), nn);
    fpop_a = zeros(nn,1);
    xpop_a(:,1:npop_r) = xpop_r;
    xpop_a(:,npop_r+1:npop_r+npop_m) = xpop_m;
    xpop_a(:,npop_r+npop_m+1:npop_r+npop_m+npop) = xpop_old;
    fpop_a(1:npop_r) = fpop_r;
    fpop_a(npop_r+1:npop_r+npop_m) = fpop_m;
    fpop_a(npop_r+npop_m+1:npop_r+npop_m+npop) = fpop_old;
    [fpop_a, ind] = sort(fpop_a);
    xpop_n = xpop_a(:,ind(1:npop));
    fpop_n = fpop_a(1:npop);
    err = norm(xpop_n - xpop_old)/sqrt(npop);
    return;
    end
```

7.4. Determination of starting points

In this section, we establish a representation formula that may be used for the determination of starting points. It is interesting to notice that the representation established is valid in general Hilbert spaces, including infinite-dimensional spaces. As a model situation, let us consider a Hilbert space V and a functional $J : V \longrightarrow \mathbb{R}$. We are interested in the determination of

$$u = \arg\min_S J, \quad i.e., u \in S, \ J(u) \leq J(v), \ \forall v \in S. \tag{7.27}$$

We assume that $S \subset V$ is closed, bounded and not empty. Then, there is a constant $\alpha \in \mathbb{R}$ such that $\|v\| \leq \alpha$, $\forall v \in S$.

Let $J : V \longrightarrow \mathbb{R}$ be a continuous functional. We assume that there is a constant $\beta \in \mathbb{R}$ such that $|J(v)| \leq \beta$, $\forall v \in S$. This assumption is verified when J is bounded from below and is coercive.

Let B_ε^* be the part of S situated in the interior of the open ball having center u and radius ε. We denote by $S_\varepsilon^* = S - B_\varepsilon^*$ its complement in S. We assume that there is $\varepsilon_0 > 0$ such that $\mu(B_\varepsilon^*) > 0, \forall \varepsilon \in (0, \varepsilon_0)$. Let χ_ε^* be the characteristic function of B_ε^* and ψ_ε^* be the characteristic function of S_ε^*. We have $\chi_\varepsilon^*(v) + \psi_\varepsilon^*(v) = 1$, $\forall v \in S$.

Let $\lambda > 0$ be a real number large enough (in the following, we let $\lambda \longrightarrow +\infty$) and $g : \mathbb{R}^2 \longrightarrow \mathbb{R}$ be a continuous function such that $g \geq 0$. We assume that there are $\varepsilon_0 > 0$ and two functions $h_1, h_2 : \mathbb{R}^2 \longrightarrow \mathbb{R}$ such that, $\forall \varepsilon \in (0, \varepsilon_0)$:

$$\frac{E(vg(\lambda, J(v)))}{E(g(\lambda, J(v)))} \xrightarrow[\lambda \to +\infty]{} u, \text{ weakly in } V. \blacksquare \tag{7.28}$$

Then

THEOREM 7.7.– Assume that [7.30] and [7.31] are satisfied. If V_d is a finite-dimensional linear subspace of V and $\P_d : V \longrightarrow V_d$ is the orthogonal projection on V_d, then

$$\frac{E(P_d(v) g(\lambda, J(v)))}{E(g(\lambda, J(v)))} \xrightarrow[\lambda \to +\infty]{} P_d(u), \text{ strongly in } V. \tag{7.29}$$

In addition, for any $\ell \in \mathcal{L}(V, \mathbb{R})$,

$$\frac{E(\ell(P_d(v))g(\lambda, J(v)))}{E(g(\lambda, J(v)))} \xrightarrow[\lambda \to +\infty]{d \to +\infty} \ell(u) \ ;$$

$$\frac{E(P_d(v)g(\lambda, J(v)))}{E(g(\lambda, J(v)))} \xrightarrow[\lambda \to +\infty]{d \to +\infty} u, \text{ weakly in } V. \blacksquare$$

COROLLARY 7.1.– Assume that, in addition to the assumptions of the preceding theorem, the Fubini–Tonnelli theorem applies. Then, $E\left((v,\varphi_n)g(\lambda,J(v))\right) = \left(E\left(vg(\lambda,J(v))\right),\varphi_n\right), \forall n \in \mathbb{N}^*$ and

$$E\left(g(\lambda,J(v))\right) \geq h_1(\lambda,\varepsilon) > 0 \,;\, E\left(\psi_\varepsilon^*(v)g(\lambda,J(v))\right) \leq h_2(\lambda,\varepsilon) \qquad [7.30]$$

$$\forall \varepsilon \in (0,\varepsilon_0) \,:\, \frac{h_2(\lambda,\varepsilon)}{h_1(\lambda,\varepsilon)} \xrightarrow[\lambda\to+\infty]{} 0 \,. \qquad [7.31]$$

These results suggest that we may consider approximations of

$$\mathbf{x}^* = \arg\min\left\{F(\mathbf{y}) : \mathbf{y} \in S\right\}$$

having the form

$$\mathbf{x}^* \approx \frac{\sum_{i=1}^{ns} \mathbf{x_i} g(\lambda, F(\mathbf{x_i}))}{\sum_{i=1}^{ns} g(\lambda, F(\mathbf{x_i}))}$$

or simply

$$\mathbf{x}^*) \approx \frac{1}{ns}\sum_{i=1}^{ns} \mathbf{x_i} g(\lambda, F(\mathbf{x_i}))$$

In practice, these approximations may be used in order to generate an initial population by using the code

Listing 7.7. *Generating an initial population*

```
function xpop = inipop_rep(f,n,npop,ntmax,alea,lambda,g)
%
% generates a population of npop elements from R^n
% by using the representation formula involving function g
% and parameter lambda
% xpop has dimensions n x npop
%
xpop = zeros(n,npop);
for i = 1: npop
    xac = zeros(n,1););
    sw = 0;
    for nt = 1: ntmax
        xt = alea(n, 1);
        ft = f(xt);
```

```
            wt = g(lambda, ft);
            xac = xac + wt*xt;
            sw = sw + wt;
      end;
      xpop(:,i) = xac/sw;
end
return;
end
```

EXAMPLE 7.12.– Let us consider $\overline{\mathbf{x}} = (1, 2, \ldots, n)$ and the Griewank's function

$$F(\mathbf{x}) = 1 + \frac{1}{200}\|\mathbf{x} - \overline{\mathbf{x}}\|^2 - \prod_{i=1}^{n} \cos\left(\frac{x_i - \overline{x}_i}{\sqrt{i}}\right).$$

We apply the method with $ns = 2{,}500$, $\lambda = 10$, $n = 5$, $g(\lambda, s) = \exp(-\lambda s)$. Random vectors are generated by variates from $N(0, \sigma)$, with $\sigma = 5$. The mean of the population generated is used as a starting point to a gradient descent with step $\mu = 0.1$. The evolution of the objective function is shown in Figure 7.12

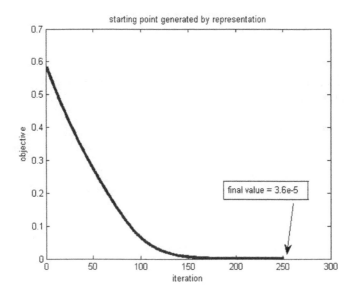

Figure 7.17. *Results for Griewank's function of dimension* 5 *with the representation*

We observe that these mathematical results remain valid in any dimension, including infinite-dimensional separable spaces. In fact, theorem 7.7 is the result of the following proposition:

PROPOSITION 7.1.– Assume that [7.30] and [7.31] are verified. Let $\ell \in \mathcal{L}(V, \mathbb{R}^d)$, $d \in \mathbb{N}^*$. Then

$$\ell(u) = \lim_{\lambda \to +\infty} \frac{E(\ell(v) g(\lambda, J(v)))}{E(g(\lambda, J(v)))}. \quad \blacksquare$$

PROOF OF THE PROPOSITION.–

1) We have

$$\frac{E(\ell(v) g(\lambda, J(v)))}{E(g(\lambda, J(v)))} = \frac{E(\ell(u) g(\lambda, J(v)))}{E(g(\lambda, J(v)))} + \frac{E(\ell(v - u) g(\lambda, J(v)))}{E(g(\lambda, J(v)))}.$$

Thus:

$$\frac{E(\ell(v) g(\lambda, J(v)))}{E(g(\lambda, J(v)))} = \ell(u) + \frac{E(\ell(v - u) g(\lambda, J(v)))}{E(g(\lambda, J(v)))}. \quad [7.32]$$

2) Let $\varepsilon \in (0, \varepsilon_0)$. We have

$$E(\ell(v - u) g(\lambda, J(v))) = E(\ell(v - u)(\chi_\varepsilon^*(v) + \psi_\varepsilon^*(v)) g(\lambda, J(v))).$$

Thus:

$$E(\ell(v - u) g(\lambda, J(v))) = E(\ell(v - u) \chi_\varepsilon^*(v) g(\lambda, J(v))) \\ + E(\ell(v - u) \psi_\varepsilon^*(v) g(\lambda, J(v))).$$

3) Since $\ell \in \mathcal{L}(V, \mathbb{R}^d)$, there exists a constant $M_p \in \mathbb{R}$ such that, for $v \in B_\varepsilon^*$,

$$|\ell(v - u)|_p \leq M_p \|v - u\| \leq M_p \varepsilon \quad [7.33]$$

while, for $v \in S_\varepsilon^*$,

$$|\ell(v - u)|_p \leq M_p \|v - u\| \leq M_p (\|v\| + \|u\|) \leq 2 M_p \alpha \quad [7.34]$$

4) On the one hand, from Jensen's inequality (Proposition 1.18),

$$|E(\ell(v - u) \chi_\varepsilon^*(v) g(\lambda, J(v)))|_p \leq E\left(|\ell(v - u)|_p \chi_\varepsilon^*(v) g(\lambda, J(v))\right).$$

and, from equation [7.33],

$$E\left(\left|\ell(v-u)\right|_p \chi_\varepsilon^*(v)g(\lambda,\ J(v))\right) \leq M_p \varepsilon E\left(\chi_\varepsilon^*(v)g(\lambda,\ J(v))\right).$$

On the other hand, Jensen's inequality (Proposition 1.18) also shows that,

$$\left|E\left(\ell(v-u)\psi_\varepsilon^*(v)g(\lambda,\ J(v))\right)\right|_p \leq E\left(\left|\ell(v-u)\right|_p \psi_\varepsilon^*(v)g(\lambda,\ J(v))\right).$$

and we have from [7.34] and [7.30]:

$$E\left(\left|\ell(v-u)\right|_p \psi_\varepsilon^*(v)g(\lambda,\ J(v))\right) \leq 2M_p \alpha h_2(\lambda,\ \varepsilon).$$

5) Thus:

$$\left|E\left(\ell(v-u)g(\lambda,\ J(v))\right)\right|_p \leq \varepsilon M_p E\left(\chi_\varepsilon^*(v)g(\lambda,\ J(v))\right) + 2M_p \alpha h_2(\lambda,\ \varepsilon),$$

and

$$\begin{aligned}\left|\frac{E(\ell(v-u)\chi_\varepsilon^*(v)g(\lambda,J(v)))}{E(g(\lambda,J(v)))}\right|_p &\leq \frac{\varepsilon M_p E(\chi_\varepsilon^*(v)g(\lambda,J(v))) + 2M_p\alpha h_2(\lambda,\varepsilon)}{E(g(\lambda,J(v)))} \\ &= \frac{\varepsilon M_p E(\chi_\varepsilon^*(v)g(\lambda,J(v)))}{E(g(\lambda,J(v)))} + \frac{2M_p\alpha h_2(\lambda,\varepsilon)}{E(g(\lambda,J(v)))}.\end{aligned}$$

6) Since

$$\frac{E\left(\chi_\varepsilon^*(v)g(\lambda,\ J(v))\right)}{E\left(g(\lambda,\ J(v))\right)} \leq 1$$

and

$$\frac{h_2(\lambda,\ \varepsilon)}{E\left(g(\lambda,\ J(v))\right)} \leq \frac{h_2(\lambda,\ \varepsilon)}{h_1(\lambda,\ \varepsilon)},$$

we have

$$\left|\frac{E\left(\ell(v-u)g(\lambda,\ J(v))\right)}{E\left(g(\lambda,\ J(v))\right)}\right|_p \leq \left(\varepsilon + 2\alpha \frac{h_2(\lambda,\ \varepsilon)}{h_1(\lambda,\ \varepsilon)}\right) M_p$$

7) This inequality, combined with equation [7.31], yields

$$\forall \varepsilon \in (0, \varepsilon_0) : \lim_{\lambda \to +\infty} \left| \frac{E\left(\ell(v-u)g(\lambda, J(v))\right)}{E\left(g(\lambda, J(v))\right)} \right|_p \leq \varepsilon M_p.$$

Thus:

$$\lim_{\lambda \to +\infty} \left| \frac{E\left(\ell(v-u)g(\lambda, J(v))\right)}{E\left(g(\lambda, J(v))\right)} \right|_p = 0$$

and we have

$$\lim_{\lambda \to +\infty} \frac{E\left(\ell(v-u)g(\lambda, J(v))\right)}{E\left(g(\lambda, J(v))\right)} = 0, \qquad [7.35]$$

8) The result is obtained by taking the limit for $\lambda \longrightarrow +\infty$ in equation [7.32] and using equation [7.35]. ∎

PROOF OF THE THEOREM.–

1) Let $\{\phi_1, ..., \phi_d\}$ be an orthonormal basis of V_d and $\ell : V \longrightarrow \mathbb{R}^d$ be given by $\ell(v) = ((v, \phi_1), ..., (v, \phi_n))$. ℓ is linear and $|\ell(v)|_2 = \|P_d(v)\| \leq \|v\|$. So, $\ell \in \mathcal{L}(V, \mathbb{R}^d)$ and, from Proposition 7.1:

$$\lim_{\lambda \to +\infty} \frac{E\left((v, \phi_n) g(\lambda, J(v))\right)}{E\left(g(\lambda, J(v))\right)} = (u, \phi_n).$$

Thus:

$$\frac{E(P_d(v)g(\lambda, J(v)))}{E(g(\lambda, J(v)))} = \sum_{n=0}^{d} \frac{E((v, \phi_n)g(\lambda, J(v)))}{E(g(\lambda, J(v)))} \phi_n$$
$$\xrightarrow[\lambda \to +\infty]{} \sum_{n=0}^{d} (u, \phi_n) \phi_n$$

strongly in V and the first claim is established.

2) Let $\ell \in \mathcal{L}(V, \mathbb{R})$, $\Phi = \{\varphi_n\}_{n \in \mathbb{N}^*} \subset V$ be an orthonormal Hilbert basis of V, $V_d = [\{\varphi_1, ..., \varphi_d\}]$, $P_d(v) = \sum_{n=1}^{d} (u, \varphi_n) \varphi_n$ the orthogonal projection of v onto V_d. Since ℓ is linear continuous, we have: $|\ell(v)| \leq M_1 \|v\|$ and $|\ell(P_d(v))| \leq M_1 \|P_d(v)\| \leq M_1 \|v\|$. Since S is bounded, there is a constant $\alpha \in \mathbb{R}$ such that $\|v\| \leq \alpha$, $\forall v \in S$ and we have $|\ell(P_d(v))| \leq$

$M_1\alpha$, $\forall v \in S$. Let

$$L(d,\lambda) = \frac{E(\ell(P_d(v))g(\lambda, J(v)))}{E(g(\lambda, J(v)))}$$

From Proposition 1.18:

$$|L(d,\lambda)| \le \frac{E(|\ell(P_d(v))|g(\lambda, J(v)))}{E(g(\lambda, J(v)))} \le M_1\alpha.$$

3) let $L = \limsup L(d,\lambda)$. Then,

$\forall \varepsilon > 0: \lambda \ge \lambda_0(\varepsilon)$ and $d \ge d_0(\varepsilon) \implies \sup L(d,\lambda) - \varepsilon \le L \le \sup L(d,\lambda)$

Since $L(d,\lambda) \longrightarrow \ell(P_d(u))$ for $\lambda \longrightarrow +\infty$:

$\forall \eta > 0: \lambda \ge \lambda_1(\eta, d) \implies \ell(P_d(u)) - \eta \le L(d,\lambda) \le \ell(P_d(u)) + \eta.$

Thus, $\lambda \ge \lambda_1(\eta, d) \implies \ell(P_d(u)) - \eta \le \sup L(d,\lambda) \le \ell(P_d(u)) + \eta$ e

$d \ge d_0(\varepsilon)$ and $\lambda \ge \lambda_1(\eta, d) \implies \ell(P_d(u)) - \eta - \varepsilon \le L \le \ell(P_d(u)) + \eta.$

For $\eta \longrightarrow 0$, we have $d \ge d_0(\varepsilon) \implies \ell(P_d(u)) - \varepsilon \le L \le \ell(P_d(u))$. Moreover, since $P_d(u) \longrightarrow u$ strongly in V, $\ell(P_d(u)) \longrightarrow \ell(u)$: we have $\ell(u) - \varepsilon \le L \le \ell(u)$. Thus, by taking the limit $\varepsilon \longrightarrow 0$, we have $L = \ell(u)$.

4) In an analogous manner, we obtain that $\liminf L(d,\lambda) = \ell(u)$. Thus, $\liminf L(d,\lambda) = \limsup L(d,\lambda) = \ell(u)$. Thus, $\ell(u) = \lim L(d,\lambda)$.

5) Let

$$u(d,\lambda) = \frac{E(P_d(v)g(\lambda, J(v)))}{E(g(\lambda, J(v)))} = \sum_{n=1}^{k} a_n(\lambda)\varphi_n;\ a_n(\lambda)$$
$$= \frac{E((v,\varphi_n)g(\lambda, J(v)))}{E(g(\lambda, J(v)))}.$$

and $\ell \in \mathcal{L}(V, \mathbb{R})$. We have $\ell(u(d,\lambda)) = \sum_{n=1}^{k} a_n(\lambda)\ell(\varphi_n) = L(d,\lambda)$. So, as previously established, $\ell(u(d,\lambda)) = L(d,\lambda) \longrightarrow \ell(u)$ for $\lambda \longrightarrow +\infty, d \longrightarrow +\infty$. Thus, $u(d,\lambda) \rightharpoonup u$, weakly in V. ∎

PROOF OF THE COROLLARY.– If $E((v, \varphi_n)g(\lambda, J(v))) = (E(vg(\lambda, J(v)), \varphi_n)$, then

$$a_n(\lambda) = \frac{(E(v), \varphi_n) g(\lambda, J(v))}{E(g(\lambda, J(v)))} \implies u(d, \lambda) = \frac{P_d(E(vg(\lambda, J(v))))}{E(g(\lambda, J(v)))}.$$

Taking $d \longrightarrow +\infty$, we have

$$u_\lambda = \sum_{n=1}^{+\infty} a_n(\lambda) \varphi_n = \frac{E(vg(\lambda, J(v)))}{E(g(\lambda, J(v)))}.$$

Since $u_\lambda \longrightarrow u$, weakly in V, for $\lambda \longrightarrow +\infty$, we obtain the result. ∎

NOTE 7.5.– In a finite-dimensional space, weak convergence implies the strong convergence. Therefore, if V is a finite-dimensional space, then

$$\frac{E(vg(\lambda, J(v)))}{E(g(\lambda, J(v)))} \xrightarrow[\lambda \to +\infty]{} u \text{ in } V. \qquad [7.36]$$

The choice of g may be guided by the following result:

PROPOSITION 7.2.– Assume that S_ε^* is weakly compact for $\varepsilon \in (0, \varepsilon_0)$ and J is weakly l.s.c. (lower semicontinuous). Let $g : \mathbb{R}^2 \longrightarrow \mathbb{R}$ be a continuous function such that $g > 0$ and $\xi \longrightarrow g(\lambda, \xi)$ is strictly decreasing for any $\lambda > 0$. Then, [7.30] and [7.31] are satisfied. ∎

The proof uses the following auxiliary result:

LEMMA 7.12.– Let $\varepsilon_0 > 0$ be small enough and $\varepsilon \in (0, \varepsilon_0)$. If S_ε^* is weakly closed and J is weakly l.s.c., then

$$\exists\, \theta = \theta(\varepsilon) > 0 \text{ such that } \max_{S_\varepsilon^*} g(\lambda, J(v)) \leq g(\lambda, J(u)) \exp(-\lambda\theta) \qquad [7.37]$$

Moreover, for any $\delta > 0$ such that $\delta < \theta$, there is $\eta = \eta(\delta) > 0$ such that:

$$\min_{B_\eta^*} g(\lambda, J(v)) \geq g(\lambda, J(u)) \exp(-\lambda\delta). \blacksquare$$

PROOF OF THE LEMMA.–

1) Since g is decreasing, equation [7.27] implies that

$$\forall x \in S : g(\lambda, J(v)) \leq g(\lambda, J(u)).$$

Thus:

$$\max_{S^*_\varepsilon} g(\lambda, J(v)) \leq g(\lambda, J(u)). \qquad [7.38]$$

2) There is a sequence $\{u_n\}_{n\geq 0} \subset S^*_\varepsilon$ such that

$$g(\lambda, J(u_n)) \xrightarrow[n\longrightarrow +\infty]{} \max_{S^*_\varepsilon} g(\lambda, J(v)).$$

Given that S^*_ε is weakly closed, this sequence has a weak cluster point $\overline{u} \in S^*_\varepsilon$.

3) Since J is weakly l.s.c., we have

$$J(\overline{u}) \leq \liminf J(u_n)$$

and

$$g(\lambda, J(\overline{u})) \geq \limsup g(\lambda, J(u_n)) = \max_{S^*_\varepsilon} g(\lambda, J(v)) . \qquad [7.39]$$

4) Equation [7.38] shows that $g(\lambda, J(\overline{u})) \leq g(\lambda, J(u))$. This inequality, combined with equation [7.39], establishes that

$$\max_{S^*_\varepsilon} g(\lambda, J(v)) = g(\lambda, J(\overline{u})) \leq g(\lambda, J(u)). \qquad [7.40]$$

5) Let us assume that

$$\max_{S^*_\varepsilon} g(\lambda, J(v)) = g(\lambda, J(u)). \qquad [7.41]$$

Equation [7.40] shows that $g(\lambda, J(\overline{u})) = g(\lambda, J(u))$. Since $\xi \longrightarrow g(\lambda, \xi)$ is strictly decreasing, we have $J(\overline{u}) = J(u)$ and the uniqueness of u shows that $\overline{u} = u$. So, $\|\overline{u} - u\| = 0$. But, $\overline{u} \in S^*_\varepsilon$ and $\|\overline{u} - u\| \geq \varepsilon$: we have $0 \geq \varepsilon > 0$, which is a contradiction.

6) Since [7.41] leads to a contradiction, quation [7.38] implies that

$$\max_{S^*_\varepsilon} g(\lambda, J(v)) < g(\lambda, J(u)) . \qquad [7.42]$$

Thus, we have [7.37].

7) Let $\eta > 0$ be small enough and $m(\eta) = \max_{B_\eta^*} J(v)$. The continuity of J shows that $m(\eta) \xrightarrow[\eta \to 0+]{} J(u)$, and the continuity of g shows that

$$\frac{g(\lambda,\ m(\eta))}{g(\lambda,\ J(u))} \xrightarrow[\eta \to 0+]{} 1. \qquad [7.43]$$

Let $\delta > 0$ be given. From equation [7.43], there exists $\eta(\delta) > 0$ such that

$$\forall\, 0 \leq \eta \leq \eta(\delta)\ :\ \frac{g(\lambda,\ m(\eta))}{g(\lambda,\ J(u))} \geq \exp(-\lambda\delta).$$

Thus, $g(\lambda,\ m(\eta)) \geq g(\lambda,\ J(u))\exp(-\lambda\delta)$. In addition,

$$\forall\, u \in B_\eta^*\ :\ m(\eta) \geq J(v).$$

Since g is strictly decreasing,

$$\forall\, x \in B_\eta^*\ :\ g(\lambda,\ m(\eta)) \leq g(\lambda,\ J(v))$$

and we have

$$\min_{B_\eta^*} g(\lambda,\ J(v)) \geq g(\lambda,\ m(\eta)) \geq g(\lambda,\ J(u))\exp(-\lambda\delta).\ \blacksquare$$

PROOF OF THE PROPOSITION.– Lemma 7.12 shows that (see equation [7.37])

$$E(\psi_\varepsilon^*(v)g(\lambda,\ J(v))) \leq g(\lambda,\ J(u))\exp(-\lambda\theta)\ E(\psi_\varepsilon^*(v)).$$

Let $h_2(\lambda, \varepsilon) = g(\lambda,\ J(u))\exp(-\lambda\theta)\ \mu(S_\varepsilon^*)$. From equation [7.37] (see Lemma 7.12):

$$E(g(\lambda,\ J(v))) \geq E(\chi_\eta^*(v)g(\lambda,\ J(v))) \geq g(\lambda,\ J(u))\exp(-\lambda\delta)E(\psi_\eta^*(v)).$$

Thus, by setting $h_1(\lambda,\ \varepsilon) = g(\lambda,\ J(u))\exp(-\lambda\delta)\ \mu(B_\eta^*)$, we have

$$E(g(\lambda,\ J(v))) \geq h_1(\lambda,\ \varepsilon) > 0$$

and, $\forall \, \varepsilon \in (0, \varepsilon_0)$:

$$\frac{h_2(\lambda, \varepsilon)}{h_1(\lambda, \varepsilon)} = \frac{\exp(-\lambda \theta) g(\lambda, J(u)) \, \mu(S_\varepsilon^*)}{\exp(-\lambda \delta) g(\lambda, J(u)) \, \mu(B_\eta^*)}$$
$$= \frac{\mu(S_\varepsilon^*)}{\mu(B_\eta^*)} \exp\left(-\lambda(\theta - \delta)\right) \xrightarrow[\lambda \to +\infty]{} 0. \blacksquare$$

These results suggest a numerical method for the determination of u, based on the generation of finite samples of the random variables considered. For instance, we may choose λ large enough and generate a sample $\mathbb{V} = (v_1, ..., v_{nr})$ of nr variates from v. Then,

$$u \approx u_a = \sum_{i=1}^{nr} v_i g(\lambda, J(v_i)) \,\Big/\, \sum_{i=1}^{nr} g(\lambda, J(v_i)), \qquad [7.44]$$

which corresponds to the approximations

$$E\left(v g(\lambda, J(v))\right) \approx \frac{1}{nr} \sum_{i=1}^{nr} v_i g(\lambda, J(v_i));$$
$$E\left(g(\lambda, J(v))\right) \approx \frac{1}{nr} \sum_{i=1}^{nr} g(\lambda, J(v_i)).$$

The sample $\mathbb{V}$ may be constructed by using the standard random number generator. For instance, a random element from $(\mathbb{N}^*)^k \times \mathbb{R}^k$ may be obtained by using, on the one hand, a sample $\boldsymbol{n} = (n_1, ..., n_k)$ of k random elements from $\mathbb{N}^*$ and, on the other hand, a random element $\boldsymbol{x} = (x_{n_1}, ..., x_{n_k})$ from $\mathbb{R}^k$. For this, it is necessary to use a random generator of integers for $\boldsymbol{n} \in (\mathbb{N}^*)^k$ and a random generator of real numbers for $\boldsymbol{x} \in \mathbb{R}^k$. Random elements of $\mathbb{R}_0^\infty$ may be obtained by a random selection of the index k followed by the generation of a random element from $(\mathbb{N}^*)^k \times \mathbb{R}^k$. From the purely mathematical standpoint, this procedure does not involve finite-dimensional approximations: the values of k and $\boldsymbol{n}$ span all the possible values and generate any value of $\mathbb{R}_0^\infty$. Of course, in practice, limitations arise from the computer (there is a maximum for the value of k) and from the generator used (some values of k may have very small probabilities and so become practically negligible).

The procedure may be illustrated by an algorithm denoted by $M1$: let $k \in \mathbb{N}^*$ be given and consider the generation of an element of $(\mathbb{N}^*)^k \times \mathbb{R}^k$. For this, we consider two strictly positive real numbers $\theta \in \mathbb{R}$ and $\sigma \in \mathbb{R}$.

$M1.1$: Set $i \longleftarrow 0$.

$M1.2$: Select k elements from $\mathbb{N}^*$ using $n = (m_1+1, ..., m_k+1)$, where $(m_1, ..., m_k)$ is a sample of k variates from Poisson's distribution $\mathcal{P}(\theta)$.

$M1.3$: Select $x = (x_{n_1}, ..., x_{n_k})$ as a sample of k values from the Gaussian distribution $\mathcal{N}(0, \sigma)$.

$M1.4$: Set $v_i = \sum_{j=1}^{k} v_{nj} \varphi_{nj}$ (or $\sum_{i=1}^{k} v_{nj} \psi_{nj}$).

$M1.5$: Increment $i : i \longleftarrow i+1$. If $i = nr$, then approximate the solution by using equation [7.44]. Otherwise, return to $M1.2$.

An alternative $M2$ is the following: let be given four strictly positive numbers $k_0 \in \mathbb{N}^*$, $\theta_0 \in \mathbb{R}$, $\theta \in \mathbb{R}$ and $\sigma \in \mathbb{R}$.

$M2.1$: Set $i \longleftarrow 0$.

$M2.2$: Select $k = k_0 + m \in \mathbb{N}^*$, where $m \in \mathbb{N}^*$ is a variate from the Poisson's distribution $\mathcal{P}(\theta_0)$.

$M2.2$: Select k elements from $\mathbb{N}^*$ using $n = (m_1+1, ..., m_k+1)$, where $(m_1, ..., m_k)$ is a sample of k variates from the Poisson's distribution $\mathcal{P}(\theta)$.

$M2.4$: Select $x = (x_{n_1}, ..., x_{n_k})$ as a sample of k variates from the Gaussian distribution $\mathcal{N}(0, \sigma)$.

$M2.5$: Set $v_i = \sum_{j=1}^{k} v_{nj} \varphi_{nj}$ (or $\sum_{i=1}^{k} v_{nj} \psi_{nj}$).

$M2.6$: Increment $i : i \longleftarrow i+1$. If $i = nr$, approximate the solution by using equation [7.44]. Otherwise, return to $M2.2$.

The readers may find examples of applications in finite-dimensional or infinite-dimensional situations in the literature (see, for instance, [DE 07, BEZ 08, BEZ 05, BEZ 10] and [ZID 13]).

8
Reliability-Based Optimization

Structural optimization looks for the best design of elements involved in a structure. Usually, the objective is cost or mass reduction and restrictions must be taken into account: for instance, a given available space imposes a restriction or failure criteria must be taken into account. Regarding practical applications, the objective and the restrictions may contain parameters affected by uncertainty: for instance, experimental data or material parameters having some variability. In addition, the external loads may appear as incertain: for instance, the design of a structure submitted to wind, sea waves or random vibration must take into account the intrinsic variability and uncertainty of the external loads.

Whenever stochastic or variable parameters are part of optimization problem, difficulties arise, such as:

– small variations of the parameters may correspond to large variations of the optimal design (lack of robustness);

– the optimal design cannot be exactly obtained in practice, since fabrication tolerances and geometrical errors arise (variability and uncertainty);

– the non-homogeneity of the materials, the variability of the external loads and of the boundary conditions may lead to situations where the optimal design does not satisfy the restrictions and becomes unsafe for some operating conditions (unreliability).

In order to take in account eventual implementation errors and variations in operating conditions, uncertainty must be taken into account in the design procedure. A complete characterization of the uncertainty of the optimal design involves the use of the approaches presented in Chapter 7, but the designer may be also simply interested in obtaining a single design having some robustness properties, such as the satisfaction of the restrictions with a prescribed probability.

A first approach consists of using safety factors, i.e. in modifying the results or the optimization problem in order to achieve the goal of probabilistic constraint satisfaction. Safety factors are generally multiplicative coefficients to be applied to loads or the variables. The main limitations in this approach are that, on the one hand, the coefficients are specific to particular situations and cannot be easily extended to other situations and, on the other hand, they are generally produced by the experience and the observation of historic failures, what makes their determination a difficulty problem for situations where the number of observed failures is insufficient.

In order to surmount these difficulties, an alternative consists of introduction of restrictions involving the probabilities of some events leading to failure. For instance, we may introduce a restriction on the maximal probability of such an event, what corresponds to a minimal reliability. This approach leads to reliability-based design optimization (RBDO), which is presented in the following.

8.1. The model situation

Reliability is characterized by the reliability index, denoted by β. The reliability index is defined for structures whose state is defined by a vector of parameters $\mathbf{x} \in \mathbb{R}^n$ – referred to as the physical variables. The failure corresponds to negative values of a real variable $Z = g(\mathbf{x})$, where g is a given function, referred to as limit curve, or limit state curve. $\mathbb{R}^n$ is split into two disjoint regions, S and D. S corresponds to a positive sign of Z, referred to as safe region – a point $x \in C$ corresponds to safe operation. D corresponds to a negative sign of Z, referred to as failure region – a point $x \in D$ corresponds to failure. The limit curve C is the boundary common to both the regions $C = \partial S = \partial D$.

$$S = \{\mathbf{x} \in \mathbb{R}^n : g(\mathbf{x}) \le 0\}, \ D = \{\mathbf{x} \in \mathbb{R}^n : g(\mathbf{x}) > 0\},$$
$$C = \{\mathbf{x} \in \mathbb{R}^n : g(\mathbf{x}) = 0\}.$$

Uncertainty is introduced by considering, for instance, the state of the structure that is not given by the nominal parameters $\mathbf{x}$ but by $\mathbf{r}(\mathbf{x}, \mathbf{v}) = \mathbf{x} + \mathbf{v}$, where $\mathbf{v}$ is a random variable. In this case, the limit curve is given by:

$$G(\mathbf{x}, \mathbf{v}) = g(\mathbf{r}(\mathbf{x}, \mathbf{v})) = 0$$

The situation is analogous if g depends on uncertain parameters $\mathbf{v}$, or $\mathbf{r}$ is a more complex function: in all these situations, the limit curve becomes a random function. Thus, the variable defining the failure becomes:

$$Z = G(\mathbf{x}, \mathbf{v}) = g(\mathbf{r}(\mathbf{x}, \mathbf{v})). \qquad [8.1]$$

Z, as defined by equation [8.1], is a random variable and the inequalities $Z \leq 0$ and $Z > 0$ define complementary events having associated probabilities. For instance, the *probability of failure* associated to a given $\overline{\mathbf{x}}$ is:

$$P_f(\overline{\mathbf{x}}) = P(Z > 0 \mid \mathbf{x} = \overline{\mathbf{x}}), \qquad [8.2]$$

while the *reliability* associated to $\overline{\mathbf{x}}$ is:

$$P_f(\overline{\mathbf{x}}) = P(Z \leq 0 \mid \mathbf{x} = \overline{\mathbf{x}}), \qquad [8.3]$$

The model RBDO problem is found below.

MODEL PROBLEM 8.1.– Let there be a target maximal failure probability P_t. Determine $\mathbf{x} \in \mathbb{R}^n$ such that:

$$\mathbf{x} = \arg\min\{F(\mathbf{x}) : \mathbf{x} \in \mathcal{A}\}, \mathcal{A} = \{\mathbf{x} \in \mathbb{R}^n : P_f(\mathbf{x}) \leq P_t\}. \blacksquare \qquad [8.4]$$

This formulation looks for the minimal value of the objective function F for a prescribed maximal failure probability P_t. Additional deterministic or stochastic restrictions may be introduced in the definition of the admissible set $\mathcal{A}$.

As an alternative, we may consider the determination of the maximal reliability for a prescribed maximal value F_t of the objective function F.

MODEL PROBLEM 8.2.– Let there be a target maximal cost F_t. Determine $\mathbf{x} \in \mathbb{R}^n$ such that:

$$\mathbf{x} = \arg\min\{P_f(\overline{\mathbf{x}}) : \mathbf{x} \in \mathcal{A}\}, \mathcal{A} = \{\mathbf{x} \in \mathbb{R}^n : F(\mathbf{x}) \leq F_t\}. \blacksquare \quad [8.5]$$

EXAMPLE 8.1.– A simple example is furnished by the situation where a structure has a maximal admissible stress equal to σ_{max}: in this case, we may take x_1 and x_2 as being the stress σ and the maximal admissible σ_{max}, respectively, i.e. $x = (\sigma, \sigma_{max})$. Then, $g(x) = x1 - x2 = \sigma - \sigma_{max}$. Other classical examples are furnished by Wohler's curves (or limit curves) – usually given by equations of the form $g(x) = 0$.

8.2. Reliability index

The standard reliability approach introduces a *reliability index* β, such that increasing β corresponds to increasing the reliability, i.e. decreasing the probability of failure. β is defined by using normalized independent variables $\mathbf{u} \in \mathbb{R}^m$, such that:

– each component of $\mathbf{u}$ has a mean equal to zero and a variance equal to one;

– $\mathbf{u}$ has a cumulative function F depending on $\|\mathbf{u}\|$: $F(\mathbf{u}) = \Phi(\|\mathbf{u}\|)$;

– $t \longrightarrow \Phi(t)$ is strictly decreasing.

For instance, the classical Nataf's transformation [NAT 62] brings $\mathbf{v}$ to a vector $\mathbf{u}$ of independent Gaussian variables and satisfies these assumptions. Thus, increasing $\|\mathbf{u}\|$ corresponds to decreasing the probability of failure P_t: this property is used in order to define the reliability index (see below).

From the mathematical standpoint, we consider a transformation $\boldsymbol{\tau} : \mathbb{R}^n \times \mathbb{R}^m \longrightarrow \mathbb{R}^n \times \mathbb{R}^k$ connecting the couples $(\mathbf{x}, \mathbf{v})$ and $(\mathbf{x}, \mathbf{u})$. Thus, there exists a function $\mathbf{T} : \mathbb{R}^n \times \mathbb{R}^m \to \mathbb{R}^k$ such that $\mathbf{v} = \mathbf{T}(\mathbf{x}, \mathbf{u})$.

In general situations, **T** is a complex transformation and its determination may involve some difficulty. In some simple situations, **T** is an affine transformation $\mathbf{T}(\mathbf{x}, \mathbf{u}) = \mathbf{A}(\mathbf{x})\mathbf{u} + \mathbf{B}(\mathbf{x})$, where $\mathbf{A}(\mathbf{x})$ and $\mathbf{B}(\mathbf{x})$ are matrices, independent from **v** and **u**. For instance, if, on the one hand, **v** is formed by independent variables and, on the other hand, the distribution of each component v_i corresponds to the cumulative function $F_i(u_i) = \Phi(|u_i|)$, having mean m_i and standard deviation s_i, we may consider normalized random variables given by $u_i = (v_i - m_i)/s_i$ and we have $v = m + diag(\sigma)u$, where $\mathbf{m} = (m_1, \ldots, m_n)$ is the vector of the means and $\sigma = (\sigma_1, \ldots, \sigma_n)$ is the vector of standard deviations.

By introducing **T**, we have:

$$Z = H(\mathbf{x}, \mathbf{u}) = G(\mathbf{x}, \mathbf{T}(\mathbf{x}, \mathbf{u})); \mathbf{R}(\mathbf{x}, \mathbf{u}) = \mathbf{r}(\mathbf{x}, \mathbf{T}(\mathbf{x}, \mathbf{u})). \qquad [8.6]$$

and it is natural to consider as working space:

$$V = \{((\mathbf{x}, \mathbf{u}) : \mathbf{x} \in \mathbb{R}^n, \mathbf{u} \in \mathbb{R}^m, \mathbf{u} \text{ vector of normalized independent variables}\}.$$

V is the *hybrid space*. We have $V = V_\mathbf{x} \times V_\mathbf{u}$, where $V_\mathbf{x} = \mathbb{R}^n$ is the physical space and $V_\mathbf{u}$ is a space of normalized random variables (mean zero, variance one), called *normalized space*. We have:

$$S = \{(\mathbf{x}, \mathbf{u}) \in V : H(\mathbf{x}, \mathbf{u}) \leq 0\}, D = \{(\mathbf{x}, \mathbf{u}) \in V : H(\mathbf{x}, \mathbf{u}) > 0\},$$
$$C = \{(\mathbf{x}, \mathbf{u}) \in V : H(\mathbf{x}, \mathbf{u}) = 0\}.$$

The reliability index β is determined in the normalized space and is defined as:

$$\beta(\mathbf{x}) = \|\mathbf{u}(\mathbf{x})\| = \min\{\|\mathbf{u}\| : H(\mathbf{x}, \mathbf{u}) = 0\}$$
$$= \min\{\|\mathbf{u}\| : H(\mathbf{x}, \mathbf{u}) \geq 0\}. \qquad [8.7]$$

$\mathbf{u}(\mathbf{x})$ is associated to the physical variables $\mathbf{v}(\mathbf{x}) = \mathbf{T}(\mathbf{x}, \mathbf{u}(\mathbf{x}))$ and corresponds to the state of the structure $\mathbf{r}(\mathbf{x}) = \mathbf{r}(\mathbf{x}, \mathbf{v}(\mathbf{x}))$ or $\mathbf{R}(\mathbf{x}) = \mathbf{R}(\mathbf{x}, \mathbf{u}(\mathbf{x}))$. The points $\mathbf{u}(\mathbf{x})$, $\mathbf{v}(\mathbf{x})$, $\mathbf{r}(\mathbf{x})$ and $\mathbf{R}(\mathbf{x})$ are usually referred to as the most probable failure points (or simply *most probable points*) associated to the state **x**. Sometimes, the expression *design point* is also used.

The determination of $\beta(\mathbf{x})$ is referred to as *reliability analysis of the point* **x**.

8.3. FORM

The determination of $\mathbf{u}(\mathbf{x})$ and $\beta(\mathbf{x})$ requests the solution of a constrained optimization problem – equation [8.7] for the given **x**, which may lead to a computationally expensive problem involving difficulties – namely for experimentally determined limit curves.

A first attempt in order to increase the efficiency by saving computational effort at this level is introduced by the First Order Reliability Method (FORM), which tends to furnish a rapid evaluation of $\mathbf{u}\left(\mathbf{x}^{(k)}\right)$ and $\beta(\mathbf{x}^{(k)})$.

When a starting point $\mathbf{u}^{(0)}$ is given, FORM generates a sequence $\mathbf{u}^{(1)}, \mathbf{u}^{(2)}, \ldots$ which is expected to converge to $\mathbf{u}(\mathbf{x})$, by using a sequence of affine approximations of the limit curve C. For a given $\mathbf{u}^{(i)}$, we determine an affine function $\mathbf{u} \longrightarrow H^{(i)}(\mathbf{u})$ such that $H(\mathbf{x},\mathbf{u}) \approx H^{(i)}(\mathbf{u})$ on the neighborhood of $\mathbf{u}^{(i)}$. Then, we determine:

$$\mathbf{u}^{\mathbf{u}^{(i+1)}} = \arg\min\{\|\mathbf{u}\| : H^{(i)}(\mathbf{x},\mathbf{u}) \geq 0\}.$$

We observe that this equation corresponds to the approximation of the failure region D by a half-space $D^{(i)}$. FORM is very popular due to its easy implementation and rapid convergence (even if the limit may be different from $\mathbf{u}(\mathbf{x})$). Breitung and Der Kiureghian [BRE 84, KIU 87, KIU 91] have proposed an analogous method, based on quadratic approximations of C – usually referred to as Second Order Reliability Method (SORM).

Both the approaches FORM/SORM involve the construction of approximations of the limit curve C. In the situations where such an approximation is not available, β may be evaluated by using a statistical approximation: a sample of points of D is generated and β is approximated by the empirical value determined on the sample:

$$\beta \approx \min\{\|\mathbf{x} - \mathbf{y_i}\| : (\mathbf{y_1}, \ldots, \mathbf{y}_N) \text{ sample of points from } D\}.$$

For Gaussian variables, this approximation is equivalent to a Monte Carlo approximation for the evaluation of the failure probability.

8.4. The bi-level or double-loop method

The model RBDO problem 8.1 reads as:

$$\mathbf{x} = \arg\min\{F(\mathbf{x}) : H(\mathbf{x}, \mathbf{u}(\mathbf{x})) \leq 0 \text{ and } \beta(\mathbf{x}) = \|\mathbf{u}(\mathbf{x})\| \geq \beta_t\}, \quad [8.8]$$

where β_t is the *target reliability index*. The alternative formulation [8.2] reads as:

$$\beta = \arg\max\{\beta(\mathbf{x}) : H(\mathbf{x}, \mathbf{u}(\mathbf{x})) \leq 0, \text{ and } F(\mathbf{x}) \leq F_t\} \quad [8.9]$$

Both the problems [8.8] and [8.9] contain difficulties which are still with us, namely those connected to their non convexity and the evaluation of $\beta(\mathbf{x}, \mathbf{u})$. Both these problems are *bi-level optimization problems*. For instance, equation [8.8] looks for $\mathbf{x}$ at the superior level, but requests the determination of $\mathbf{u}(\mathbf{x})$ at the inferior level. Analogous, equation [8.9] looks for β at the superior level, but requests the determination of $\mathbf{x}$ at the inferior level. Typical iterations for the solution consist of generating a sequence $\mathbf{x}^{(0)}, \mathbf{x}^{(1)}, \mathbf{x}^{(2)}, \ldots$ as follows:

Algorithm 8.1. Bi-level or double-loop RBDO

Require: $kmax > 0, precmin > 0$;
Require: an admissible starting point $\mathbf{x}^{(0)}$;
Require: a method for the determination of $\mathbf{u}(\mathbf{x})$ for a given $\mathbf{x}$;
Require: a method for updating $\mathbf{x}$ for a known $\mathbf{u}(\mathbf{x})$;
 1. $k = 0$;
 2. **inferior level (internal loop):** determine $\mathbf{u}^{(k)} = \mathbf{u}\left(\mathbf{x}^{(k)}\right)$;
 3. **superior level (external loop):** determine $\mathbf{x}^{(k+1)}$ by updating $\mathbf{x}^{(k)}$ for $\mathbf{u}\left(\mathbf{x}^{(k)}\right)$ known;
 4. $k = k+1$;
 5. Test the stopping conditions: $\|\mathbf{x}^k - \mathbf{x}^{k-1}\| \leq precmin$ or $k \geq kmax$. If the stopping conditions are not satisfied, go to 2.
 return $\mathbf{x}^{(k)}, \mathbf{u}\left(\mathbf{x}^{(k)}\right)$

Examples of methods for updating $\mathbf{x}^{(k)}$, may be found in the literature. The determination of $\mathbf{u}^{(k)}$ is referred to as being the *reliability level*, while the determination of $\mathbf{x}^{(k+1)}$ is referred to as the *optimization level*. The resolution may request a large number of calls to each level. Thus, methods of simplification have been proposed in the literature – for instance, the use of FORM at the inferior level. We examine other approaches in the following.

An alternative approach consists of generating $\mathbf{x}^{(k+1)}$ by determining successively:

$$\mathbf{u}^{(k)} = \arg\min\{\|\mathbf{u}\| : H\left(\mathbf{x}^{(k)}, \mathbf{u}\right) \geq 0 \text{ and } \|\mathbf{u}\| \geq \beta_t\}$$

and

$$\mathbf{x}^{(k+1)} = \arg\min\{F(\mathbf{x}) : H\left(\mathbf{x}, \mathbf{u}^{(k)}\right) \leq 0\}.$$

In this case, $\mathbf{u}^{(k)} \neq \mathbf{u}\left(\mathbf{x}^{(k)}\right)$.

8.5. One-level or single-loop approach

A way in order to save computational effort may be the reduction of the cost associated to the internal level (reliability analysis). FORM/SORM approximations may be considered, but an alternative approach may be introduced by transforming the bi-level problem into a one-level problem, where the variables $\mathbf{x}$ and $\mathbf{u}$ are considered as independent and the equality $\|\mathbf{u}\| = \beta(\mathbf{x})$ is achieved only at the limit of the iterations, i.e. at the optimal point. These ideas lead to the introduction of a new objective function J_d, where a convenient "penalty" term is introduced in order to ensure that the optimal point $(\mathbf{x}, \mathbf{u})$ satisfies $\|\mathbf{u}\| = \beta(\mathbf{x})$. For instance, we may consider a positive function $d : \mathbb{R}^2 \longrightarrow \mathbb{R}$ such that, on the one hand, $a \longrightarrow d(a, b)$ is increasing for any $b > 0$ and, on the other hand, $b \longrightarrow d(a, b)$ is decreasing for any $a > 0$. Then, we set:

$$J_d(\mathbf{x}, \mathbf{u}) = J(\mathbf{x}) d(\|\mathbf{u}\|, \beta(\mathbf{x})),$$

and we consider the problem:

$$(\mathbf{x}, \mathbf{u}) = \arg\min\{J_d(\mathbf{x}, \mathbf{u}) : (\mathbf{x}, \mathbf{u}) \in \mathcal{A}_d\};$$
$$\mathcal{A}_d = \{(\mathbf{x}, \mathbf{u}) : H(\mathbf{x}, \mathbf{u}) = 0, \|\mathbf{u}\| \geq \beta_t\}. \qquad [8.10]$$

Taking into account the properties of d, it is expected that optimization procedures will decrease $\|\mathbf{u}\|$ while increasing $\beta(\mathbf{x})$. It is expected to get at the limit, a point verifying $H(\mathbf{x},\mathbf{u}) \geq 0, \|\mathbf{u}\| \geq \beta_t$, minimizing $\|\mathbf{u}\|$ and corresponding to the largest $\beta(\mathbf{x})$ compatible with the restrictions: the set of all these properties achieves the equality $\|\mathbf{u}\| = \beta(\mathbf{x}) \geq \beta_t$, since the minimality of $\|\mathbf{u}\|$ joined to $H(\mathbf{x},\mathbf{u}) \leq 0$ (i.e. to the admissibility of $(\mathbf{x},\mathbf{u})$) lead to $\mathbf{u} = \mathbf{u}(\mathbf{x})$. This new formulation is expected to decrease the computational cost, since we do not need to perform a reliability analysis at each iteration – this method generates a sequence $(\mathbf{x}^{(0)}, \mathbf{u}^{(0)}), (\mathbf{x}^{(1)}, \mathbf{u}^{(1)}), (\mathbf{x}^{(2)}, \mathbf{u}^{(2)}), \ldots$ such that the condition $\mathbf{u}^{(k)} = \mathbf{u}(\mathbf{x}^{(k)})$ is not satisfied at each iteration number k. The problem in equation [8.10] has to be solved by an appropriate descent method and may be implemented by the following algorithm:

Algorithm 8.2. One-level or single-loop RBDO

Require: $kmax > 0, precmin > 0$;
Require: an admissible starting point $(\mathbf{x}^{(0)}, \mathbf{u}^{(0)})$;
Require: a descent method for the minimization of J_d;
  1. $k = 0$;
  2. Determine $(\mathbf{x}^{(k+1)}, \mathbf{u}^{(k+1)}) \in \mathcal{A}_d$ by one iteration of the descent method;
  3. $k = k + 1$;
  4. Test the stopping conditions: $\|\mathbf{x}^k - \mathbf{x}^{k-1}\| \leq precmin$ or $k \geq kmax$. If the stopping conditions are not satisfied, go to 2.
  return $\mathbf{x}^{(k)}, \mathbf{u}(\mathbf{x}^{(k)})$

The reader will find in the literature examples of d. A simple convenient choice is $d(a,b) = a$, which corresponds to:

$$J_d(\mathbf{x},\mathbf{u}) = J(\mathbf{x})\|\mathbf{u}\|.$$

This choice has shown to be effective to calculate, namely for situations where the response of the mechanical system may be analytically determined (i.e. the systems where a model explicitly connecting the design variables and the response is available). Analogous to the bi-level approach, equation [8.10] has difficulties which are still with us, namely those connected to non-convex optimization – the iterations may converge to local minima. In addition, this approach tends to give an identical relative importance to the values of $\|\mathbf{u}\|$

and J, which may be considered as inconvenient. Attempts to generate more equilibrated objective functions may be found in the literature, such as

$$J_{d,R}(\mathbf{x}, \mathbf{u}) = J_d(\mathbf{x}, \mathbf{u}) J(E(\mathbf{R}(\mathbf{x}, \mathbf{u}))),$$

where $E(\mathbf{R}(\mathbf{x}, \mathbf{u}))$ is the mean of $\mathbf{R}(\mathbf{x}, \mathbf{u})$ – the variables defining the state of the system. In this case, we determine:

$$(\mathbf{x}, \mathbf{u}) = \arg\min\{J_{d,R}(\mathbf{x}, \mathbf{u}) : (\mathbf{x}, \mathbf{u}) \in \mathcal{A}_{d,R}\};$$
$$\mathcal{A}_{d,R} = \{(\mathbf{x}, \mathbf{u}) : H(\mathbf{R}(\mathbf{x}, \mathbf{u}), \mathbf{u}) = 0, \|\mathbf{u}\| \geq \beta_t\}.$$

Choosing a modified and more equilibrated objective function such as $J_{d,R}$ often generates a significant increase of the computational time, which makes this approach unsuitable. As an alternative, semi-analytic methods, such as safety factors, may be used.

8.6. Safety factors

A safety factor is usually a real number associated to a variable and destined – generally by a multiplication – to increase the reliability. Safety factors are generally introduced for the critical parameters, such as, for instance, external loads, maximal stresses, displacements, deformations. They are generally produced by the observation of preceding failures and are connected to empirical design rules. Their empirical determination often involves experimentation, inverse problems and calibration. However, the basic idea of determining a correction to be applied to the variables may be exploited into another way: we show in the following that corrections may be generated by using optimality conditions – in such a situation, the corrections are usually referred to as *optimal safety factors* (OSF), but the reader must keep in mind that these are not empirical safety factors, but only numerical corrections destined to improve the preceding approaches.

In order to generate a correction, let us recall that $\mathbf{u}(\mathbf{x})$ verifies:

$$\|\mathbf{u}(\mathbf{x})\| = \min\{\|\mathbf{u}\| : H(\mathbf{x}, \mathbf{u}) \geq 0\}.$$

This problem may be rewritten under the following equivalent form:

$$\|\mathbf{u}(\mathbf{x})\|^2 = \min\{\|\mathbf{u}\|^2 : H(\mathbf{x}, \mathbf{u}) \geq 0\}.$$

Let us introduce the Lagrange's multiplier $\lambda \geq 0$ associated to the restriction $H(\mathbf{x}, \mathbf{u}) \geq 0$. The Lagrangian associated to this problem is:

$$L(\mathbf{u}, \lambda) = \|\mathbf{u}\|^2 - \lambda H(\mathbf{x}, \mathbf{u}).$$

The stationary points of L satisfy:

$$\nabla_\mathbf{u} L(\mathbf{u}(\mathbf{x}), \lambda) = 0, \ \lambda \geq 0, \ H(\mathbf{x}, \mathbf{u}(\mathbf{x})) \geq 0, \ \lambda H(\mathbf{x}, \mathbf{u}(\mathbf{x})) = 0.$$

i.e.

$$2\mathbf{u}(\mathbf{x}) - \lambda \nabla_\mathbf{u} H(\mathbf{x}, \mathbf{u}(\mathbf{x})) = 0 \Longrightarrow \mathbf{u}(\mathbf{x}) = \frac{\lambda}{2} \nabla_\mathbf{u} H(\mathbf{x}, \mathbf{u}(\mathbf{x}))$$

and

$$\beta(\mathbf{x})^2 = \|\mathbf{u}(\mathbf{x})\|^2 = \frac{\lambda^2}{4} \|\nabla_\mathbf{u} H(\mathbf{x}, \mathbf{u}(\mathbf{x}))\|^2.$$

Assume that $\beta(\mathbf{x}) > 0$. Then,

$$\frac{\lambda^2}{4} \|\nabla_\mathbf{u} H(\mathbf{x}, \mathbf{u}(\mathbf{x}))\|^2 \neq 0 \Longrightarrow \lambda \neq 0 \text{ and } \nabla_\mathbf{u} H(\mathbf{x}, \mathbf{u}(\mathbf{x})) \neq \mathbf{0}.$$

Thus,

$$\frac{\lambda}{2} = \left|\frac{\lambda}{2}\right| = \frac{\beta(\mathbf{x})}{\|\nabla_\mathbf{u} H(\mathbf{x}, \mathbf{u}(\mathbf{x}))\|}$$

and we have:

$$\mathbf{u}(\mathbf{x}) = \beta(\mathbf{x}) \frac{\nabla_\mathbf{u} H(\mathbf{x}, \mathbf{u}(\mathbf{x}))}{\|\nabla_\mathbf{u} H(\mathbf{x}, \mathbf{u}(\mathbf{x}))\|}$$

If $\beta(\mathbf{x}) = \beta_t$, we have:

$$\mathbf{u}(\mathbf{x}) = \beta_t \frac{\nabla_\mathbf{u} H(\mathbf{x}, \mathbf{u}(\mathbf{x}))}{\|\nabla_\mathbf{u} H(\mathbf{x}, \mathbf{u}(\mathbf{x}))\|} \qquad [8.11]$$

Equation [8.11] may be used into different ways: on the one hand, it defines a set of nonlinear equations which may be solved in order to determine $\mathbf{u}(\mathbf{x})$; on the other hand, it may be also used in order to update $\mathbf{u}^{(k)}$. For instance, we may choose a number of sub-iterations $i_{max} > 0$ and set $\mathbf{u}^{(k+1)} = \mathbf{u}^{(k+1,i_{max})}$, where:

$$\mathbf{u}^{(k+1,i)} = \beta_t \frac{\nabla_{\mathbf{u}} H(\mathbf{x}, \mathbf{u}^{(k+1,i-1)})}{\|\nabla_{\mathbf{u}} H(\mathbf{x}, \mathbf{u}^{(k+1,i-1)})\|}, i = 1, ..., i_{max};$$

$$\mathbf{u}^{(k+1,0)} = \mathbf{u}^{(k)}.$$
[8.12]

An alternative correction, involving a single sub-iteration, is furnished by:

$$\mathbf{u}^{(k+1)} = \mathbf{u}\left(\mathbf{x}^{(k)}\right) - \left(\beta_t - \beta(\mathbf{x}^{(k)})\right) \frac{\nabla_{\mathbf{u}} H(\mathbf{x}^{(k)}, \mathbf{u}(\mathbf{x}^{(k)}))}{\|\nabla_{\mathbf{u}} H(\mathbf{x}^{(k)}, \mathbf{u}(\mathbf{x}^{(k)}))\|},$$

$$i = 0, ..., i_{max}$$
[8.13]

These corrections may be interpreted as the use of multiplicative safety factors in the situations where $\mathbf{v} = \mathbf{x} + diag(\boldsymbol{\sigma})\mathbf{u}$. Recalling that $H(\mathbf{x}, \mathbf{u}) = G(\mathbf{x}, \mathbf{T}(\mathbf{x}, \mathbf{u}))$, we have:

$$\nabla_{\mathbf{u}} H(\mathbf{x}, \mathbf{u}) = [\nabla_{\mathbf{u}} T(\mathbf{x}, \mathbf{u})]^t \nabla_{\mathbf{v}} G(\mathbf{x}, T(\mathbf{x}, \mathbf{u}))$$

Thus,

$$\|\nabla_{\mathbf{u}} H(\mathbf{x}, \mathbf{u})\|^2 = [\nabla_{\mathbf{v}} G(\mathbf{x}, T(\mathbf{x}, \mathbf{u}))]^t \nabla_{\mathbf{u}} T(\mathbf{x}, \mathbf{u}) [\nabla_{\mathbf{u}} T(\mathbf{x}, \mathbf{u})]^t$$
$$\times \nabla_{\mathbf{v}} G(\mathbf{x}, T(\mathbf{x}, \mathbf{u}))$$

and we have:

$$\mathbf{u}(\mathbf{x}) = \beta(\mathbf{x}) \frac{[\nabla_{\mathbf{u}} T(\mathbf{x}, \mathbf{u})]^t \nabla_{\mathbf{v}} G(\mathbf{x}, T(\mathbf{x}, \mathbf{u}))}{\sqrt{[\nabla_{\mathbf{v}} G(\mathbf{x}, T(\mathbf{x}, \mathbf{u}))]^t \nabla_{\mathbf{u}} T(\mathbf{x}, \mathbf{u}) [\nabla_{\mathbf{u}} T(\mathbf{x}, \mathbf{u})]^t \nabla_{\mathbf{v}} G(\mathbf{x}, T(\mathbf{x}, \mathbf{u}))}}$$

Taking into account that $T(\mathbf{x}, \mathbf{u}) = \mathbf{x} + diag(\boldsymbol{\sigma})\mathbf{u}$, we have $\nabla_{\mathbf{u}} T(\mathbf{x}, \mathbf{u}) = diag(\boldsymbol{\sigma})$ and

$$u_i(\mathbf{x}) = \sigma_i \eta_i, \eta_i = \beta(\mathbf{x}) \frac{\frac{\partial G}{\partial v_i}(\mathbf{x}, T(\mathbf{x}, \mathbf{u}))}{\sqrt{\sum \sigma_i^2 \left(\frac{\partial G}{\partial v_i}(\mathbf{x}, T(\mathbf{x}, \mathbf{u}))\right)^2}}.$$

Let $\gamma_i = \sigma_i/x_i$ be the coefficient of variation of v_i: we have $u_i(\mathbf{x}) = \eta_i \gamma_i x_i$ and $v_i(\mathbf{x}) = (1 + \eta_i \gamma_i) x_i$. Thus, the factor $S_i = 1 + \eta_i \gamma_i$ appears as a multiplicative factor to be applied to the deterministic variables $\mathbf{x}$ in order to ensure reliability. In such a situation, S_i is called the *OSF associated to variable i*.

Equation [8.11] may be exploited by the preceding algorithms in order to update $\mathbf{u}^{(k)}$. For instance, we may use the following algorithm:

Algorithm 8.3. RBDO using OSF

Require: $kmax > 0, precmin > 0, \beta_t > 0$;
Require: an admissible starting point $\left(\mathbf{x}^{(0)}, \mathbf{u}^{(0)}\right)$;
 1. $k = 0$;
 2. determine $\mathbf{u}^{(k+1)}$ by updating $\mathbf{u}^{(k)}$ (for instance, use equation [8.12] or equation [8.13]);
 3. determine $\mathbf{x}^{(k+1)} = \arg\min \left\{ J(\mathbf{x}) : H(\mathbf{x}, \mathbf{u}^{(k+1)}) \leq 0 \right\}$ (or other updating method)
 4. $k = k + 1$;
 5. Test the stopping conditions: $\left\| \mathbf{x}^k - \mathbf{x}^{k-1} \right\| \leq precmin$ or $k \geq kmax$. If the stopping conditions are not satisfied, go to 2.
 return $\mathbf{x}^{(k)}, \mathbf{u}^{(k)}$

Bibliography

[AUT 97] AUTRIQUE L., DE CURSI J.E.S., "On stochastic modification for global optimization problems: an efficient implementation for the control of the vulcanization process", *International Journal of Control*, vol. 67, no. 1, pp. 1–22, 1997.

[AZE 88] AZENCOTT R., "Simulated annealing", *Séminaire Bourbaki*, 1987–1988.

[BAD 70] BADRIKIAN A., *Séminaire sur les fonctions aléatoires et les mesures cylindriques*, vol. 139 of Lecture Notes in Mathematics, Springer, 1970.

[BAD 74] BADRIKIAN A., CHEVET S., *Mesures cylindriques, espaces de Wiener et fonctions aléatoires gaussiennes*, vol. 379 of Lecture Notes in Mathematics, Springer, 1974.

[BEL 06] BELLIZZI S., SAMPAIO R., "POM analysis of randomly vibrating systems obtained from Karhunen-Loève", *Journal of Sound and Vibration*, vol. 303, pp. 774–793, 2006.

[BEL 09] BELLIZZI S., SAMPAIO R., "Karhunen-Loève modes obtained from displacement and velocity fields: assessments and comparisons", *Mechanical Systems and Signal Processing*, vol. 23, no. 4, pp. 1218–1222, 2009.

[BEZ 05] BEZ E.T., DE CURSI J.E.S., GONÇALVES M.B., "A hybrid method for continuous global optimization involving the representation of the solution", *6th World Congress on Structural and Multidisciplinary Optimization*, Rio de Janeiro, Brazil, 2005.

[BEZ 08] BEZ E.T., DE CURSI J.E.S., GONÇALVES M.B., "A hybrid method for continuous global optimization involving the representation of the solution", *Proceedings of EngOpt 2008*, Rio de Janeiro, Brazil, 2008.

[BEZ 10] BEZ E.T., GONÇALVES M.B., DE CURSI J.E.S., "A procedure of global optimization and its application to estimate parameters in interaction spatial models", *International Journal for Simulation and Multidisciplinary Design Optimization*, vol. 4, no. 2, pp. 85–100, 2010.

[BHA 10] BHAI Z., SILVERSTEIN J., *Spectral Analysis of Large Dimensional Random Matrices*, Springer, New York, 2010.

[BOC 33] BOCHNER S., "Integration von Funktionen, deren Werte die Elemente eines Vectorraumes sind", *Fundamenta Mathematicae*, vol. 20, pp. 262–276, 1933.

[BRE 84] BREITUNG K., "Asymptotic approximation for multi-normal integrals", *Journal of Engineering Mechanics, ASCE*, vol. 110, no. 3, pp. 357–366, 1984.

[CAM 47] CAMERON R., MARTIN W., "The orthogonal development of nonlinear functionals in series of Fourier-Hermite functionals", *Annals of Mathematics*, vol. 48, no. 2, p. 385, 1947.

[CAT 09] CATALDO E., SOIZE C., SAMPAIO R., *et al.*, "Probabilistic modeling of a nonlinear dynamical system used for producing voice", *Computational Mechanics*, vol. 43, no. 2, pp. 265–275, 2009.

[CHO 62] CHOQUET G., "Le probléme des moments", *Séminaire Choquet: Initiation à l'analyse*, vol. 1, no. 4, pp. 1–10, 1962.

[CHO 09] CHOUP L.N., "Edgeworth expansion of the largest eigenvalue distribution function of Gaussian orthogonal ensemble", *Journal of Mathematical Physics*, vol. 50, p. 013512, 2009.

[CRO 10] CROQUET R., DE CURSI J.E.S., "Statistics of uncertain dynamical systems", TOPPING B., ADAM J., PALLARES F., BRU R., *et al.* (eds.), *Proceedings of the 10th International Conference on Computational Structures Technology*, Civil-Comp Press, Stirlingshire, UK, pp. 541–561, 2010.

[DAT 09] DATTA K.B., *Matrix and Linear Algebra: Aided with Matlab*, Prentice-Hall, India, 2009.

[DAU 89] DAUTRAY R., *Méthodes probabilistes pour les équations de la physique*, Dunod, Paris, France, 1989.

[DE 92] DE CURSI J.E.S., *Introduction aux probabilités et statistiques*, Ecole Centrale Nantes, Nantes, France, 1992.

[DE 04a] DE CURSI J.E.S., ELLAIA R., BOUHADI M., "Global optimization under nonlinear restrictions by using stochastic perturbations of the projected gradient", FLOUDAS C.A., PARDALOS P.M. (eds.), *Frontiers in Global Optimization*, Kluwer Academic Press, Dordrecht, the Netherlands, pp. 541–561, 2004.

[DE 04b] DE CURSI J.E.S., ELLAIA R., BOUHADI M., "Stochastic perturbation methods for affine restrictions", FLOUDAS C.A., PARDALOS P.M. (eds.), *Advances in Convex Analysis and Global Optimization*, Kluwer Academic Press, Dordrecht, the Netherlands, pp. 487–499, 2004.

[DE 07] DE CURSI J.E.S., "Representation of solutions in variational calculus", TAROCCO E., DE SOUZA NETO E.A., NOVOTNY A.A., (eds.), *Variational Formulations in Mechanics: Theory and Applications*, International Center for Numerical Methods in Engineering (CIMNE), Barcelona, Spain, pp. 87–106, 2007.

[DE 08] DE CURSI J.E.S., SAMPAIO R., *Modélisation et convexité*, Hermès-Lavoisier, Paris, France, 2008.

[DER 28] DER POL B.V., DER MARK J.V., "The heartbeat considered as a relaxation oscillation, and an electrical model of the heart", *Philosophical Magazine Supplement*, vol. 6, pp. 763–775, 1928.

[DEV 86] DEVROYE L., KROESE D.P., *Non-Uniform Random Variate Generation*, Springer-Verlag, New York, 1986.

[DIE 77] DIESTEL J., UHL J.-J., *Vector Measure*, American Mathematical Society, 1977.

[DOR 12] DORINI F., SAMPAIO R., "Some results on the random wear coefficient of the Archard model", *Journal of Applied Mechanics, ASME*, vol. 79, no. 5, pp. 051008–051014, 2012.

[ESS 09] ES-SADEK M., ELLAIA R., DE CURSI J.E.S., "Application of an hybrid algorithm in a logistic problem", *Journal of Advanced Research in Applied Mathematics*, vol. 1, no. 1, pp. 34–52, 2009.

[FOR 09] FORRESTER P.J., KRISHNAPUR M., "Derivation of an eigenvalue probability density function relating to the Poincare disk", *Journal of Physics A: Mathematical and Theoretical*, vol. 38, no. 42, p. 385204, 2009.

[GAV 03] GAVRILIADIS P.N., ATHANASSOULIS G.A., "Moment data can be analytically completed", *Probabilistic Engineering Mechanics*, vol. 18, no. 4, pp. 329–338, 2003.

[GAV 08] GAVRILIADIS P.N., ATHANASSOULIS G.A., "Moment information for probability distributions, without solving the moment problem. I: where is the mode?", *Communications in Statistics – Theory and Methods*, vol. 37, no. 5, pp. 671–681, 2008.

[GAV 09] GAVRILIADIS P.N., ATHANASSOULIS G.A., "Moment information for probability distributions, without solving the moment problem. II: Main-mass, tails and shape approximation", *Journal of Computational and Applied Mathematics*, vol. 229, no. 1, pp. 7–15, 2009.

[GEM 86] GEMAN S., HWANG C.-R., "Diffusions for global optimization", *SIAM Journal on Control and Optimization*, vol. 24, no. 5, pp. 1031–1043, 1986.

[GHA 91] GHANEM R., SPANOS P., *Stochastic Finite Elements: A Spectral Approach*, Springer, 1991.

[GRI 02] GRIFFEL D.H., *Applied Functional Analysis*, Dover, Mineola, New York, 2002.

[HAL 64] HALMOS P., *Measure Theory*, Van Nostrand, 1964.

[HAN 78] HANDELMAN M., "A generalized eigenvalue distribution", *Journal of Mathematical Physics*, vol. 19, no. 12, pp. 2509–2513, 1978.

[JAY 57a] JAYNES E., "Information theory and statistical mechanics", *The Physical Review*, vol. 106, no. 4, pp. 620–630, 1957.

[JAY 57b] JAYNES E., "Information theory and statistical mechanics II", *The Physical Review*, vol. 108, pp. 171–190, 1957.

[KAP 92] KAPUR J.K.H., *Entropy Optimization Principles with Applications*, Academic Press, Inc., USA, 1992.

[KIU 87] KIUREGHIAN A.D., LIN H.Z., HWANG S.J., "Second-order reliability approximations", *Journal of Engineering Mechanics, ASCE*, vol. 113, no. 8, pp. 1208–1225, 1987.

[KIU 91] KIUREGHIAN A.D., STEFANO M.D., "Efficient algorithm for second-order reliability analysis", *Journal of Engineering Mechanics, ASCE*, vol. 117, no. 12, pp. 2904–2923, 1991.

[KLE 01] KLEYWEGT A.J., SHAPIRO A., "Stochastic optimization", *Handbook of Industrial Engineering*, 3rd ed., John Wiley, New York, 2001.

[KNU 98] KNUTH D., *Seminumerical Algorithms: The Art of Computer Programming*, vol. 2, 3rd ed., Addison Wesley, USA, 1998.

[LOP 11] LOPEZ R.H., DE CURSI J.E.S., LEMOSSE D., "Approximating the probability density function of the optimal point of an optimization problem", *Engineering Optimization*, vol. 43, no. 3, pp. 281–303, 2011.

[MAN 93a] MANOHAR C.S., KEANE A.J., "Axial vibrations of stochastic rod", *Journal of Sound and Vibration*, vol. 165, no. 2, pp. 341–359, 1993.

[MAN 93b] MANOHAR C.S., LYENGAR R.N., "Probability distribution of the eigenvalues of systems governed by the stochastic wave equation", *Probabilistic Engineering Mechanics*, vol. 8, pp. 57–64, 1993.

[MAU 12] MAUPRIVEZ J., CATALDO E., SAMPAIO R., "Estimation of random variables associated to a model for the vocal folds using ANNs", *Inverse Problems in Science & Engineering*, vol. 20, no. 2, pp. 209–225, 2012.

[MOU 06] MOUATASIM A.E., ELLAIA R., DE CURSI J.E.S., "Random perturbation of the variable metric method for unconstrained nonsmooth nonconvex optimization", *Applied Mathematics and Computer Science*, vol. 16, no. 4, pp. 463–474, 2006.

[NAT 62] NATAF A., "Détermination des distributions de probabilités dont les marges sont données", *Comptes rendus de l'Académie des Sciences*, vol. 225, pp. 42–43, 1962.

[POG 94] POGU M., DE CURSI J.E.S., "Global optimization by random perturbation of the gradient method with a fixed parameter", *Journal of Global Optimization*, vol. 5, no. 2, pp. 159–180, 1994.

[RAH 06] RAHMAN S., "A solution of the random eigenvalue problem by a dimensional decomposition method", *International Journal for Numerical Methods in Engineering*, vol. 67, pp. 1318–1340, 2006.

[RAH 07] RAHMAN S., "Stochastic dynamic systems with complex-valued eigensolutions", *International Journal for Numerical Methods in Engineering*, vol. 71, pp. 963–986, 2007.

[RAH 09] RAHMAN S., "Probability distributions of natural frequencies of uncertain dynamic systems", *AIAA Journal*, vol. 47, no. 6, pp. 1579–1589, 2009.

[RAH 11] RAHMAN S., YADAV V., "Orthogonal polynomial expansions for solving random eigenvalue problem", *International Journal for Uncertainty Quantification*, vol. 1, no. 2, pp. 163–187, 2011.

[RIT 08] RITTO T.G., SAMPAIO R., CATALDO E., "Timoshenko beam with uncertainty on the boundary conditions", *Journal of the Brazilian Society of Mechanical Sciences and Engineering*, vol. 30, no. 4, pp. 291–299, 2008.

[RIT 09] RITTO T.G., SOIZE C., SAMPAIO R., "Nonlinear dynamics of a drill-string with uncertain model of the bit-rock interaction", *International Journal of Non-Linear Mechanics*, vol. 44, no. 8, pp. 865–876, 2009.

[RIT 10a] RITTO T.G., SOIZE C., SAMPAIO R., "Probabilistic model identification of the bit-rock-interaction-model uncertainties in nonlinear dynamics of a drill-string", *Mechanics Research Communications*, vol. 37, no. 6, pp. 584–589, 2010.

[RIT 10b] RITTO T.G., SOIZE C., SAMPAIO R., "Robust optimization of the rate of penetration of a drill-string using a stochastic nonlinear dynamical model", *Computational Mechanics*, vol. 45, pp. 415–427, 2010.

[RIT 12] RITTO T.G., SAMPAIO R., "Stochastic drill-string dynamics with uncertainty on the imposed speed and on the bit-rock parameters", *International Journal for Uncertainty Quantification*, vol. 2, no. 2, pp. 111–124, 2012.

[ROS 06] ROSS S.M., *Simulation*, Elsevier, 2006.

[RUB 08] RUBINSTEIN R.Y., *Simulation and the Monte Carlo Method*, John Wiley & Sons, 2008.

[SAM 07] SAMPAIO R., SOIZE C., "On measures of nonlinearity effects for uncertain dynamical systems – application to a vibro-impact system", *Journal of Sound and Vibration*, vol. 303, pp. 659–674, 2007.

[SAM 08] SAMPAIO R., RITTO T.G., Short course on dynamics of flexible structures – deterministic and stochastic analysis, 2008.

[SAM 10] SAMPAIO R., CATALDO E., "Two strategies to model uncertainties in structural dynamics", *Shock and Vibration*, vol. 17, no. 2, pp. 171–186, 2010.

[SCH 69] SCHWARTZ L., "Probabilités cylindriques et applications radonifiantes", *Comptes Rendus de l'Académie des Sciences de Paris*, vol. 265, pp. 646–648, 1969.

[SHA 48] SHANNON C., "A mathematical theory of communication", *Bell System Technical Journal*, vol. 27, pp. 379-423 and 623-659, 1948.

[SHO 09] SHONKWILLER R., MENDIVIL F., *Explorations in Monte Carlo Methods*, Springer, New York, 2009.

[SOI 00] SOIZE C., "A nonparametric model of random uncertities for reduced matrix models in structural dynamics", *Probabilistic Engineering Mechanics*, vol. 15, pp. 277–294, 2000.

[TAO 09] TAO T., VU V., "Universality of the spectral distribution of random matrices", *Bulletin of the American Mathematical Society*, vol. 46, pp. 377–396, 2009.

[TAO 10] TAO T., VU V., "Random matrices: the distribution of the smallest singular values", *Geometric And Functional Analysis*, vol. 20, no. 1, pp. 260–297, 2010.

[TIE 84] TIEL J.V., *Convex Analysis, An Introductory Text*, John Wiley and Sons, New York, 1984.

[TRI 05] TRINDADE M.A., WOLTER C., SAMPAIO R., "Karhunen–Loève decomposition of coupled axial/bending of beams subjected to impacts", *Journal of Sound and Vibration*, vol. 279, pp. 1015–1036, 2005.

[WIE 38] WIENER N., "The homogeneous chaos", *American Journal of Mathematics*, vol. 60, no. 4, pp. 897–936, 1938.

[WIR 95] WIRSCHING P., ORTIZ K., *Random Vibrations: Theory and Practice*, John Wiley & Sons, 1995.

[XIU 10] XIU D., *Numerical Methods for Stochastic Computations*, Princeton University Press, 2010.

[ZID 13] ZIDANI H., Representation de solutions en optimisation continue, multiobjectif et applications, PhD Thesis, INSA Rouen, 2013.

Index

A, B, C

algebraic equation
  linear, 227
  nonlinear, 265
Brownian motion, 112, 113, 115
Borel, 37, 42, 58, 65
central limit, 93, 99, 100
characteristic function, 23, 44, 45, 52, 73, 83, 87, 88, 91–94, 96, 97, 103, 216, 408
conditional
  distribution, 32, 86, 81
  expectation, 31, 86, 181, 318
  mean, 31, 33, 86, 181
convergence
  almost sure, 90
  in distribution, 90, 216, 217, 375, 387, 398
  in probability, 89
  relations, 91
  quadratic mean, 88
couples, 11–13, 76, 77, 87, 348, 424
covariance, 13, 17, 18, 27–29, 78, 80, 94, 97, 98, 107, 135, 156
cumulative distribution function, 6, 13, 184, 217, 246, 247

D, E, F

deterministic, 1, 133, 134, 159, 160, 227, 235, 260, 288, 289, 319, 342, 346, 364, 370, 391, 392, 404, 423, 433
differential equation
  linear, 298, 302
  nonlinear, 306, 307
distribution
  Gaussian, 10, 92, 93, 156, 419
  Poisson, 36, 92, 138, 145, 150, 156, 240
  uniform, 36, 92, 138, 145, 150, 156, 240
elliptic, 325, 327
empirical
  distribution, 93, 99, 100
  mean, 99, 100, 197, 237, 351, 360–363
expectation, 4, 25, 31, 71, 77, 86, 181, 184, 318, 351, 362, 363
Euclidean, 378, 388, 399
finite population, 1, 3–5, 13, 36, 72, 73, 76, 78, 80, 82–85, 99
Frobenius, 157

G, H, I

Gaussian
  sample, 101
  vector, 21, 28–30, 94, 96, 98, 101, 103, 104, 168, 170, 191
Glivenko-Cantelli, 99, 101
Hamiltonian system, 314–319
Hilbert-Schmidt, 157
hybrid, 160, 370, 391, 425
inequality, 63, 67, 69, 228, 293, 294, 370, 382, 395, 397, 413, 416
infinite-dimensional, 53, 57, 58, 62, 63, 161, 297, 336, 340, 407, 411, 419
integral, 43, 45, 47, 48, 50–53, 62–64, 73, 78, 116–119, 127, 144, 340
Ito
  calculus, 122
  diffusion, 127, 130, 324, 332
  integral, 114, 122, 127

J, L, M

Jensen's inequality, 63, 64, 411, 412
law of large numbers, 99, 100
Lebesgue measure, 41, 42, 50, 52, 181, 185, 187, 191, 199, 207, 213, 218, 234, 340, 379, 380, 389, 400, 401
Levy's theorem, 91, 215, 217
linear correlation, 26–28, 84, 186, 187
measure
  external, 37, 38, 42
  internal, 37, 38, 42
  finite, 34, 36, 57, 60, 61–63
measurable function, 43, 44, 47, 52, 63, 65
multidimensional, 80, 113, 127, 130, 325, 327, 329
multivariate, 39, 30, 113

N, O, P

norm, 2, 22, 25, 55, 82, 88, 116, 157, 162, 163, 168, 178, 181, 217, 324
normalized, 162, 163, 165, 168, 254, 424, 425
numerical characteristic, 1, 3–5, 11, 15, 21, 23, 25–27, 31–33, 65, 76, 80, 85
ordinary, 51, 298, 319
orthogonal
  transformation, 102, 103
  projection, 22 -26, 82–85, 102–105, 178, 341, 408, 413
  family, 204–207
orthonormal, 55, 102, 103, 164, 194, 195, 206, 207, 255, 257, 258, 413
parabolic, 228, 330
partial, 67, 69, 70, 93, 310, 325, 328, 364
partition, 46, 108, 115, 118
principle, 109, 117, 120, 126, 134, 135, 364

R, S, U

rectangular, 45, 46
relative frequency, 3, 11
scalar product, 55, 82, 88, 108, 109, 111, 116, 154, 178, 195, 197, 199, 231, 238, 283, 290, 301, 302
simple function, 44–47, 108, 340
standard deviation, 4, 7, 10, 25, 71, 74, 75, 99, 100, 184, 425
unidimensional, 112, 113, 234
univariate, 27, 122

Lightning Source UK Ltd.
Milton Keynes UK
UKOW06n0328020515

250743UK00001B/3/P